Vitamins in Human Health and Disease

Vitamins in Human Health and Disease

Tapan K. Basu, PhD, FACN, FICN
Professor of Nutritional Biochemistry
Department of Agricultural, Food and Nutritional Science
University of Alberta
Edmonton, Alberta, Canada

and

John W.T. Dickerson, PhD, FI Biol, FIFST, FRSH
Professor Emeritus of Nutrition
School of Biological Science
University of Surrey, Guildford, UK

CAB INTERNATIONAL

CAB INTERNATIONAL
Wallingford
Oxon OX10 8DE
UK

Tel: +44 (0)1491 832111
Fax: +44 (0)1491 833508
E-mail: cabi@cabi.org
Telex: 847964 (COMAGG G)

A catalogue record for this book is available from the British Library.

ISBN 0 85198 986 1

Typeset in Photina by AMA Graphics Ltd
Printed and bound in the UK by Biddles Ltd, Guildford

Contents

Foreword by Elsie M. Widdowson xv

Preface xvii

1 General Introduction 1
History 2
Factors Affecting Vitamin Requirements 4
Factors Affecting Vitamin Status 6
Recommended Daily Allowances (RDAs) 8
References 9

2 Thiamin (Vitamin B_1) 11
Occurrence in Foods 11
Thiamin Losses 12
Absorption and Metabolism 13
Thiamin Activation 13
Role of Thiamin Pyrophosphate (TPP) 14
Coenzyme function 14
Pyruvate decarboxylase 14
α-Ketoglutarate decarboxylase 15
Branched-chain α-ketoacid decarboxylase 15
Transketolase 15
Other functions 18
Thiamin Deficiency 19
Situations Where Thiamin Deficiency Occurs 20
Thiamin-associated Genetic Disorders 24
Deficiency of pyruvate decarboxylase 24
Deficiency of branched chain α-ketoacid dehydrogenase 24
Leigh's disease (subacute necrotizing encephalomyelopathy) 24
Deficiency of transketolase enzyme 24

Requirements 25
Assessment of Thiamin Status 25
References 26

3 Riboflavin (Vitamin B_2) 28
Chemistry 28
Sources 29
Absorption 29
Transport 30
Metabolism 30
Metabolic Functions 31
Deficiency and its Consequences 33
Interactions with Other Nutrients 34
Iron 34
Vitamin B_6 34
Folic acid 35
Requirements 35
Biochemical Assessment 35
References 37

4 Niacin (Vitamin B_3) 39
Chemistry 39
Sources 40
Absorption 42
Transport 42
Tissue Storage 43
Metabolism 43
Metabolic Functions 45
Deficiency 46
Requirements 48
Measurement of Niacin Status 48
References 49

5 Pyridoxine (Vitamin B_6) 51
Chemistry 51
Sources 53
Bioavailability 53
Absorption and Transport 56
Metabolism 56
Metabolic Functions 57
Transaminases (aminotransferases) 58
Decarboxylases 60
Enzymes involved in cysteine synthesis 60
Kynureninase 61

Other vitamin B_6-dependent enzymes 61
Vitamin B_6 Deficiency 64
Requirements 65
Assessment of Vitamin B_6 Status 65
References 66

6 Biotin 68
Chemistry 68
Sources 68
Absorption and Transport 69
Activation of Biotin as a Coenzyme 69
Metabolic Role 70
Pyruvate carboxylase 70
Acetyl CoA carboxylase 71
Propionyl CoA carboxylase 72
β-Methylcrotonyl CoA carboxylase 72
Desaturases 73
Deficiency 73
Avidin-induced deficiency 74
Deficiency of carboxylases 75
Other contributory factors to biotin deficiency 75
Sudden infant death syndrome (SIDS) 75
Requirements 77
Assessment of Biotin Status 77
References 78

7 Pantothenic Acid 79
Chemistry 79
Sources 80
Absorption 81
Excretion 81
Metabolism 82
Metabolic Functions 83
Deficiency 83
Requirements 84
References 85

8 Pteroylglutamic Acid (Folic Acid, Folacin) 86
Chemistry 86
Sources 88
Absorption and Transport 89
Metabolic Role 90
Methionine synthesis 91
Biosynthesis of nucleic acids 92

Amino acid interconversions 94
Signs of Folate Deficiency 95
Factors Affecting Folate Status 96
Dietary factors 96
Malabsorption 97
Increased requirement 97
Pregnancy 97
Infancy and Childhood 98
Adolescents 98
Elderly population 98
Haemolytic anaemia 99
Disturbances of folate metabolism 99
Drug-induced disturbances 99
Genetical disturbances 99
Requirements 99
Assessment of Folate Status 100
Serum folate 100
RBC folate 101
FIGLU test 101
References 103

9 Vitamin B_{12} (Cobalamins) 106
Structure 106
Sources 107
Absorption and Transport 108
Metabolic Role 111
Methylcobalamin 111
5′-Adenosylcobalamin 113
Signs of Deficiency 113
Causes of Deficiency 115
Strict vegetarians (vegans) 116
Gastric factor 116
Intestinal factor 117
Abnormality of vitamin B_{12} transport 118
Derangements of metabolism 118
Requirements 119
Assessment of Vitamin B_{12} Status 119
Serum concentrations 119
Methylmalonic acid excretion 120
Schilling test 121
Deoxyuridine (dU) suppression test 122
References 122

10 Vitamin C (Ascorbic Acid) 125
Biosynthesis 125
Occurrence in Foods 126
Vitamin Losses 127
Absorption 129
Distribution in the Body 129
Metabolism and Excretion 130
Biochemical Role 131
Free-radical scavenger 132
Role as a cofactor 132
Mono-oxygenase 133
Dioxygenases 134
Mixed-function oxygenase 136
Other reducing roles 136
Cyclic nucleotides 136
Iron absorption and metabolism 137
Activation of folic acid 138
Nitrite scavenger 138
Immune function 138
Deficiency 139
Requirements 141
Elderly people 142
Alcoholics 142
Smokers 142
Stress 144
Exposure to foreign compounds 144
Assessment of Vitamin C Status 145
Biological tissues and fluids 145
Analytical techniques 146
References 146

11 Vitamin A 148
Chemistry 148
Sources 149
Vitamin A Activity 150
Metabolism 150
Absorption 150
Storage and transport 151
Distribution within the cell 153
Bio-transformation and excretion 154
Factors Affecting Circulatory Vitamin A and Holo-RBP (Retinol–RBP Complex) 156
Dietary Factors 156
Vitamin A 156

Lipid 158
Protein 158
Zinc 159
Age 159
Stress 160
Disease 160
Pathophysiology of Vitamin A 161
Somatic (systemic) function 161
Vitamin A and the immune system 163
Vitamin A and the visual cycle 163
Vitamin A and reproduction 165
Occurrence of Vitamin A Deficiency 165
Involving eyes 165
Involving epithelial cells 168
Requirements 168
Assessment of Vitamin A Status 170
Retinal Function Tests 170
Rapid dark adaptation (RDA) 170
Conjunctival impression cytology (CIC) 171
Rose bengal staining test (RBST) 171
Biochemical Tests 172
Serum vitamin A and its carrier proteins 172
Relative dose response test (RDR) 172
Methods for analysing serum vitamin A 173
References 173

12 β-Carotene and Related Substances 178
Biosynthesis of Carotenes 178
Structure of Carotenes 179
Properties of β-Carotene 181
Dietary Sources of β-Carotene 181
Absorption, Transport and Storage of β-Carotene 181
Bio-potency of β-Carotene 183
Provitamin A activity 183
Antioxidant property 184
LDL oxidation 185
Arachidonic acid oxidation 187
Immunomodulatory effect 187
Requirements 188
Factors that Influence β-Carotene Status 188
Assessment of Carotene Status 190
References 190

13 Vitamin D 193
Sources 195
Food sources 195
Cutaneous synthesis 195
Absorption 196
Metabolism 197
Regulation of Metabolism 198
Functions 199
Recommended Intake 202
Vitamin D and Diseases of Man 203
Rickets 203
Clinical features 203
Osteomalacia 204
Drug-induced osteomalacia 207
Osteomalacia after gastrectomy 207
Osteomalacia and intestinal malabsorption 207
Osteomalacia and liver disease 207
Osteomalacia and renal disease 208
Osteoporosis and ageing 208
Hereditary rickets 209
Biochemical Assessment of Vitamin D Status 210
References 211

14 Vitamin E 214
Chemistry 214
Sources 215
Absorption and Transport 216
Metabolism 218
Function 219
Deficiency 221
Requirements 224
Assessment of Vitamin E Status 225
References 226

15 Vitamin K 228
Sources 228
Absorption and Metabolism 229
Functions 230
The clotting of blood 230
Other potential roles 233
Deficiency 234
Malabsorptive states 234
Hepatic insufficiency 235
Antagonistic effects of vitamins A and E 236

Newborn infants 236
Hospitalized patients 237
Requirements 238
Assessment of Vitamin K Status 238
References 239

16 Vitamin-like Substances – Pseudovitamins 240
Amygdalin 240
Pangamic Acid 242
Bioflavonoids 243
Carnitine 244
Choline 247
Inositol 248
p-Aminobenzoic Acid 249
Coenzyme Q (CoQ) 249
References 249

17 Vitamins and Cancer 252
Vitamin A 253
β-Carotene 255
Vitamin E 258
Vitamin C 259
Vitamin D 261
Folic Acid 261
Conclusions 261
References 262

18 Interactions of Drugs and Vitamins 267
Drug-induced Vitamin Deficiency 268
Drug-induced malabsorption 268
Drugs affecting synthesis 271
Drugs affecting utilization 271
Drug-induced oxidation and excretion 275
Alcohol-induced vitamin deficiency 277
Cytotoxic drugs affecting vitamin status 279
Vitamin Deficiency and Drug Biotransformation 279
Drug metabolism 279
Vitamin deficiency and drug metabolism 280
Risk Populations 282
Maternal vitamin deficiency and the fetal outcome 282
Vitamin deficiency and the elderly 283
Consequences of drug–vitamin interactions 284
References 286

19 Therapeutic Potential of Vitamins 288
Vitamin A 290
Acne vulgaris 290
Bronchopulmonary dysplasia 290
Infectious disease 291
β-Carotene 292
Erythropoietic protoporphyria 292
Vitamin D 293
Nutritional rickets 293
Congenital rickets 293
Renal osteodystrophy 294
Hypertension 294
Osteoporosis 295
Vitamin E 295
Abetalipoproteinaemia 295
Biliary atresia 296
Cystic fibrosis 296
Intermittent claudication 297
Platelet function 297
Ischaemic heart disease 297
Cataracts 297
Anaemias 298
Ageing 299
Vitamin C 299
Immune function 299
Wound healing 299
Cardiovascular disease 300
Blood pressure 301
Platelets 302
Stroke 302
Cataracts 302
The 'common cold' 302
Riboflavin 303
Niacin 303
Hyperlipidaemias 303
Hartnup's disease 304
Pyridoxine 304
Dependency conditions 304
Carpal tunnel syndrome 306
Premenstual tension syndrome 306
Conclusions 306
References 307

20 Vitamin Abuse 313
Megavitamin Use 314
Are Megavitamin Supplements Harmful? 315
The Use of High Doses of Vitamins 316
Concluding Comments 317
References 318

21 Safety Considerations of Excess Vitamin Intakes 319
Lipid-soluble Vitamins 319
Retinoids 320
Acute toxicity 320
Chronic toxicity 320
Teratogenicity 321
Mechanism of action of vitamin A toxicity 322
Carotenoids 323
Vitamin D 324
Toxicity 324
Mechanism of action of vitamin D toxicity 325
Vitamin E 325
Vitamin K 326
Water-soluble Vitamins 326
Nicotinic Acid 327
Vitamin B_6 327
Ascorbic Acid 328
Gastrointestinal disturbances 328
Renal stones 329
Conditioning effect 330
Impaired copper status 331
Lysis of erythrocytes 331
Iron overload 331
Interference with glucose assay 332
Folic Acid 332
Other Water-soluble Vitamins 333
Conclusions 333
References 333

Index 337

Foreword

The authors of this book are both experienced teachers of nutrition to students taking both scientific and medically oriented degrees; the biochemistry and biological functions of the vitamins form an important part of the courses. Moreover, in their own research over many years, Professors Basu and Dickerson have made studies on the applied aspects of some of the vitamins. Their teaching and research made them realize the need for a comprehensive and up-to-date book on the subject, and they courageously decided to write one themselves. This volume is the result. In the Preface they state that their aim was to present a 'concise account of up-to-date knowledge of vitamins which will meet the needs of students reading for degrees in nutrition.' This is too modest. The book will be of value to a much wider readership. Biochemists, physiologists and doctors will find much to help over problems concerning vitamins.

The book is written in a clear, readable style, and covers the biochemistry and structure of the vitamins, the amounts needed and the diseases that arise when they are lacking in the diet or from some other source. The effects of excesses are also covered. There has been increasing interest in recent years in the more subtle effects of mild deficiencies, commonly referred to as subclinical deficiencies. This subject is also discussed.

As the authors explain, scientific interest in vitamins has waxed and waned over the years, and besides this, the exact mode of function of some of them has taken a long time to be unravelled. An example is vitamin D. In 1920, Mellanby established that there were two fat-soluble vitamins, one of which prevented and cured xerophalmia, the other which prevented and cured rickets. Both were present in cod liver oil. A few years later Dame Harriette Chick established that sunlight acting on the skin also prevented and cured rickets. It was not until 50 years later, however, that Kodicek and colleagues in Cambridge, and De Luca and co-workers in Madison, discovered the many biological mechanisms involved.

Knowledge about vitamins is still increasing, and there is still much to discover. If this book inspires some students to make an aspect of vitamins in medicine the subject of their research, the authors will be doubly rewarded.

Elsie M. Widdowson,
CH, CBE, FRS, DSc

Preface

The history of our knowledge of the substances that we now call 'vitamins' is like a detective story, the unravelling of which has involved people with different backgrounds – dietitians, nutritionists, pharmacists, clinicians, chemists and biochemists, to name but a few. Diseases caused by deficiencies of these micronutrients have been known from ancient times and have played a part in shaping human civilization. They have decimated armies and ships' crews and have been responsible for the deaths of explorers. They are still a major cause of morbidity and mortality in the poorer nations where there is a high dependency on a few 'staple' foods. Overt deficiency diseases are found today in western societies only in certain vulnerable groups, such as ethnic minorities, old people, and among alcohol abusers and clinical patients. However, some nutritionists have long contended that subclinical deficiencies of some vitamins are more common and have a potentially harmful effect on health by lowering resistance to infections or increasing the toxicity of drugs and other xenobiotics. In contrast, it has been argued by some that the requirements of individuals vary widely and only by consuming large amounts of some vitamins can a state of 'optimum' health be achieved. The large intakes recommended, and the amounts required to rectify or prevent deficiency in seriously vulnerable people, such as those at risk of vitamin A deficiency in Africa, Asia and the Western Pacific, cannot be met from dietary sources. Knowledge of the chemistry of the vitamins has made it possible to synthesize them and to make them available in concentrated forms.

We have known for some time that excesses of the fat-soluble vitamins A and D are injurious to health and that excess vitamin A, in the mother, can interfere with the development of the fetus. But starting some 20 years ago there have been important and exciting developments that have revived interest in these minor nutrients. We now know that maternal deficiency of folic acid at the time of conception is associated with the birth of a fetus with a neural tube defect. At the other end of the life-span, we now know that oxidative damage caused by free oxygen radicals may be involved in the aetiology of cardiovascular disease and cancer. This has focused interest on the part that could be played in preventive medicine by those vitamins with antioxidant properties – ascorbic acid, α-tocopheral and β-carotene.

Our aim in writing this book has been to present a concise account of up-to-date knowledge of vitamins which will meet the needs of students reading for degrees in nutritional, pharmaceutical and medical sciences. Sufficient biochemistry has been included to provide a satisfactory outline of the mechanisms of action of the various vitamins. We are conscious that our knowledge of vitamins is at present in a dynamic state. We hope that students using this text will catch something of the present excitement in relation to what were first called 'accessory food factors'.

The contribution of Darcey Abel in drawing the structural figures is recognized with thanks and gratitude. We are also indebted to both Linda Callan and Bev Cote for typing the manuscript with care and patience.

General Introduction 1

Vitamins are a heterogeneous group of organic substances which are essential food factors of very high biological activity and required in exceedingly small amounts for the growth and maintenance of normal cell and organ functions. They need to be supplied in the diet either because the body cannot make them, or cannot do so in the amounts that are required for health. Vitamins are generally classified into two groups, the fat-soluble vitamins including vitamins A, D, E, and K and the water-soluble vitamins including vitamin C and a series of B-vitamins, which include B_1, B_2, B_3, B_6, pantothenic acid, biotin, folic acid and B_{12}. Broadly speaking, the water-soluble vitamins, owing to their chemical constitution and configuration, are able to act as coenzymes accelerating enzymatic reactions, often as carriers of some particular chemical grouping. On the other hand, fat-soluble vitamins appear to function as integral parts of cell membranes, and these vitamins are more akin to the steroid hormones than to the vitamins that function as enzymic cofactors. Diseases due to vitamin deficiencies have probably been known from antiquity. Some, such as scurvy, pellagra and beriberi, have decimated armies, ships' crews, Arctic explorers and even nations. It has been known since the 1800s that these diseases can be cured by diet, but isolation and identification of the metabolically essential nutrients occurred only with the development of biochemistry. The first 40 years or so of this century has been described as the 'vitamin era.'

Although we know that deficiencies of particular vitamins are associated with certain clinical manifestations, the definitive biochemical aetiology of some of these manifestations is still incomplete. The reason for this is that the natural occurrence in man of disease due to deficiency of a single nutrient is rare. In addition, the results of deficiency may be non-specific, with a similar clinical presentation resulting from deficiency of more than one vitamin.

Diseases due to vitamin deficiencies do not develop quickly. Transition from adequacy to deficiency is a gradual rather than an abrupt phenomenon (Table 1.1). Dietary deficiency first causes a decrease in the amount of vitamin in transit, that is, in the circulation. As this deficiency persists, the level in the cells

Table 1.1. Stages in the development of a vitamin deficiency.

I	Deficiency: primary (low dietary intake) and secondary (malabsorption)
II	Reduced plasma levels: consequent on I or on failure to synthesize (or release) a carrier protein
III	Depletion of tissue levels and body stores: may reflect decreased availability of vitamin in circulation, increased utilization or increased losses
IV	Metabolic lesions: may occur as a result of reduced availability of vitamin (for whatever reason) or as a response to a metabolic challenge
V	Covert deficiency: clinical signs become evident only as a result of stress of some kind or when dietary intake is only of threshold adequacy level
VI	Overt clinical deficiency: associated with breakdown of metabolic processes resulting in pathological changes and the occurrence of signs and symptoms

Modified from Dickerson and Williams, 1990.

falls and metabolic processes involving the vitamin are ultimately impaired. However, these changes do not occur at a uniform rate throughout all the tissues of the body because some retain particular vitamins more strongly, whilst other tissues, by virtue of their metabolic peculiarities, are sensitive to quite small changes in vitamin availability. The sensitivity of the brain to thiamin (vitamin B_1) deficiency by virtue of its high consumption of glucose is an example of this latter phenomenon.

History

Animal experimentation has made a considerable contribution to the elucidation of vitamin biochemistry and nutrition. Similarities between diseases in animals and man provided the stimulus and key to their scientific investigation. In general, biological and nutritional observation preceded the isolation and identification of the chemical nature of vitamins. The existence of 'accessory food factors' was postulated by Gowland Hopkins (Hopkins, 1912) and in the same year Casimir Funk isolated and identified the substance that prevented human beriberi. Funk called the substance 'beriberi vitamine'. It later became known as vitamin B_1, and still later, when its chemical constitution had been determined, as thiamin. The name 'vitamine' (i.e. vital amine) was applied by Funk to other substances, e.g. the scurvy vitamine, that were later shown not to be amines. It was because of this that Drummond suggested in 1920 that the accessory food factors should be called 'vitamins' and given a letter until such time as their chemical constitution had been determined. Rudolph Peters and his colleagues later described the biochemical role of the pyrophosphate form of thiamin in carbohydrate metabolism (Peters and Thompson, 1934).

Table 1.2. Landmarks in the isolation and characterization of some vitamins.

1912	Vitamin B_1 – Thiamin (Funk)
1922	Vitamin D – Cholecalciferol (McCollum *et al.*)
1930	Vitamin A – Retinol (Karrer*)
1932	Vitamin C – Ascorbic acid (Svirbely and Szent-Gyorgy*; King* and Waugh)
1933	Vitamin B_2 – Riboflavin (Kuhn *et al.*)
1936	Biotin (Kogl and Tonnis)
1937	Vitamin B_3 – Niacin (Elvehjem *et al.*)
1938	Vitamin B_6 – Pyridoxine (Lepkovsky; Keresztesy and Stevens; Gyorgy)
1939	Vitamin K – Phylloquinone (Doisy* *et al.*; Dam* *et al.*)
1939	Pantothenic acid (Williams *et al.*)
1948	Vitamin B_{12} – Cyanocobalamin (Rickes *et al.*; Smith and Parker)

*Nobel Prize awardees.

Table 1.3. Coenzyme forms of vitamins of the B group.

Vitamin	Coenzyme Form
Thiamin	Thiamin pyrophosphate
Riboflavin	Riboflavin 5′ – monophosphate (FMP) Flavin–adenine dinucleotide (FAD)
Nicotinic acid	Nicotinamide adenine dinucleotide (NAD) Nicotinamide adenine dinucleotide phosphate (NADP)
Pyridoxine	Pyridoxal phosphate Pyridoxamine phosphate
Vitamin B_{12}	5′-Deoxyadenosylcobalamin and analogous derivatives
Folic acid	A number of derivatives of folinic acid, N^5-formyl-tetrahydrofolic acid: leucovorin (IX)
Pantothenic acid	Coenzyme A (CoA)
Biotin	Biotinyl enzyme. Three types known: Carboxybiotinyl carboxylases, enzyme transcarboxylases and decarboxylases

Research on vitamins captivated the attention of scientists in different parts of the world and there was considerable competition in respect of different stages of their history. It is not unusual therefore to find similar work being published almost simultaneously in at least two places. An example of this from more recent times was the isolation of vitamin B_{12} which was reported by Rickes and co-workers from the US (1948) and in the UK by Smith and Parker (1948). Table 1.2 shows some other milestones in the history of vitamins.

Table 1.4. Alternative names for folic acid and vitamin K.

	IUNS	IUPAC-IUB
Folic acid	Folic acid Folic acid glutamate Folic acid diglutamate	Pteroylglutamic acid Pteroyldiglutamic acid Pteroyltriglutamic acid
Vitamin K_1	Phytylmenaquinone Phenylmenaquinone-6 (or 7,8)	Phylloquinone Menaquinone-6 (or 7,8)

The chemical form in which a vitamin is usually described is often not the form in which it is metabolically active, and some vitamins are active in a variety of forms depending on the particular reaction in which they are involved. Thus, the active form of thiamin is thiamin pyrophosphate; pyridoxine functions as a coenzyme as pyridoxal phosphate and as pyridoxamine phosphate (Table 1.3). The different forms in which a vitamin such as vitamin B_6 is biologically active are known as 'vitamers' and the overall name, i.e. vitamin B_6, is called the 'generic descriptor'.

Two separate bodies – the International Union of Nutritional Sciences (IUNS) and the International Union of Pure and Applied Chemistry – International Union of Biochemistry Commission on Biological Nomenclature (IUPAC–IUB) – have been involved in deciding the systematic nomenclature of vitamins. Although it seems likely that the IUPAC–IUB nomenclature is accepted most widely, the differing nomenclature for folic acid and vitamin K (Table 1.4) may still be seen.

Factors Affecting Vitamin Requirements

The amounts of energy, nitrogen and the various minerals needed to maintain the metabolic processes of the human body in a steady state can be determined with some degree of precision in both health and sickness. The amounts of these materials entering the body in the diet, and the amounts leaving by the various routes, can be measured and the difference between intake and output, expressed as a 'balance', is the change in the body content of the particular constituent. Providing that time has been allowed for adaptation to occur to any changes in the composition of the diet, that the diet is nutritionally adequate in all respects and, further, that the measurements are made for a sufficient length of time, an estimate can be made of the body's 'requirements' of the corresponding nutrient to maintain 'balance.'

The determination of man's requirements for the different vitamins is a much more difficult undertaking. The main reason is that a vitamin can be broken down to many metabolic products which escape in the urine, faeces, and sweat, and that these metabolites are often undetectable by the conventional

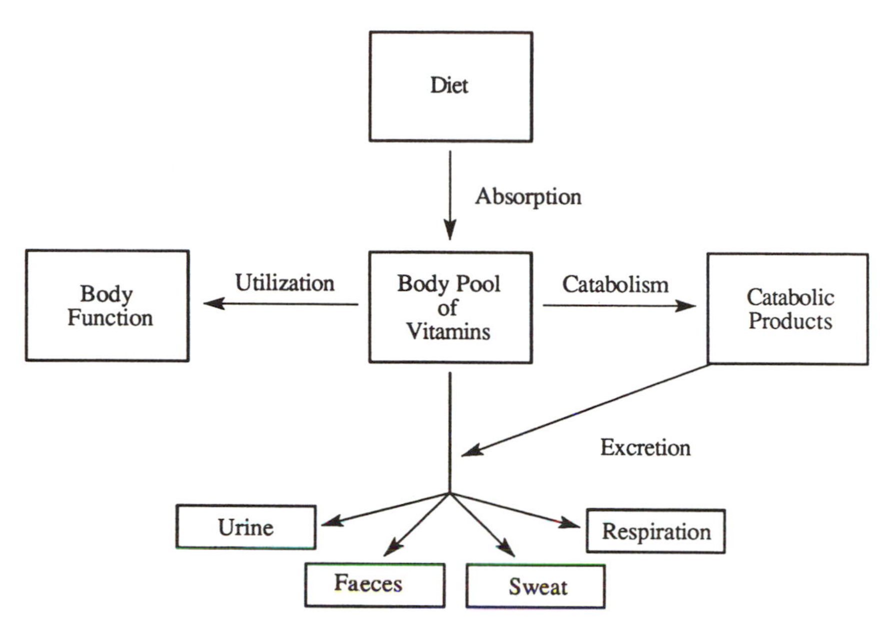

Fig. 1.1. Factors which determine the body pool size of a vitamin.

assay procedures (Fig. 1.1). One approach to the problem would be to measure the body content of a particular vitamin and then to assess, with a radioactively labelled tracer, the amount of the vitamin necessary to maintain the body pool size. This has been done for vitamin C (Kallner *et al.*, 1979). When adjustment was made for the proportion of the intake absorbed, the method suggested a requirement of about 100 mg day^{-1}. Earlier estimates in the UK were based on the experimental depletion of human volunteers with the production of clinical scurvy. Subsequent supplementation showed that 10 mg of the vitamin per day would cure the disease and addition of a further 20 mg as a 'safety factor' was judged to give a recommended allowance of 30 mg day^{-1}. This method can be criticized on the grounds that it places an over-emphasis on the role of vitamin C in the metabolism of connective tissue and does not take into account evidence derived from studies involving other metabolic pathways. However, it seems that the evidence from these studies is not sufficiently in agreement for it to be used as a basis for estimating requirements. The UK Committee on Medical Aspects of Food Policy (Department of Health, 1991) has recently decided on a Reference Nutrient Intake of 40 mg day^{-1} for vitamin C.

Vitamins circulate in the plasma and are also present in cells. Blood measurements can therefore be made directly, or indirectly, by measuring the stimulation of the activity of an appropriate enzyme for which the vitamin acts as a cofactor. Examples of vitamins whose status is measured in this way are thiamin and riboflavin using red cell transketolase and glutathione reductase respectively (see Chapters 2 and 3). Status with respect to other vitamins can be assessed indirectly by measuring the excretion of a metabolite after giving a loading dose of a substance whose metabolism is facilitated by the vitamin. Examples of this technique are the assessment of pyridoxine status by measurement of tryptophan metabolites after a loading dose of the amino acid (Chapter 5), and of folic acid by determination of the excretion of formiminoglutamic acid (FIGLU) after a loading dose of histidine (Chapter 8). Such measurements can then form the basis of assessments of requirements and of recommended allowances.

Factors Affecting Vitamin Status

The requirements for all vitamins rise during growth to maturity. In the mature organism requirements tend to be slightly higher in men than in women due to their greater cell mass and greater intakes of energy and protein. Pregnancy, and particularly lactation, impose additional demands. There is no good evidence that requirements change substantially in old age, although many old people are exposed to the risk of deficiency for a variety of reasons. Insufficient food, poor food choice, conditions of storage, preparation and cooking of food all contribute to the likelihood of vitamin depletion. In addition, a number of the factors discussed below which are connected with disease are also more likely to occur in older people. In the elderly, vitamin deficiencies may be both a cause and a consequence of ill-health.

Nutrients, and for our present discussion, vitamins, are not consumed as pure chemicals, but as constituents of foods. If the variety of foods is restricted, food choice becomes an important determinant of nutritional adequacy. This is particularly true for those nutrients, like some vitamins, that are present in high concentrations in relatively few foods. A diet lacking green vegetables and based on chicken or hamburgers and chips may eventually cause megaloblastic anaemia due to folic acid deficiency. A proportion of the niacin present in maize is in a bound form which is not readily available for absorption. Thus, a dietary staple of maize is inclined to lead to niacin deficiency (pellagra) unless steps are taken during processing to liberate the bound vitamin or to supplement the diet with an alternative source of niacin.

Status with respect to some vitamins is affected by the intake of other nutrients. Vitamin A status, as judged by the concentration of retinol in the plasma, is affected by the intake of energy and protein. Retinol circulates in the plasma bound to a specific globulin, retinol binding protein (RBP), which in turn is bound in a complex with transthyretin (see Chapter 11). RBP has a short half-life and its synthesis in the liver is therefore very sensitive to protein–energy nutrition. A low food intake results in decreased RBP synthesis and a low plasma retinol concentration, regardless of how much retinol may be present in hepatic stores. Furthermore, the release of RBP from the liver is affected by the availability of zinc (see Chapter 11). It follows therefore that a low plasma retinol concentration alone does not indicate a low vitamin A status unless protein–energy malnutrition and zinc deficiency have been excluded.

Vitamin C has a sparing effect on a number of vitamins, including thiamin, riboflavin, pantothenic acid, biotin, folic acid, cyanocobalamin, retinol and α-tocopherol. Most of these observations have been made in rats, but a thiamin-sparing effect has been demonstrated in female geriatric patients who had evidence of thiamin deficiency as determined by the thiamin pyrophosphate (TPP) stimulating effect on red cell transketolase activity (Basu *et al.*, 1976). Oral administration of vitamin C (1 g day^{-1}) for one week resulted in a reduction of the TPP effect, that is an improvement in thiamin status (see Chapter 2).

Drugs may decrease the absorption of vitamins, increase requirements or influence tissue uptake. Some oral contraceptive agents reduce riboflavin status, increase the requirement for vitamin B_6 and reduce status with respect to folic acid, vitamin B_{12} and ascorbic acid. Some anticonvulsants, particularly phenytoin (dilantin) and phenobarbitone, increase the requirement for folic acid by stimulating the synthesis of cytochrome P450, a haemo-protein involved in drug metabolism, in which folic acid plays a part. These same drugs also increase the requirement for vitamin D and, in the absence of sufficient exposure to sunlight for endogenous synthesis, may induce rickets in children and osteomalacia in adults. Phenothiazines may decrease riboflavin utilization because some of these drugs can act as riboflavin antagonists. Laxatives decrease the absorption of fat-soluble vitamins and β-carotene. Intestinal antibiotics decrease the synthesis of vitamin K and the utilization of folic acid. Anti-inflammatory agents, such as aspirin, reduce the intestinal absorption of vitamin C and its entry into leucocytes. Drug–vitamin interactions are treated in more detail in Chapter 18.

In medical conditions in which there is malabsorption of fat (i.e. steatorrhoea) the absorption of the fat-soluble vitamins A, D, E and K is reduced to a degree which roughly correlates with the severity of the steatorrhoea. The nature of the disease responsible seems of less importance than the severity of the steatorrhoea. In contrast, the malabsorption of folic acid and vitamin B_{12} varies with the aetiology of the steatorrhoea (see Chapters 8 and 9, respectively).

Any disease that induces anorexia will cause reduced vitamin status because of an overall effect on food intake. Other diseases, such as cancer, may cause deficiency because they increase the need for some vitamins (see Chapter 17).

There are conditions where symptoms of vitamin deficiency can occur even when the dietary intake of a vitamin is adequate, at least in terms of a recommended intake. In these disorders there is an inherited, or induced, defect of vitamin utilization such that abnormally large amounts of a vitamin are necessary to prevent the occurrence of symptoms or to cure an existing disorder. These diseases are collectively known as 'vitamin dependency disorders' and often involve an abnormality in the structure of the apoenzyme such that its affinity for the coenzyme is reduced. There are a number of these disorders in which there is a dependency on pyridoxine, and in which symptoms are caused by the low activity of a specific enzyme (see Chapter 19). Recommended daily allowances for vitamins are designed to cover the requirements of about 95% of the population. The optimal requirements of individuals may vary quite considerably, and it is likely that a substantial amount of this variability is due to variations in enzyme structure and hence in the affinity of apoenzymes for coenzymes.

Recommended Daily Allowances (RDAs)

The RDA has been defined as 'the average amount of a nutrient which should be provided per head in a group of people if the needs of practically all members of the group are to be met'. The tables of RDAs published periodically by government departments in different countries are decided by committees. This not only gives scope for varying opinions but also, in some countries, for the values to be subject to political considerations. These factors together go some way to account for the sometimes quite large differences in the values for RDAs published by different countries (Table 1.5). The ideal intake for any nutrient is

Table 1.5. Daily recommended intakes of vitamins for males in some European countries, the USA, Canada and by the World Health Organization (WHO).[1]

	Vitamin A (retinol equiv. μg)	Thiamin (mg)	Riboflavin (mg)	Niacin (mg equiv.)	Ascorbic acid (mg)	Vitamin D[2] (μg)
Denmark	1000	1.4	1.6	18	45	2.5
Germany (FRG)	850	1.6	2.0	15	75	2.5
Netherlands	850	1.2	1.7	–	50	–
Sweden	900	1.4	1.7	18	60	10
UK (1991)	700	1.0	1.3	17	40	2.5[4]
US (1980)	1000	1.4	1.6	18	60	5
Canada	1000	1.2[3]	1.5	21[3]	40	2.5[5]
WHO (1990)	600	–	–	–	30	2.5
USSR	1500	1.8	2.4	20	75	12.5

[1] Modified from Williams and Dickerson (1986).
[2] Cholecalciferol.
[3] Assuming the daily total energy intake is 2000 kcal.
[4] 2.5 μg for elderly housebound persons.
[5] 5.0 μg for elderly.

the ‘optimum’ amount for ‘health’. This latter term is difficult to define. It is clearly not synonymous with the absence of disease. ‘Health’ is a condition in which all the body systems are capable of working at optimal levels of efficiency. We are not able to define the optimal intake of any vitamin. In no other aspect of nutrition has so much profit been made out of ignorance.

RDAs are often wrongly used as a basis for assessing the adequacy of the calculated intakes of individuals. If they are used in this way it is probable that only intakes less than two-thirds of the RDA suggest a low intake. RDAs are to be used for quite specific purposes: the assessment of dietary surveys, the planning of diets, planning food supplies, nutritional labelling and the calculation of nutrient density.

References

Basu, T.K., Jenner, M. and Williams, D.C. (1976) The ‘Thiamin-sparing’ effect of ascorbic acid. *Nutrition and Metabolism* 20, 425–431.

Dam, von H., Geiger, A., Glavind, J., Karrer, P., Karrer, W., Rothschild, E. and Salomon, H. (1939) Isolierung des Vitamins K in hoch gereinigter Form. *Helvetica Chimica Acta* 22, 310–312.

Department of Health (1991) Dietary reference values for food energy and nutrients for the United Kingdom. *Report on Health and Social Subjects, No. 41*. HMSO, London.

Dickerson, J.W.T., and Williams, C.M. (1990) Vitamin-related disorders. In: Cohen, R.D., Lewis, B., Alberta, K.G.M.M. and Denman, A.M. (eds) *The Metabolic and Molecular Basis of Acquired Disease*. Baillière Tindall, London, pp. 634–669.

Doisy, E.A., MacCorquodale, D.W., Thayer, S.A., Binkley, S.B. and McKee, R.W. (1939) The isolation and synthesis of vitamin K1. *Science* 90, 407–409.

Drummond, J.C. (1920) The nomenclature of the so-called accessory food factors (vitamins). *Biochemical Journal* 14, 660.

Elvehjem, C.A., Madden, R.J., Strong, F.M. and Woolley, D.W. (1937) Relation of nicotinic acid and nicotinic acid amide to canine black tongue. *Journal of the American Chemical Society* 59, 1767–1768.

Funk, C. (1912) The etiology of the deficiency diseases. *Journal of State Medicine* 20, 341.

Gyorgy, P. (1938) Crystalline vitamin B_6. *Journal of the American Chemical Society* 60, 983–984.

Hopkins, F.G. (1912) Feeding experiments illustrating the importance of accessary factors in normal dietaries. *Journal of Physiology* 44, 425–560.

Kallner, A., Hartman, D. and Hornig, D. (1979) Steady state turnover and body pool of ascorbic acid in man. *American Journal of Clinical Nutrition* 32, 530–539.

Karrer, P. (1930) Uber die Konstitition des Lycopens und Carotins. *Helvetica Chimica Acta* 13, 1084.

Keresztesy, J.C. and Stevens, J.R. (1938) Crystalline vitamin B_6. *Proceedings of the Society of Experimental Biology and Medicine* 38, 64–65.

King, C.G. and Waugh, W.A. (1932) The chemical nature of vitamin C. *Science* 75, 357–358.

Kogl, F. and Tonnis, B. (1936) Uber das Bios-Prolem. Darstellung von krystallisiertem Biotin aus Ergelb. 20. Mitteilung uber pflanzliche. Wachstumsstoffe. *Zeitschrift Physiologie Chemie* 242, 43–73.

Kuhn, R., Gyorgy, P. and Wagner-Jauregg, T. (1933) Uber eine neue Klasse von Naturfarbstoffen (Vorlaufige Mitteilung). *Berlin* 66, 317–320.

Lepkovsky, S. (1938) Crystalline Factor 1. *Science* 87, 169–170.

McCollum, E.V., Simonds, N., Becker, J.E. and Shipley, P.G. (1922) Studies on experimental rickets XX1. An experimental demonstration of the existence of a vitamin which promotes calcium deposition. *Journal of Biological Chemistry* 53, 293–312.

Peters, R.A. and Thompson, R.H.S. (1934) Pyruvic acid as an intermediary metabolite in the brain tissue of avitamous and normal pigeons. *Biochemical Journal* 28, 916–925.

Rickes, E.L., Brink, N.G., Koniuszy, F.R., Wood, T.R. and Folders, K. (1948) Crystalline Vitamin B_{12}. *Science* 107, 396–397.

Smith, E.L. and Parker, L.F.J. (1948) Purification of the anti-pernicious anaemia factor. *Biochemical Journal* 43, viii–ix.

Svirbely, J.L. and Szent-Gyorgy, A. (1932) Hexuronic acid as the antiscorbutic factor. *Nature* 129, 576.

Williams, C.A. and Dickerson, J.W.T. (1986) Energy and nutrient requirements. In: Heatley, R.V., Losowsky, M.S. and Kelleher, J. (eds) *Clinical Nutrition in Gastroenterology*. Churchill Livingstone, Edinburgh, pp. 18–33.

Williams, R.J., Weinstock, H.H., Rohrmann, E., Truesdail, J.H., Mitchell, H.K. and Meter, C.E. (1939) Pantothenic acid III. Analysis and determination of constituent group. *Journal of the American Chemical Society* 61, 454–457.

Thiamin (Vitamin B_1) 2

The thiamin molecule consists of a pyrimidine ring and a thiazole moiety linked by a methylene bridge (Fig. 2.1). The most common commercially available form is thiamin hydrochloride (mol. wt 337). It is a colourless crystalline substance, readily soluble in water (1 g ml^{-1}), and has a faintly yeasty odour with a slight bitter taste. In the absence of ultraviolet light and moisture the thiamin salt is stable towards atmospheric oxygen even when warm. Acid solutions are also fairly stable but in alkaline or neutral solution it decomposes.

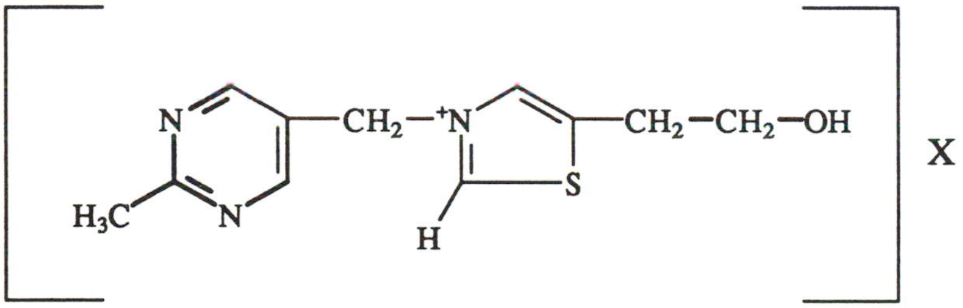

$X = Cl^- \text{ or } HCl$

Fig. 2.1. Structure of thiamin hydrochloride.

Occurrence in Foods

Thiamin is present in most plant and animal tissues, but its ultimate source is the plant which, unlike mammals, has the ability to synthesize thiamin. In plants the pyrimidine and thiazole residues of thiamin are formed independently, and then joined together to form the thiamin molecule as shown in Fig. 2.2.

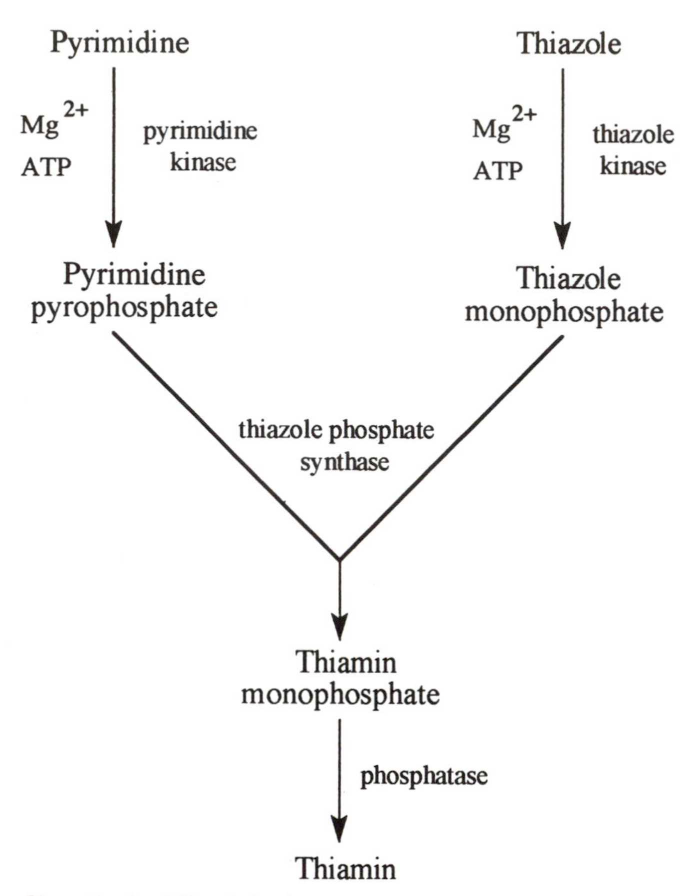

Fig. 2.2. Biosynthesis of thiamin in plants.

The primary dietary sources of thiamin are unrefined cereal grains, nuts, legumes, organ meats and lean cuts of pork. Certain foods, such as flour, bread, corn and macaroni products, are enriched with the synthetic vitamin. This has been a significant source of thiamin in the diet for people especially of the developed countries, and has helped eliminate thiamin deficiency. Generally speaking, fruits, chicken, and milk are poor sources. Thiamin occurs in a phosphorylated form in animal foods while it is present in its free form in plant foods and enriched products.

Thiamin Losses

Because of its water solubility, thiamin is readily leached out of foods during boiling and blanching. Any method of preparation that minimizes the length of time a food is in contact with water and the amount of surface area exposed will decrease thiamin loss. The use of rapidly boiling water in cooking vegetables will increase the amount of oxygen in contact with the food; this will enhance the loss of thiamin. In addition, the practice of adding baking soda to the water used for cooking green vegetables to make them a brighter green destroys thiamin. It is also destroyed by ultraviolet irradiation. Thiamin-degrading enzymes, thiaminases, present in uncooked freshwater and shell fish destroy thiamin by displacing the pyrimidine methylene group with a nitrogenous base or SH-compound. In addition, heat-stable thiamin antagonists may occur in tea, coffee, betel nuts, red cabbage and Brussels sprouts. These include *O*- and *p*-hydroxy polyphenols, which oxidize the thiazole ring to yield thiamin disulphide, the form in which thiamin is not absorbed.

Absorption and Metabolism

Absorption of thiamine can be both active and passive, depending upon the concentration of the vitamin (Thompson and Leevy, 1972). At low concentrations ordinarily found in food, thiamin is absorbed by a Na^+- dependent active process. This absorption is greatest in the jejunum and ileum. At high concentrations of thiamin attained by supplemental intake, it is absorbed by passive diffusion. Almost no absorption of thiamin occurs in the stomach and distal small intestine regions. The vitamin formed by the flora of the large intestine is not absorbed.

Orally administered thiamin is rapidly absorbed, distributed, and excreted with a half-life of one hour. An adult human body contains approximately 30 mg of thiamin with high concentrations found in the heart, liver, kidney, and brain. Fifty per cent of the body's thiamin is distributed throughout the skeletal muscle. The main excretion products are thiamin itself, thiamin carboxylic acid, 4-methyl-thiazole-5-acetic acid, 2-methyl-4-amino-5-formyl-amino-pyrimidine, and a number of hitherto unidentified metabolites. The higher the doses of thiamin administered, the more unchanged thiamin is excreted in the urine within 4–6 hours. Since thiamin has a high metabolic turnover rate and can only be stored in limited amounts in the human organism, it must be taken daily in quantities sufficient to maintain the tissue saturation state.

Thiamin Activation

The thiamin molecule is not biologically active. To be active as a coenzyme, it must be converted to its pyrophosphate (diphosphate) form. This conversion is catalysed by Mg^{2+} and ATP-dependent thiamin pyrophosphokinase, an enzyme found especially in the liver and brain. In thiamin pyrophosphate (TPP) the OH

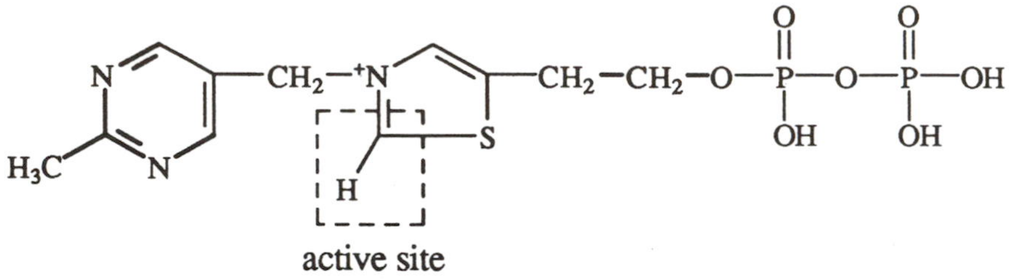

Fig. 2.3. Thiamin pyrophosphate (TPP) showing active site in the thiazole ring.

group in the thiazole ring side-chain is replaced by -O-PO(OH)-O-PO$(OH)_2$ (Fig. 2.3). The action of TPP is thought to occur at a chemically active site on the thiazole ring where a carbon bears a negative charge (carbon-2). Although the thiazole ring is the centre of phosphorylation and active site of the thiamin molecule, it should be remembered that the pyrimidine ring of the molecule is just as necessary for TPP to be active as a coenzyme.

Role of Thiamin Pyrophosphate (TPP)

Coenzyme function

TPP functions in several metabolic reactions in mammalian cells; it is actually thought to be a coenzyme in over 24 enzymes (Sauberlich, 1967). One of these reactions is oxidative decarboxylation where TPP is the coenzyme for the decarboxylases, the group of enzymes that remove a carboxyl group. These enzymes include pyruvate decarboxylase, α-ketoglutarate decarboxylase, and branched-chain amino acid decarboxylase. Because of the function of TPP to release CO_2 from the keto acids, it is commonly called co-carboxylase. Another known coenzyme function for TPP is the transfer of ketones (—C=O) from an α-keto sugar in a transketolase reaction in the hexose monophosphate shunt pathway of glucose metabolism.

Pyruvate decarboxylase

Pyruvate, a three-carbon keto acid, is formed in the cytosol during glycolysis, the anaerobic stage of glucose metabolism. To enter the aerobic stage of the metabolism in mitochondria, pyruvate must lose a carbon to form acetyl CoA, a two-carbon compound. This oxidative decarboxylation is accomplished by a multienzyme complex bound to the mitochondrial membrane called pyruvate dehydrogenase complex. It is composed of three enzymes which include pyruvate decarboxylase, dihydrolipoyl transacetylase and dihydrolipoyl dehydrogenase; these are critical to the extraction of energy from foods.

In TPP the carbon atom at 2 position, the active site between the nitrogen and sulphur atoms in the thiazole ring, is highly acidic. It ionizes to form a carbanion, which is stabilized by the positively charged nitrogen. The carbanion combines with the carbonyl group of keto acids, such as pyruvic acid. The removal of CO_2 is followed rapidly by reactions catalysed by pyruvate decarboxylase. The overall reaction requires lipoic acid, FAD, NAD^+ and, finally, coenzyme A to which the two carbons become attached to form acetyl CoA, which then enters into the tricarboxylic acid (TCA) cycle for complete combustion by an aerobic system. The overall reaction may be summarized as:

$$\text{Pyruvate} + \text{CoASH} + \text{NAD}^+ \xrightarrow[\text{lipoic acid}]{\text{TPP}} \text{Acetyl CoA} + \text{CO}_2 + \text{NADH H}^+ \qquad (2.1)$$

More detailed steps for the oxidative decarboxylation of pyruvate involving pyruvate dehydrogenase complex are illustrated in Fig. 2.4. Thiamin in its coenzyme form catalyses the production of acetyl CoA, the link between carbohydrate, fat, and protein metabolism.

α-Ketoglutarate decarboxylase

Another TPP dependent enzyme, α-ketoglutarate decarboxylase, acts within the TCA cycle to convert α-ketoglutarate to succinyl CoA. This enzyme functions similarly to pyruvate decarboxylase and removes a carbon from the five-carbon compound, α-ketoglutarate. The overall reaction is shown in Fig. 2.5.

Branched-chain α-ketoacid decarboxylase

TPP is also needed as the coenzyme for branched chain α-keto acid decarboxylase which prepares the branched chain amino acids, leucine, isoleucine and valine, for synthesis of fatty acids if there is a surplus of these amino acids or for use as energy if there is a shortage of carbohydrate. In this process, these amino acids are first converted to their respective α-keto acids following deamination (Fig. 2.6). Like pyruvate and α-ketoglutarate these keto acids are then changed to acyl CoA thioesters following decarboxylation under the influence of α-keto acid decarboxylase. The thioester from leucine enters the TCA cycle in acetyl CoA, and those from the isoleucine and valine enter in succinyl CoA.

A genetically defective version of the branched-chain α-keto acid dehydrogenase complex is associated with maple syrup urine disease, so named because of the odour of the branched chain α-keto acids that accumulate. Sometimes individuals with this inborn error of metabolism can be helped by supplements of thiamin. The binding of TPP is believed to change the shape of the complex and increase its stability (Haas, 1988).

Transketolase

Another coenzymatic role of TPP is in activating transketolase, a key enzyme in the hexose monophosphate shunt pathway. This pathway is not so much for energy production, as for production of pentoses for RNA and DNA synthesis and for NADPH for biosynthesis of fatty acids. In transketolase reactions a

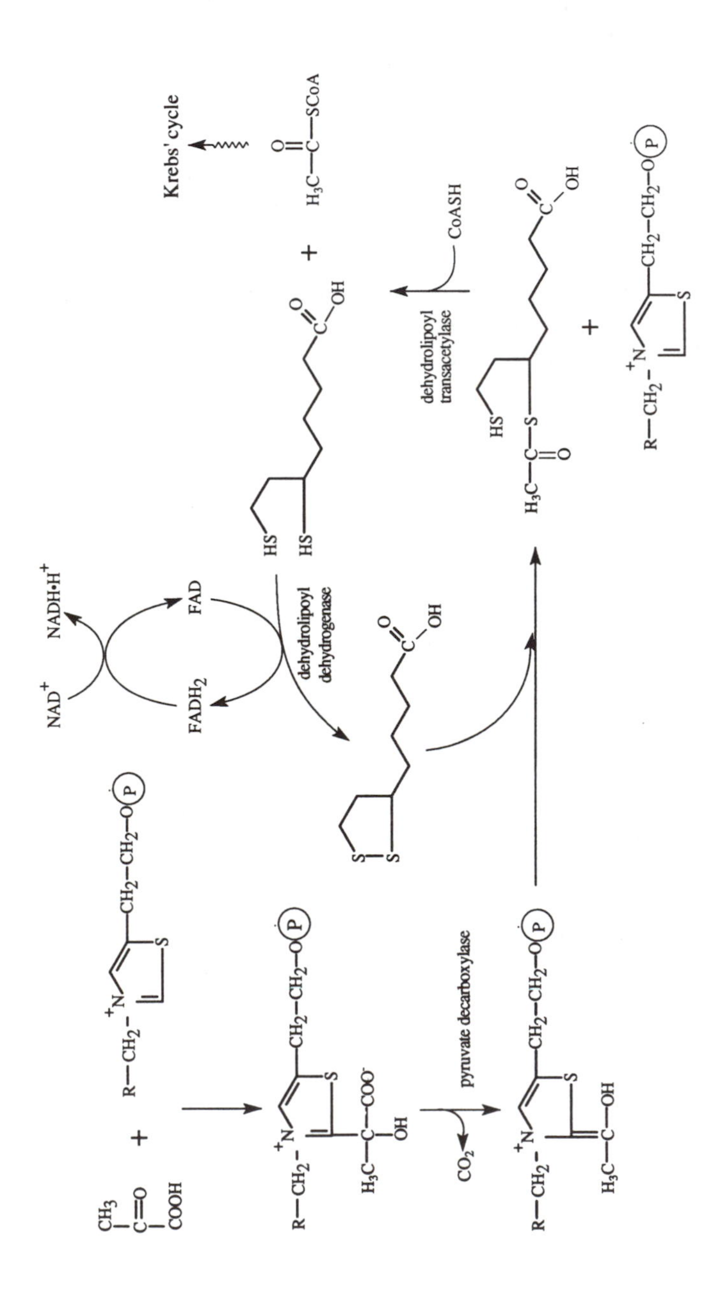

Fig. 2.4. Oxidative decarboxylation of pyruvate.

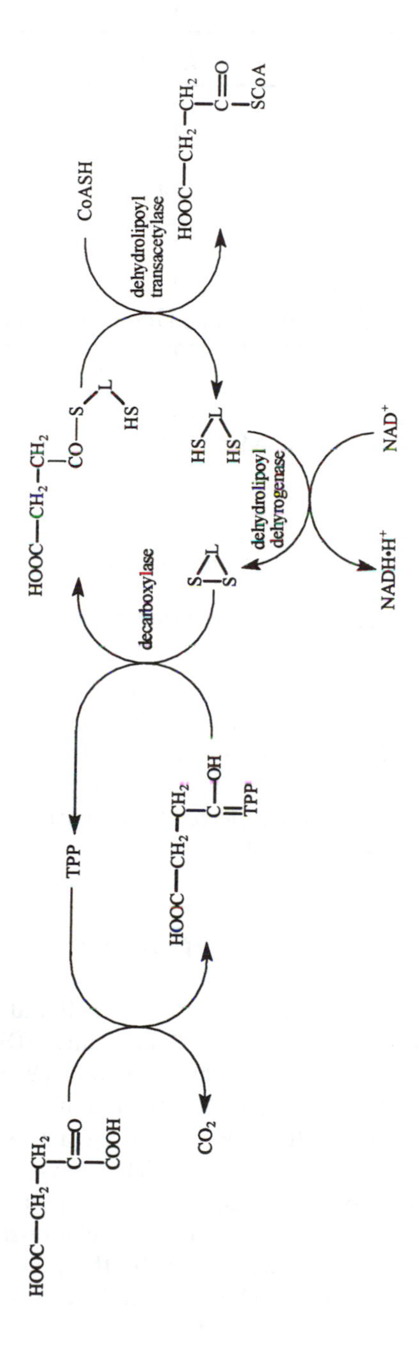

Fig. 2.5. Oxidative decarboxylation of α-ketoglutarate (TPP, thiamin pyrophosphate; S–S L, lipoic acid).

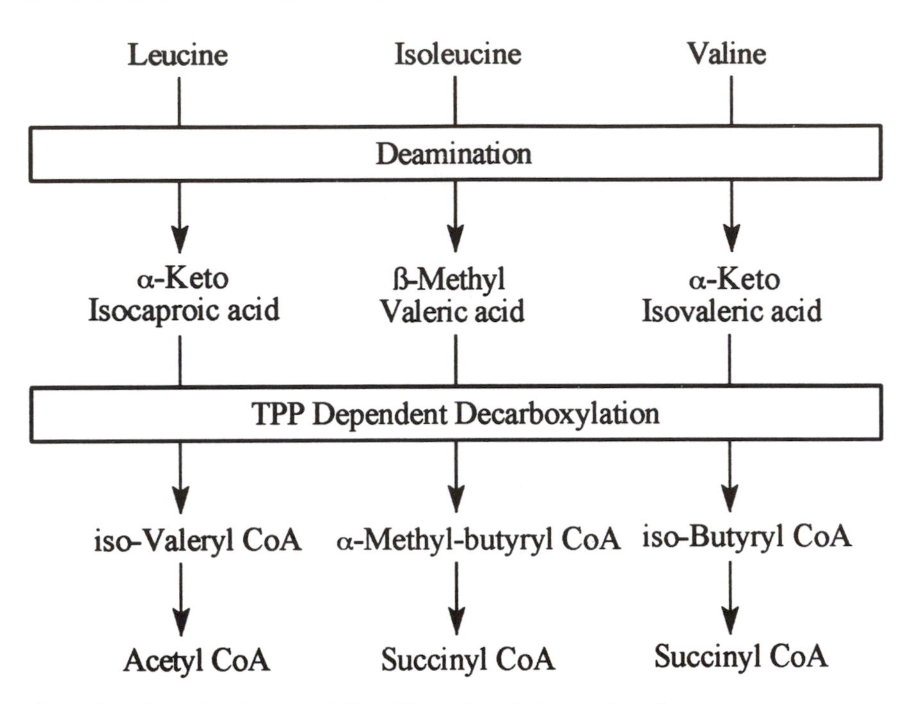

Fig. 2.6. Oxidative decarboxylation of branched-chain α-ketoacids.

two-carbon unit of a ketose is transferred to the aldehyde carbon of an aldose sugar. Thus, xylulose-5-phosphate is cleaved between carbons 2 and 3 to form glyceraldehyde-3-phosphate and an intermediate, 2-(1,2-dihydroxyethyl)-TPP (Fig. 2.7). Subsequently the 1,2-dihydroxyethyl group is set free from its complex with TPP, and is then transferred to the carbon 1 of ribose-5-phosphate yielding sedoheptulose-7-phosphate.

Other functions

Apart from its coenzyme function thiamin is believed to have an independent role in nerve conduction. Thiamin as diphosphate (TDP) or triphosphate (TPP) appears to be present in neural membranes (Haas, 1988); these phosphorylated thiamins disappear from the brains of patients suffering from subacute necrotizing encephalomyelopathy (Cooper and Pincus, 1976). In addition to reduced nerve impulse transmission, thiamin deficiency-associated impaired nerve function may be due to a lack of energy and a decreased amount of acetylcholine, the synthesis of which requires TPP. The relationship between thiamin and function of the brain is further evident by the prevalence of cerebrocortical necrosis (CCN) in ruminants where availability of thiamin is blocked by the action of a thiaminase with anti-thiamin activity (Edwin and Lewis, 1971).

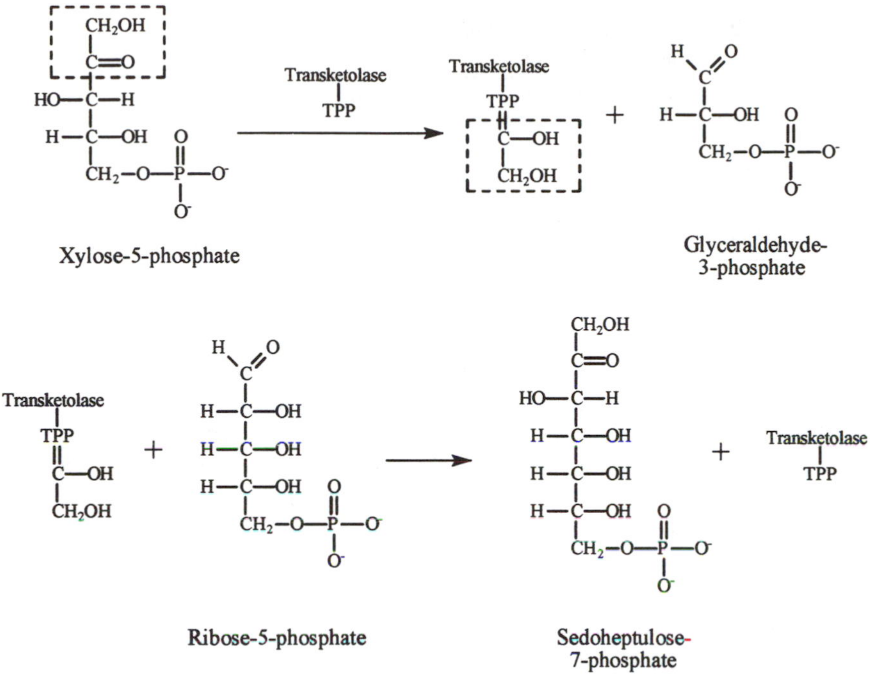

Xylose-5-phosphate

Transketolase
TPP

Glyceraldehyde-3-phosphate

Ribose-5-phosphate

Sedoheptulose-7-phosphate

Fig. 2.7. Hexose monophosphate shunt pathway involving thiamin pyrophosphate (TPP).

Thiamin Deficiency

An early stage of thiamin deficiency is characterized by reduced urinary excretion of the vitamin. A biochemical phase is seen within 10 days when erythrocyte transketolase activity (Table 2.1) is depressed with a positive 'TPP effect' of about 15%.

Thiamin deficiency usually occurs along with deficiencies of other B-vitamins. It is therefore difficult to attribute specific symptoms of early clinical signs to thiamin deficiency. However, in about 21–28 days of dietary inadequacies, symptoms such as tiredness, emotional instability, irritability, depression, and even fall in body weight, have been observed. One explanation for these manifestations may be that a lack of thiamin causes a complete failure to provide energy to the body cells from food (Fig. 2.8). In early days of thiamin deficiency, the elasticity of the wall of the gastrointestinal tract is also decreased to the point that normal gastric mobility is diminished, the colon becomes distended, and constipation results. Consequently, there is a tendency to suffer from loss of appetite, sometimes associated with nausea and vomiting.

Prolonged intake of diets low in thiamin will eventually lead to clinical signs referred to as the disease beriberi. It has been suggested that often the active,

Table 2.1. Thiamin pyrophosphate stimulating effect of transketolase activity (TPP effect) and thiamin nutritional state.

Thiamin status coefficient[1]	TPP effect[1]	Activity
Adequate	15	1.00–1.15
Marginal	16–20	1.16–1.20
Deficient	> 20	> 1.20

[1]Interdepartmental Committee on Nutrition for National Defence (ICNND), 1963.

stronger, or supposedly slightly better nourished members of a poor community, or the infants who are overfed, are generally the victims of beriberi (Thurnham, 1978). The underlying cause appears to be the greater energy intake from a food deficient in thiamin leading to a thiamin–energy ratio below requirements.

Beriberi is characterized by peripheral neuropathy (called dry beriberi) and/or cardiovascular problems (called wet beriberi). Dry beriberi is often accompanied by tingling, numbness, tenderness, weakness, and paralysis of the arms. Victims are generally thin and emaciated. In thiamin deficiency, less pyruvate is converted to acetyl CoA, which is a precursor for acetylcholine, a constituent of myelin sheaths. A deficiency of this important neurotransmitter causing myelin degeneration may be a responsible factor for the neuropathies in dry beriberi. Wet beriberi, on the other hand, is associated with oedema and cardiac disturbances. The accumulation of pyruvic acid (and lactic acid) resulting from failure of its oxidative decarboxylation, is believed to cause dilatation of peripheral blood vessels leading to vasodilatation. In this state, fluid may leak out through capillaries producing oedema. In the presence of vasodilatation, cardiac output is increased in order to maintain the circulation. Sudden death among patients with wet beriberi may result from myocardial exhaustion.

Overall, the symptoms of gross thiamin deficiency are peripheral neuropathies with sensitivity disorders, muscular weakness, and coordination disturbances. These signs are conditioned by the central nervous system, ataxia and pareses, together with psychic, gastrointestinal, and cardiovascular disorders accompanied by extensive oedema.

Situations Where Thiamin Deficiency Occurs

Thiamin intake is lowest in developing parts of the world where polished rice is the main staple food. The germ and bran fractions of the cereal grain are the richest sources of thiamin. The use of highly milled rice has traditionally been associated with thiamin deficiency; the underlying basis for this may be a combination of high carbohydrate intake from rice low in fat and protein, and a loss of thiamin with the husk due to the milling process.

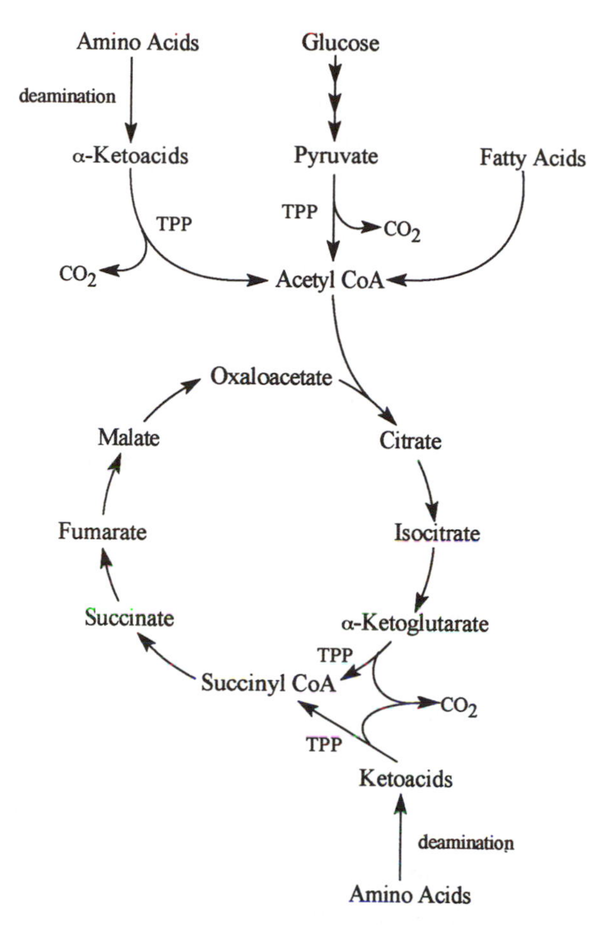

Fig. 2.8. Thiamin pyrophosphate and its active sites in energy metabolism.

In 1988, there was an outbreak of beriberi in the Gambia, killing 22 people (Tang *et al.*, 1989). The outbreak occurred three weeks before the beginning of the rainy season. There had been a recent trend away from local grains and toward imported milled rice, which was particularly low in thiamin. The affected individuals responded within a few days to 10 mg day^{-1} of thiamin. This was the first report of an outbreak of beriberi in Western Africa unrelated to alcoholism or to institutional feeding. Generally speaking, because of the improved social and economic factors as well as the availability of synthetic thiamin for treatment and prophylaxis, an overt deficiency of thiamin in rice-eating nations has greatly declined in recent decades.

In North America, Europe, and other industrialized nations, alcoholics are the only group in which thiamin deficiency with clinical signs is fairly commonly seen (Bonjour, 1980). The underlying basis is that an excessive intake of alcohol, owing to its high calorie content (7.0 kcal g^{-1}), increases the requirement of thiamin, while the food intake is considerably reduced. In addition, alcohol induces gastritis interfering with thiamin absorption, and damages hepatic functions resulting in an inhibition of thiamin activation (to TPP), which in turn affects its metabolic availability. In chronic alcoholism, thiamin deficiency may reach a point where it can contribute towards cardiomyopathy with dilatation of the right ventricle, and to polyneuropathy, Wernicke's encephalopathy and Korsakoff's psychosis (Leevy, 1982). Most alcohol-induced thiamin deficiency symptoms may disappear with thiamin supplementation. However, there may be evidence of irreversible cerebral damage, depending upon the extent of its deficiency and its duration. Wernicke's syndrome is a neurological disorder of acute onset, characterized by features that include double vision due to damaged cranial nerve (diplopia), rapid movement of eyeballs (nystagmus), and uncoordinated gait (ataxia). Korsakoff's psychosis, on the other hand, refers to an abnormal mental state in which the individual shows confabulation, experiences a loss of memory and an inability to learn new information.

Although an overt deficiency of thiamin is relatively less common except in chronic alcoholics, there is considerable evidence of subclinical deficiency of this vitamin in all communities across the world. This could be attributed to a number of factors which affect absorption, utilization, and requirement of thiamin (Sinclair, 1982).

Biochemical evidence of thiamin deficiency has been reported to exist in aged populations living at home. Thus 3–25% of normal, free-living elderly people (Iber *et al.*, 1982) appear to be associated with an increased erythrocyte transketolase-activation coefficient (Eq. 2.2). It is not clear if such an increase is a normal event in ageing or represents nutritional inadequacy. Probably the cause of deficiency may be related to inadequate intake due to eating habits, poor appetite, and low income accompanied by alcoholism. Furthermore, it is generally believed that elderly persons have an increased requirement for thiamin while their energy requirement is reduced. The latter is thought to be due to the fact that both the absorptive function and the ability to activate thiamin in the liver are considerably diminished in older people. Subclinical thiamin deficiency could result in symptoms such as general malaise, insomnia and increased irritability which are frequently noted in older people. Abnormal thiamin status in the elderly has also been found to be associated with infections (Puxty, 1985). The increase in 'TPP effect' was greatest in those patients in whom the infection was associated with an acute confusional state. Post-operative confusion in the elderly is also associated with a rise in the 'TPP effect' (Older and Dickerson, 1982).

Subclinical thiamin deficiency is also frequently seen in children, especially among those who subsist largely on snack foods and unbalanced diets. The requirement of this vitamin is already relatively high during the period of growth. This requirement is further increased if snacks, such as chocolates and potato chips, are excessively consumed as a substitute for balanced foods. The greater energy intake from these snacks, which are devoid of thiamin, tends to lead to a low thiamin–energy ratio. This may account for the subclinical symptoms, such as loss of appetite, out-of-sort feeling, irritability, and depression that are often experienced by many children.

Gastrointestinal disorders, such as dyspepsia, peptic ulcer, gall bladder and ulcerative colitis are associated with failure of absorption of thiamin. The non-specific signs of deficiency of B-vitamins in general, including soreness of the tongue, mouth and lips, fatigue, neuritis, and depression, are commonly seen among patients with these disorders. Thiamin requirement is also related to metabolic rate. High rates of metabolism and sweat losses in athletic training appear to increase the need for the vitamin. The biochemical evidence of thiamin deficiency has been shown to be not uncommon among athletes of high performance. A pathological state, thyrotoxicosis, is associated with excessive secretion of thyroxine leading to an increased metabolic rate. As a result the requirement of thiamin for the subjects with this disorder is increased. Thiamin deficiency has been observed in association with thyrotoxicosis.

The neurons in the brain depend upon glucose for energy unless ketones are available in sufficient concentrations to meet their metabolic needs. During total parenteral nutrition (TPN), large carbohydrate loads are normally given, and the continuous infusion of insulin suppresses ketogenesis. This may explain why the thiamin requirement is considerably increased in patients receiving TPN with or without insulin (Kishi *et al.*, 1979). The deficiency of thiamin, ranging from asymptomatic to symptomatic phase, can also occur in a variety of other situations, including starvation diet, haemodialysis, diabetic acidosis, and severe or acute liver function disturbances.

Thiamin deficiency appears to occur not only by utilization, decreased metabolic availability, or increased requirement, but also through exposure to a variety of thiamin antagonists. These compounds may be structurally similar to thiamin. They act as antimetabolites by competing with the vitamin for its coenzymatic functions. The most common thiamin antagonists are pyrithiamin, oxythiamin and amprolium. In addition, the thiamin content of foods can be reduced due to the presence of antithiamin factors, such as the thiaminases in raw fish. These enzymes are, however, thermolabile, and hence cooking of fish renders the enzymes inactive. On the other hand, other antithiamin factors that are thermostable may be found in foods such as tea, blackcurrants, Brussels sprouts, and red cabbage (Sauberlich, 1985).

Thiamin-associated Genetic Disorders

Deficiency of pyruvate decarboxylase

This rare hereditary disorder is associated with a defect in the TPP-dependent oxidative decarboxylation of pyruvic acid. Biochemically, there appear to be elevated urinary levels of lactic acid, pyruvic acid, and alanine, as characteristics of this condition. Clinical manifestations include intermittent attacks of cerebellar ataxia. The synthesis of acetylcholine, and hence the oxidation of pyruvate, is especially important in the cerebellum. This condition does not respond to thiamin supplements, but a low-carbohydrate ketogenic diet may be beneficial (Koike and Koike, 1982).

Deficiency of branched chain α-ketoacid dehydrogenase

This is an inborn error of metabolism in which the infant eats poorly, vomits, and fails to thrive. The biochemical basis of this condition has been described on p. 15. Some patients appear to respond to very large dose levels (100–200 mg day^{-1}) of thiamin. The extent of response, however, is determined by the level of mutancy of the enzyme involved. Dietary restriction of branched-chain amino acids is important in controlling the disease.

Leigh's disease (subacute necrotizing encephalomyelopathy)

This is an autosomal recessive disease, associated with weakness, ataxia, convulsions, and peripheral neuropathy (Cooper and Pincus, 1976). The condition occurs before the age of two years. At post mortem examination, patients have been found to have very little or no thiamin triphosphate (TTP) in their brain and other tissues. There appears to be an inhibitor of TTP-ATP phosphoryltransferase that synthesizes TTP from the TPP. This condition responds to large doses of thiamin but the patients become refractory to this therapy within three months to two years.

Deficiency of transketolase enzyme

Wernicke–Korsakoff syndrome is associated with reduced affinity of transketolase enzyme for TPP, and is believed to be of genetic origin. These patients are particularly susceptible to development of the syndrome upon abuse of alcohol (Blass and Gibson, 1977). They appear to respond fairly well to large doses of thiamin.

Requirements

Recommendations for thiamin for all age groups are related to the total energy intake. The relationship is based on the fact that the vitamin is involved in the metabolism of carbohydrates, lipids, and protein, and hence yielding energy. Therefore the total daily requirement of thiamin depends upon the factors that influence one's energy requirement. These include age, body weight/size, degree of activity, high carbohydrate intake (including sweets), increased energy metabolism (during the period of growth, pregnancy, and lactation), and alcohol consumption.

In adults the current recommendation of the Health and Welfare Canada (1990) for thiamin is 0.4 mg per 1000 kcal day^{-1}. With a dietary intake of less than 2000 kcal day^{-1}, however, the vitamin supply should not fall below 1.0 mg.

Assessment of Thiamin Status

Various approaches have been described for the laboratory evaluation of thiamin status in humans. Urinary excretion of thiamin in a timed collection has been the most popular approach. The widely used assay for the vitamin in urine is the fluorimetric method, which is based on the oxidation of thiamin by potassium ferricyanide in the presence of alkali to the strongly fluorescent derivative, thiochrome. Urinary thiamin levels can provide information as to the adequacy of the dietary intake but do not necessarily evaluate accurately a marginal deficiency.

There are, however, a great number of other techniques that are suitable for the determination of thiamin status. For instance, blood pyruvic acid and lactic acid are both elevated in thiamin deficiency. Glucose loading is helpful as it accentuates the further elevation of pyruvic acid, and it stays high for longer than in normal subjects. The blood pyruvate level is, however, not sensitive enough to identify early deficiency of thiamin and is non-specific, although it can provide evidence of thiamin insufficiency, particularly when clinical signs are evident.

Adequacy of thiamin nutriture is most accurately determined by a functional test, that is by measurement of TPP stimulating effect of transketolase enzyme activity (or TPP effect) in haemolysed whole blood or erythrocytes (Smeets *et al.*, 1971). Transketolase occurs in a variety of tissues including red blood cells. The enzyme is associated with the glucose oxidative pathway. Although the synthesis of this enzyme is not affected by thiamin status, its catalytic activity depends on its binding TPP (Fig. 2.7). Haemolysates from subjects who are thiamin deficient thus demonstrate a relatively greater stimulation from the added TPP *in vitro* than do subjects who have adequate thiamin

stores. An activity coefficient (AC) is calculated as the ratio of the change in absorbance over time (absorbance) of the stimulated to the non-stimulated haemolysate.

$$AC = \frac{\text{Absorbance (stimulated)}}{\text{Absorbance (non-stimulated)}} \quad (2.2)$$

An activity coefficient in excess of 1.2 is generally regarded as deficient (Table 2.1).

Some variants of this procedure calculate the 'percent stimulation' of transketolase activity following an *in vitro* addition of TPP to an incubation mixture containing haemolysed erythrocytes (Eq. 2.3). According to this, there has been a guideline indicating the degree of deficiency state (Table 2.1).

$$\text{Percent stimulation} = (AC \times 100) - 100 \quad (2.3)$$

References

Blass, J.P. and Gibson, G.E. (1977) Abnormality of a thiamin-requiring enzyme in patients with Wernicke–Korsakoff syndrome. *New England Journal of Medicine* 297, 1367–1370.

Bonjour, J.P. (1980) Vitamins and alcoholism. IV. Thiamin. *International Journal of Vitamin and Nutrition Research* 50, 321–338.

Cooper, J.R. and Pincus, J.H. (1976) Subacute necrotizing encephalomyelopathy. In: C.J. Gubler, M. Fujiwara, P.M. Dreyfus (eds) *Thiamin.* John Wiley and Sons, New York.

Edwin, E.E. and Lewis, G. (1971) Thiamin deficiency with particular reference to cerebrocortical necrosis – a review and discussion. *Journal Dairy Research* 38, 79–90.

Haas, R.H. (1988) Thiamin and the brain. *Annual Review of Nutrition* 8, 483–515.

Health and Welfare Canada (1990) In: *Nutrition Recommendations.* Canadian Government Publishing Centre, Ottawa.

Iber, F.L., Blass, J.P., Brin, M. and Leevy, C.M. (1982) Thiamin in the elderly – relation to alcoholism and to neurological degenerative disease. *American Journal of Clinical Nutrition* 36, 1067–1082.

Interdepartmental Committee on Nutrition for National Defence (ICNND) (1963) *Manual for Nutrition Surveys*, 2nd edn, US Government Printing Office, Washington, DC.

Kishi, H., Nishii, S., Ono, T., Yamagi, A., Kasahara, N., Hiraoka, E., Okada, A., Itakura, T. and Tagaki, Y. (1979) Thiamin and pyridoxine requirements during intravenous hyperalimentation. *American Journal of Clinical Nutrition* 32, 332–338.

Koike, M. and Koike, K. (1982) Biochemical properties of mammalian 2-oxoacid dehydrogenase multienzyme complexes and clinical relevancy with chronic lactic acidosis. *Annals New York Academy of Sciences* 378, 225–235.

Leevy, C.M. (1982) Thiamin deficiency and alcoholism. *Annals New York Academy of Sciences* 378, 316–326.

Older, M.J.W. and Dickerson, J.W.T. (1982) Thiamin and the elderly orthopaedic patient. *Age and Ageing* 11, 101–107.

Puxty, J.H. (1985) Infections, vitamins, and confusion in the elderly. In: Kemm, J.R. and Ancill, R.J. (eds) *Vitamin Deficiency in the Elderly – Prevalence, Clinical Significance and Effects on Brain Function*, Blackwell Scientific Publications, Oxford, pp. 103–106.

Sauberlich, H.E. (1967) Biochemical alterations in thiamin deficiency – their interpretation. *American Journal of Clinical Nutrition* 20, 528–542.

Sauberlich, H.E. (1985) Bioavailability of vitamins. *Progress in Food and Nutrition Science* 9, 1–33.

Sinclair, H.M. (1982) Thiamin. In: Barker, B.M. and Bender, D.A. (eds) *Vitamins in Medicine*, Vol. 2., William Heinemann, London.

Smeets, E.H.J., Muller, H. and De Wael, J. (1971) A NADPH-dependent transketolase assay in erythrocyte hemolysates. *Clinica Chimica Acta* 33, 379–386.

Tang, C.M., Wells, J.C., Rolfe, M. and Chan, K. (1989) Outbreak of beriberi in the Gambia. *The Lancet* 2, 206–207.

Thompson, A.D. and Leevy, C.M. (1972) Observations on the mechanism of thiamin hydrochloride absorption in man. *Clinical Science* 43, 153–163.

Thurnham, D.I. (1978) Thiamin. In: Rechcigl, M. (ed.) *CRC Handbook Series in Nutrition and Food*, Vol. 3, CRC Press, Florida.

Riboflavin (Vitamin B_2) 3

Chemistry

Flavins isolated from different sources (e.g. lactoflavin, hepato-flavin, and ovo-flavin) were subsequently all shown to be chemically identical. The crystalline yellow fluorescent needles were designated 'riboflavin' in 1937. It consists of flavin (isoalloxazine ring), to which is attached a ribityl side chain (mol. wt 376). The structure of riboflavin along with its two coenzyme derivatives is shown in Fig. 3.1. The compound is heat stable and described as water soluble although its solubility in water is actually very low (120 mg l^{-1} at 27.5°C). Its solubility is, however, increased by the addition of urea and by formation of a complex

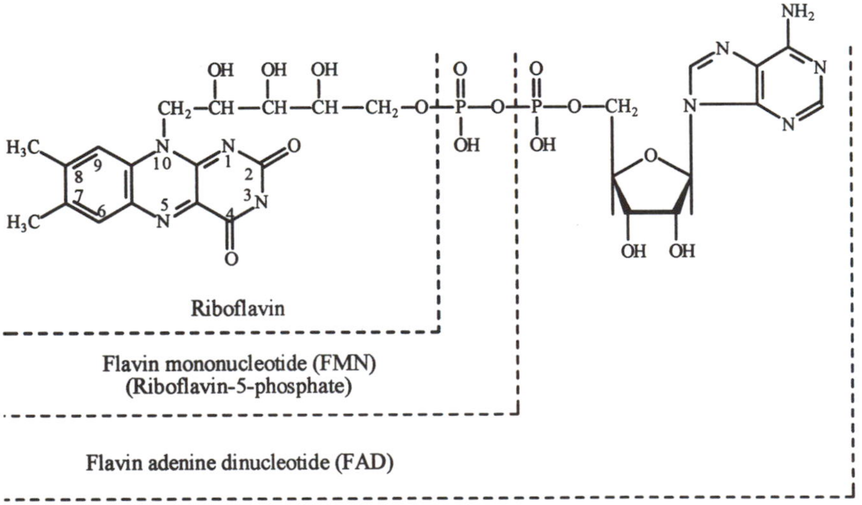

Riboflavin

Flavin mononucleotide (FMN)
(Riboflavin-5-phosphate)

Flavin adenine dinucleotide (FAD)

Fig. 3.1. Riboflavin (7,8-dimethyl-10-(1-D-ribityl)-isoalloxazine) and the flavin coenzymes. Source: Matts, 1880).

with boron (Combs, 1992). It is insoluble in lipid solvents, stable in acid but not in alkaline solutions, and unstable in the presence of UV light which causes irreversible decomposition. When exposed to daylight in neutral solution the ribose chain is split off to give lumichrome which has a greenish-yellow fluorescence, and consequently the vitamin loses its activity.

Sources

Riboflavin differs from other components of what is commonly known as the 'vitamin B complex' in that it occurs in good amounts in dairy products. Milk (2.0 mg l^{-1}) is, in fact, one of the main sources in the average UK and North American diets. Other good sources are liver and kidney (2.0–4.0 mg/100 g), egg (0.47 mg/100 g), cheeses (0.25–0.80 mg/100 g) and some green leafy vegetables (0.03–0.3 mg/100 g). Very good sources include yeast extract such as Marmite (11.0 mg/100 g), and meat extract such as Bovril (7.4 mg/100 g).

Absorption

Riboflavin is absorbed in the free form by an active process which is specific and saturable and is located in the proximal small intestine. Absorption is enhanced by bile salts. Thus children with biliary atresia (a congenital condition in which the bile duct is either absent or is pathologically closed) have reduced absorption of riboflavin (Combs, 1992).

Riboflavin occurs in most foods as coenzyme forms of the nucleotides flavin mononucleotide (FMN) and flavin adenine dinucleotide (FAD) complexed with proteins and absorption depends on the hydrolysis of these forms to yield the free vitamin (Fig. 3.2). This involves a two stage process in which the riboflavin coenzymes are released from the protein complexes by the proteolytic intestinal

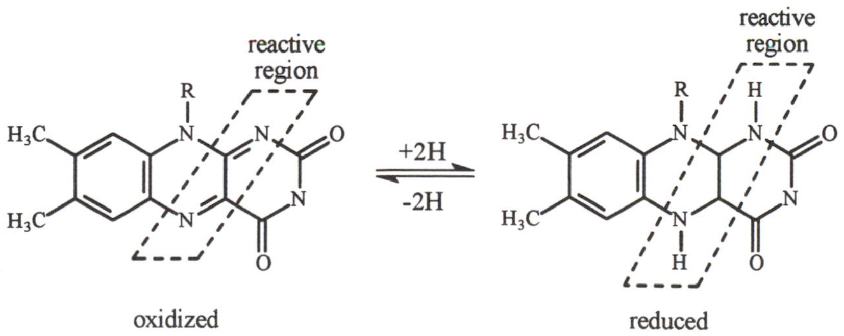

Fig. 3.2. Oxidized and reduced forms of FAD or FMN.

enzymes and the subsequent hydrolysis of the coenzymes by brush-border phosphatases to liberate free riboflavin. These phosphatases include the relatively non-specific alkaline phosphatase, an FAD pyrophosphatase which converts FAD to FMN, and an FMN phosphatase which converts FMN to free riboflavin. Free riboflavin enters the mucosal cell but much of it is then converted back to FMN by an enzyme, flavokinase, using ATP. This enters the portal circulation as a mixture of a mononucleotide and free riboflavin.

Transport

Riboflavin is transported in the plasma both as free riboflavin and as FMN. About half of each form is bound to plasma proteins. Of these, the most important is albumen but some is also found bound to globulins and fibrinogen. Since this binding is by weak hydrogen bonding, the vitamin is easily displaced from binding sites by several drugs, for example ouabain, theophylline and penicillin which can thus inhibit transport of the vitamin to the peripheral tissues (see Chapter 18). In some species, including the laying hen, the pregnant cow and pregnant rats and mice, specific riboflavin-binding proteins have been identified. These are synthesized in the liver under the stimulus of oestrogens and are believed to be involved in the transplacental/transovarian movement of riboflavin.

Riboflavin enters tissue cells in its free form and is converted predominantly into FMN (60–95%) with the remainder being converted to FAD. These nucleotides are bound almost exclusively to specific flavoproteins. Highest concentrations of the vitamin occur in the liver, kidney and heart. Significant amounts of free riboflavin are found only in the retina, the urine and cow's milk. The latter differs from human milk in that it contains 1.16–2.0 mg l^{-1} compared with 0.2–0.48 mg l^{-1} respectively. The riboflavin in cow's milk occurs mostly as the free vitamin whereas in human milk it occurs mainly as FAD and FMN. It has been estimated that the amount of riboflavin in the adult human body would meet metabolic demands for 2–6 weeks. The riboflavin content of the brain is not high but is turned over rapidly and is resistant to gross changes in riboflavin intake.

Metabolism

Riboflavin enters cells in the free state and is converted to FMN and FAD (Fig. 3.3). Both of these steps are regulated by thyroid hormones. The thyroid gland regulates the activity of the enzyme, riboflavin flavokinase. Hypothyroidism reduces the activity of the enzyme and this results in reduced tissue levels of FMN and FAD. By contrast, hyperthyroidism raises the activity of the enzyme but the levels of FMN and FAD do not rise. It appears that the levels of the

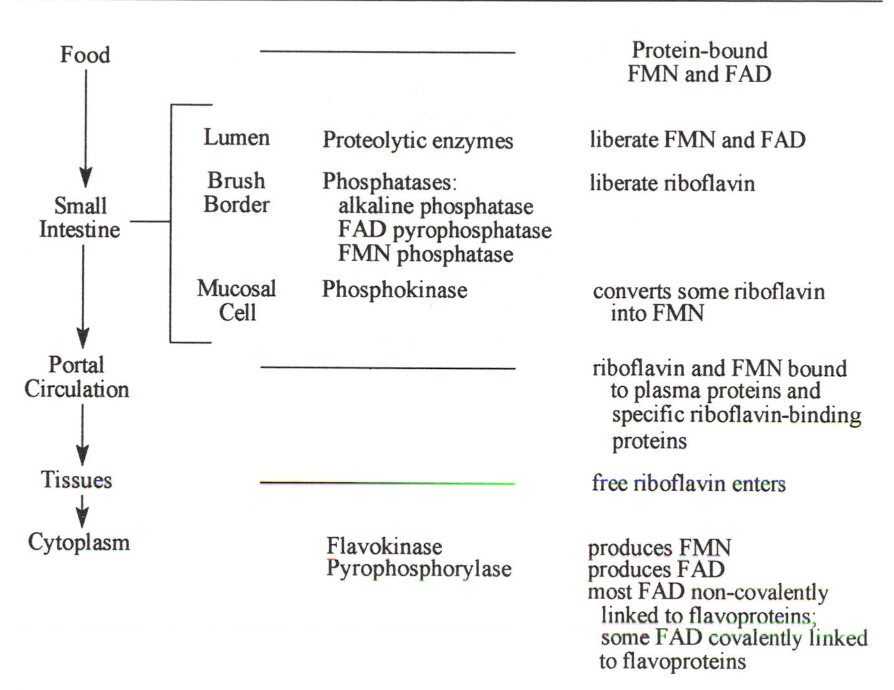

Fig. 3.3. Absorption and tissue distribution of riboflavin.

nucleotides are regulated by degradation. Tissue riboflavin is mostly in the form of flavoproteins (Fig. 3.3). Flavins bound to proteins in this way are resistant to degradation; the unbound ones are subject to catabolism in ways similar to those by which food riboflavin is broken down during its absorption. Chemically, riboflavin is 7,8-dimethyl-10-(1-D-ribityl)-iso-alloxazine (Fig. 3.1). Degradation of riboflavin involves its hydroxylation at the 7- and 8-positions of the isoalloxazine ring by hepatic microsomal cytochrome P450-dependent processes. Subsequent stages are oxidation and then removal of the methyl groups.

Riboflavin is excreted rapidly in the urine mainly as the free vitamin with smaller amounts of riboflavin degradation products including riboflavin 1-α-D-glucoside and 10-formylmethylflavin; it contributes the yellow colour of the urine. Small amounts of riboflavin degradation products are excreted in the faeces and are mostly thought to be of microbial origin.

Metabolic Functions

Metabolically, riboflavin is the essential component of FMN and FAD which are involved in the transfer of electrons in oxidation–reduction reactions. The reactive region in the riboflavin molecule is shown in Fig. 3.2. FAD or FMN undergoes oxidation and reduction states with a change in the number of hydrogens and a shift in the placement of double bonds in the riboflavin molecule. Both FAD and NAD^+ are two-hydrogen carriers but unlike NAD^+, FAD accepts both hydrogens from the substrates and becomes $FADH_2$ (Fig. 3.2). FMN and FAD function as coenzymes for more than 100 enzymes called flavoproteins or flavoenzymes. These are oxidases, which function aerobically, and dehydrogenases, which function anaerobically. One-electron and two-electron transfers allow flavoproteins to serve as two-electron donors or one-electron acceptors. Many flavoproteins contain a metal, for example Fe, Mo or Zn, and this combination of flavin and metal ion increases the versatility of the enzyme in transfers between single- and double-electron donors.

Flavoproteins are essential for the metabolism of carbohydrates, amino acids and lipids and for the conversion of the vitamins, such as pyridoxine and folate, to their respective coenzyme forms (Matts, 1980; Combs, 1992). Some examples of flavoproteins are given in Table 3.1.

Table 3.1. Some important flavoproteins in animal tissues.

Type of reaction	Flavoprotein	Flavin involved
1-Electron transfers	Mitochondrial electron transfer-flavoprotein (ETF)	FAD
	Ubiquinone reductase	FAD
	NADH-cytochrome P450 reductase[1]	FMN
Pyridine-linked dehydrogenases	NADH-cytochrome P450 reductase[1]	FAD
	NADH dehydrogenase	FMN
Non-pyridine nucleotide-dependent dehydrogenases	Succinate dehydrogenase	FAD
	Acyl-CoA dehydrogenases	FAD
Pyridine nucleotide oxidoreductases	Glutathione reductase	FAD
	Lipoamide dehydrogenase	FAD
Reactions of reduced flavoproteins with O_2	D-Amino acid oxidase	FAD
	L-Amino acid oxidase	FMN
	Monoamine oxidase	FAD
	Xanthine oxidase	FAD
	L-Gulonolactone oxidase	FAD
Flavoprotein monooxygenase	Microsomal flavoprotein mono-oxygenase	FAD

Modified from Combs, 1992.

[1]A component of microsomal cytochrome P450; it contains one molecule each of FMN and FAD.

Deficiency and its Consequences

Riboflavin deficiency has been reported to be common in China, India, Africa and Thailand. In North America, deficiency has been reported to exist in hospital patients, in patients with tuberculosis and among alcoholics. Phototherapy given to infants with hyperbilirubinaemia may also lead to deficiency of riboflavin by splitting the ribityl moiety of the vitamin. It often coexists with nicotinic acid deficiency. In the UK, the biochemical evidence of riboflavin deficiency as determined by a glutathione reductase coefficient (for details see later) was found in about a third of elderly people with a higher proportion in the lower socioeconomic classes included in two Government DHSS surveys (Thurnham, 1985). An abnormal biochemical index on its own may have little clinical significance, and in hospital patients such evidence of low riboflavin status has been found not to have any prognostic significance (Kemm and Alcock, 1984). The high values for the glutathione reductase activation coefficient seem to reflect the fact that elderly people have a lower than recommended intake of riboflavin. The clinical signs of riboflavin deficiency occur mostly in the mucocutaneous surfaces of the mouth. These include angular stomatitis (cracks at the corners of the mouth), cheilosis (inflammation of the lips), atrophied lingual papillae, glossitis (inflammation of the tongue), magenta tongue and corneal vascularization. Seborrhoeic skin lesions occur on the side of the nose; skin lesions also occur around the genitalia. Most lesions are only moderately specific and do not always respond to riboflavin (Bates, 1987). Unlike some other B-vitamin deficiencies, there is no one disease associated with riboflavin deficiency.

It is not easy to establish a link between a biochemical abnormality and the lesions produced by riboflavin deficiency. Riboflavin deficiency can be complicated by pyridoxine deficiency and it has been suggested, on the basis of studies on rats (Prasad *et al.*, 1983), that the skin lesions of riboflavin deficiency might result from a localized pyridoxine deficiency. Pyridoxal phosphate acts as a cofactor for the enzyme lysyl oxidase, which is involved in the formation of cross-links in the collagen molecule. Reduced activity of this enzyme would result in skin breakdown as a consequence of impaired collagen synthesis (Bamji, 1985). In Thai school children riboflavin supplements were found to stimulate hydroxyproline excretion, indicating that they were affecting collagen metabolism, but the change did not correlate with biochemical deficiency (Bates, 1987).

In communities where riboflavin deficiency is common, maternal riboflavin status falls during pregnancy. For reasons that are not clear, the erythrocyte glutathione reductase activation ratio was found to fall after parturition in spite of the secretion of riboflavin in breast milk (Bates *et al.*, 1981; Bamji *et al.*, 1986). In the latter study in India the activation levels rose again after 30 days. In Gambian women it seems that an intake of 2.5 mg day^{-1} is required to maintain a normal biochemical index (Bates *et al.*, 1984). Although angular stomatitis and seborrhoeic dermatitis have not been reported in premature infants, biochemical ariboflavinosis has been seen in unsupplemented babies, and this is exacerbated by phototherapy (Sisson, 1987). Although the functional significance of this is not clear, riboflavin has been implicated in the maintenance of normal intestinal function and this is of particular importance in the neonate (Powers *et al.*, 1993).

Reduced glutathione levels have been reported in all forms of human cataract and FAD is a coenzyme for glutathione reductase. Studies in rats have shown that riboflavin deficiency predisposes to cataracts but the common assumption that this is due to impaired glutathione synthesis may not be correct (Bates, 1991). There is only limited evidence for an association between riboflavin deficiency and increased cataract risk in humans. It remains to be established whether the degree of riboflavin deficiency seen in the elderly is cataractogenic (Skalka and Prehal, 1981). A recent review (Bates, 1993) has discussed the possible role of a number of vitamins in cataract formation. Two recent studies of human populations in the USA and Canada have suggested that certain vitamins may exert protective effects against senile cataracts (Jacques and Chylack, 1991; Robertson *et al.*, 1991).

Interactions with Other Nutrients

Iron

Riboflavin deficiency is sometimes accompanied by microcytic anaemia. The anaemia responds more rapidly to iron and riboflavin than to iron alone (Powers *et al.*, 1985). It seems that riboflavin plays a role in the mobilization from tissue stores, predominantly ferritin, and in transport across membrane barriers, including the intestinal wall (Powers, 1986). The biochemical involvement of riboflavin in this process seems to be by virtue of the fact that reduced flavins are most efficient biological reductants and the absorption and its mobilization require reduction of ferric to ferrous iron. Possible enzymes involved are NADPH–FMN oxidoreductase which exists in a soluble form and also possibly in microsomes, and mitochondral ubiquinone-linked ferriductase which is also flavin dependent and in the rat highly sensitive to riboflavin deficiency.

Vitamin B_6

Pyridoxine exists as a number of interconvertible vitamers (see Chapter 5). Conversion of pyridoxine phosphate or pyridoxamine to pyridoxal phosphate is catalysed by an enzyme, pyridoxine oxidase, which requires FMN as a cofactor.

Riboflavin deficiency may also affect other enzymes involved in pyridoxine metabolism such as kinases and phosphatases. Reference has already been made to this in connection with the effect of riboflavin deficiency.

Folic acid

The final step in the conversion of oxidized folate to 5-methyl tetrahydrofolate is catalysed by a flavin-dependent enzyme, 5,10-methyl tetrahydrofolate reductase, which is highly sensitive to mild riboflavin deficiency (Bates and Fuller, 1986).

Requirements

It is difficult to determine minimum daily dietary requirements of riboflavin, because the clinical signs of deficiency are relatively non-specific. Moreover, requirements are dependent on the intakes of energy and a number of other nutrients including protein. Indeed, the recommended dietary intakes of the vitamin have been related to the intakes of energy and protein and also to the metabolic body size. Studies of groups of subjects indicate that adult men and women require 0.5–0.8 mg day^{-1} (Department of Health, 1991); adolescent youths require a similar amount (Lo, 1984). Experimental studies carried out by Horwitt *et al.* (1949, 1950) in which subjects fed 0.55 mg day^{-1} for four months developed symptoms of deficiency, agree with these amounts.

Values for the requirements and recommended dietary allowances for riboflavin differ for different countries according to the evidence that has been used to arrive at the amounts and the relative weight given to evidence from the different sources. Values for the UK, USA and Canada are shown in Table 3.2.

Biochemical Assessment

Urinary excretion of riboflavin has been widely used in population studies, but has limited accuracy. Its serum concentration is also considered variable and indicative only of current intake. Adequacy of riboflavin status is effectively determined by a functional test, that is by measurement of the activity coefficient of glutathione reductase in erythrocytes (EGRAC). Glutathione reductase is a NADPH- and FAD-dependent enzyme, and is the major flavoprotein in erythrocytes. This enzyme catalyses the reduction of oxidized glutathione (GSSG) in the following expression:

$$\text{GSSG} + \text{NADPH·H}^+ \longrightarrow 2\ \text{GSH} + \text{NADP}^+ \qquad (3.1)$$

The EGRAC is defined as the ratio of FAD-stimulated to unstimulated activity of erythrocyte glutathione reductase (Becker *et al.*, 1991) (Eq. 3.2).

$$\text{EGRAC} = \frac{\text{Glutathione reductase activity (+FAD)}}{\text{Glutathione reductase (–FAD)}} \tag{3.2}$$

The degree of stimulation of glutathione reductase in erythrocytes depends on the FAD saturation of the apoenzyme, which in turn depends on the availability of riboflavin. When there is a riboflavin deficiency, the erythrocyte glutathione reductase activity falls: the activity rises when FAD is added *in vitro* to the erythrocytes. An activity coefficient (EGRAC) in excess of 1.2 is generally regarded as deficient (Table 3.3).

Table 3.2. Dietary reference values for riboflavin (mg day^{-1}).

	EAR[1]	RNI[2]	RDA[3]
Children			
0–12 months	0.3	0.3–0.5	0.4–0.6
1–6 years	0.5–0.6	0.5–0.7	0.8–1.0
7–10 years	0.8	1.0	1.4
Males			
11–14 years	1.0	1.0	1.6
15–50+ years	1.0	1.2	1.4–1.7
Females			
11–14 years	0.9	0.9	1.3
15–50+ years	0.9	0.8	1.2–1.3
Pregnancy		+0.1	+0.3
Lactation		+0.2	+0.5

[1]Estimated average requirement (Department of Health, UK, 1991).
[2]Report of the Scientific Review Committee (Health and Welfare, Canada, 1990).
[3]Committee on Dietary Allowances (Food and Nutrition Board, USA, 1980).

Table 3.3. FAD-stimulating effect of glutathione reductase activity (EGRAC) and riboflavin nutritional status.

Riboflavin[1]	EGRAC
Adequate	< 1.2
Marginal	1.2–1.4
Deficient	> 1.4

[1] McCormic, 1985.

References

Bamji, M.S. (1985) Vitamin B responsive lesions–molecular basis. *Nutrition News (India)* 6, 1–3.

Bamji, M.S., Prema, K., Jacob, C.M., Ramalakshmi, B.A. and Madhavapeddi, R. (1986) Relationship between vitamins B_2 and B_6 status and the levels of these vitamins in milk at different stages of lactation. *Human Nutrition Clinical Nutrition* 40c, 119–124.

Bates. C.J. (1987) Human riboflavin requirements, and metabolic consequences of deficiency in man and animals. *World Review of Nutrition and Dietetics* 50, 215–265.

Bates, C.J. (1991) Glutathione and related indices in rat lenses, liver and red cells during riboflavin deficiency and its correction. *Experimental Eye Research* 53, 123–130.

Bates, C.J. (1993) Vitamin undernutrition. *Proceedings of the Nutrition Society* 52, 143–154.

Bates, C.J. and Fuller, J.N. (1986) The effect of riboflavin deficiency on 5,10-methylene tetrahydrofolate reductase (NADPH) (EC1.5.1.20) and folate metabolism in the rat. *British Journal of Nutrition* 55, 455–464.

Bates, C.J., Prentice, A.M., Paul, A.A., Sutcliffe, B.A., Watkinson, M. and Whitehead, R.G. (1981) Riboflavin status in Gambian pregnant and lactating women and its implications for recommended dietary allowances. *American Journal of Clinical Nutrition* 34, 928–935.

Bates, C.J., Prentice, A.M., Watkinson, M., Morrell, P., Foord, F.A., Watkinson, A. and Whitehead, R.G. (1984) Efficacy of a food supplement in correcting riboflavin deficiency in pregnant Gambian women. *Human Nutrition Clinical Nutrition* 38c, 363–374.

Becker, K., Krebs, B. and Schirmer, R.H. (1991) Protein-chemical standardization of the erythrocyte glutathione reductase activation test (EGRAC test): Application to hypothyroidism. *International Journal of Vitamin and Nutrition Research* 61, 180–187.

Combs, G.F. (1992) In: *The Vitamins*. Academic Press, New York, pp. 277–287.

Committee on Dietary Allowances, Food and Nutrition Board (1980) *Recommended Dietary Allowances*, 9th revised edition. National Academy of Sciences, Washington, DC.

Department of Health (1991) *Dietary Reference Values for Food Energy and Nutrients for the United Kingdom*. Report on Health and social Subjects, No. 41. H.M. Stationery Office, London.

Health and Welfare Canada (1990) *The Report of the Scientific Review Committee*. Canadian Government Publishing Centre, Ottawa.

Horwitt, M.K., Hills, O.W., Harvey, C.C., Liebert, E. and Steinberg, D.L. (1949) Effects of dietary depletion of riboflavin. *Journal of Nutrition* 39, 357–373.

Horwitt, M.K., Harvey, C.C., Hills, O.W. and Liebert, E. (1950) Correlation of urinary excretion with dietary intake and symptoms of ariboflavinosis. *Journal of Nutrition* 41, 247–264.

Jacques, P.F. and Chylack, L.T. (1991) Epidemiologic evidence of a role for the antioxidant vitamins and carotenoids in cataract prevention. *American Journal of Clinical Nutrition* 53, 352S–355S.

Kemm, J.R. and Alcock, J. (1984) The distribution of supposed indicators of nutritional status in elderly patients. *Age and Ageing* 13, 21–28.

Lo, C.S. (1984) Riboflavin status of adolescents in Southern China: average intake of riboflavin and clinical findings. *Medical Journal of Australia* 141, 635–637.

Matts, S.G.F. (1980) Riboflavin. In: Barker, B.M. and Bender, D.A. (eds) *Vitamins in Medicine*, Vol.1. Heinemann Medical Books, London, pp. 398–443.

McCormic, D.B. (1985) Vitamins. In: *Textbook of Clinical Chemistry*. W.B. Saunders, Philadelphia.

Powers, H.J. (1986) Investigation into the relative effects of riboflavin deprivation on iron economy in the weanling rat and the adult. *Annals of Nutrition and Metabolism* 30, 308–315.

Powers, H.J., Bates, C.J., Lamb, W.H., Singh, T., Gelman, W. and Webb, E. (1985) Effects of a multivitamin and iron supplement on running performance in Gambian children. *Human Nutrition Clinical Nutrition* 39c, 427–437.

Powers, H.J., Weaver, L.T., Austin, S. and Beresford, J.K. (1993) A proposed intestinal mechanism for the effect of riboflavin deficiency on iron loss in the rat. *British Journal of Nutrition* 69, 553–561.

Prasad, R., Lakshmi, A.V. and Bamshi, M.S. (1983) Impaired collagen maturity in vitamin B_2 and B_6 deficiency – probable molecular basis of skin lesions. *Biochemical Medicine* 30, 333–341.

Robertson, J.Mc., Donner, A.P. and Trevithick, J.R. (1991) A possible role for vitamin C and E in cataract prevention. *American Journal of Clinical Nutrition* 53, 346S–351S.

Sisson, T.R.C. (1987) Photodegradation of riboflavin in neonates. *Federation Proceedings* 46, 1883–1885.

Skalka, H.W. and Prehal, J.T. (1981) Cataracts and riboflavin deficiency. *American Journal of Clinical Nutrition* 34, 861–863.

Thurnham, D.I. (1985) The interpretation of biochemical measurements of vitamin status in the elderly. In: Kemm, J.R. and Ancill, R.J. (eds) *Vitamin Deficiency in the Elderly*. Blackwell Scientific Publications, Oxford, pp. 46–67.

Niacin (Vitamin B_3) 4

Chemistry

'Niacin' is the generic term for two vitamers, nicotinic acid and its amide, nicotinamide. In North America, niacin is often referred to exclusively as nicotinic acid; this may cause some confusion between the generic and specific terms. Nicotinamide (Fig. 4.1) is the active form, which functions as a constituent of two coenzymes: nicotinamide adenine dinucleotide (NAD) and nicotinamide adenine dinucleotide phosphate (NADP). These coenzymes in their reduced stages (NADH/NADPH) are the principal forms of niacin that exist in animal tissues. Chemically, niacin is pyridine 3-carboxylic acid (mol. wt 123). It is a non-hygroscopic, stable white crystalline solid which sublimes without decomposition at about 230°C. It is soluble in water (1 g in 60 ml at 25°C) and in alcohol (1 g in 80 ml at 25°C) but is insoluble in ether. It is easily converted to the amide which is more readily soluble in water (1 g in 1 ml) and in alcohol (1 g in 1.5 ml) but is also soluble in ether.

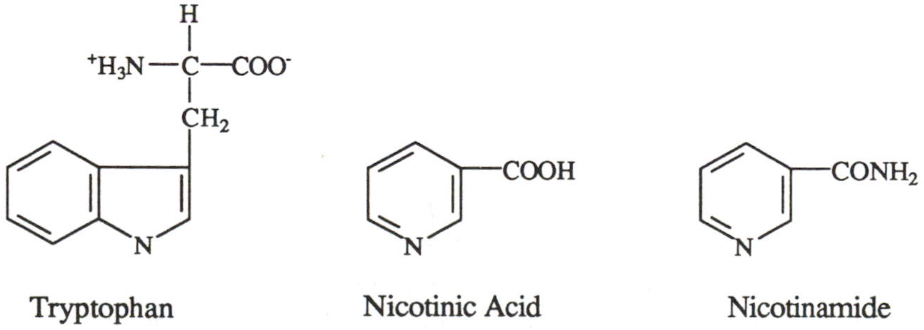

H
^+H_3N—C—COO^-
CH_2
N
COOH
N
$CONH_2$
N

Tryptophan | Nicotinic Acid | Nicotinamide

Fig. 4.1. Tryptophan and niacin vitamers.

Sources

Humans are able to synthesize niacin from tryptophan (Henderson, 1983) (Fig. 4.2). Indeed, it has been claimed that normal intakes of tryptophan are adequate to meet niacin requirements without the need for any preformed niacin in the diet, assuming unimpaired oxidation metabolism of tryptophan (Bender, 1992). Because of this interrelationship, the niacin content of foods is usually expressed as niacin equivalents (NE). This is calculated from the amount of niacin present plus 1/60th of the amount of tryptophan. The factor of 1/60th was derived from the amount of tryptophan chosen to be equivalent to 1 mg of nicotinic acid. The ratio of 60 mg tryptophan to 1.0 mg nicotinic acid was, in fact, a compromise based upon studies of the amounts of tryptophan converted to N-methylnicotinamide (Horwitt *et al.*, 1956).

Although niacin is endogenously synthesized in the body, it is categorized as a vitamin. This is because its precursor, tryptophan, is an essential amino acid. The ultimate synthesis of niacin is thus dependent upon diets.

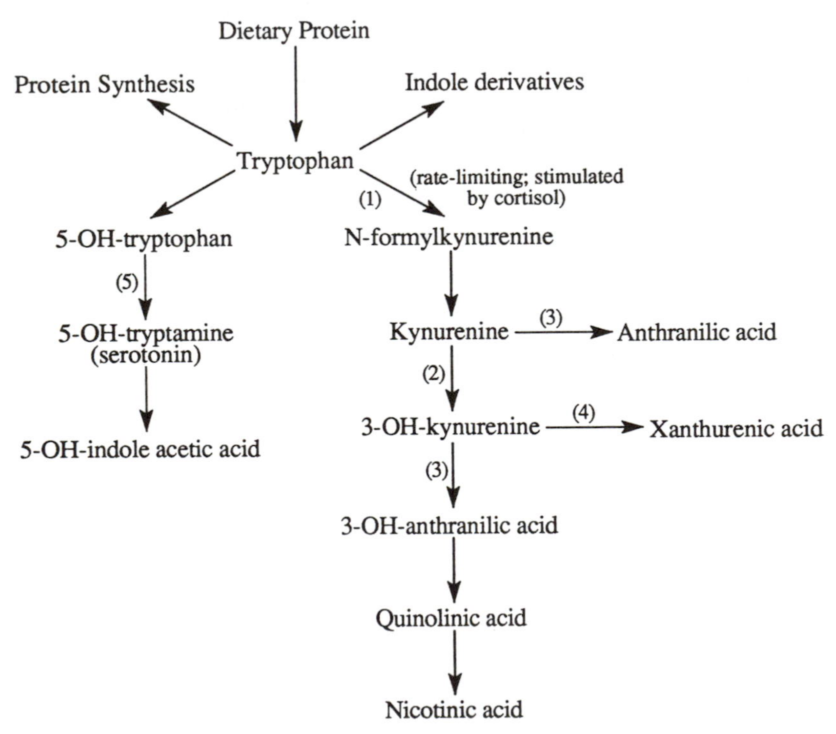

Fig. 4.2. Tryptophan metabolism – synthesis of nicotinic acid.
Enzymes and vitamins involved: (1) tryptophan pyrrolase (thiamin); (2) kynurenine hydroxylase (riboflavin); (3) kynureninase (pyridoxine); (4) amino transferase (pyridoxine); (5) tryptophan decarboxylase (pyridoxine). Source: Dickerson and Williams, 1990.

Preformed niacin is widely distributed in foods of both plant and animal origin. However, the vitamin is not released by digestion with equal facility from all foods and some is present in forms that render it unavailable to the consumer. In typical western diets the most important sources of preformed niacin are meat and meat products, cereals, dairy products, beverages and eggs. In cereals, niacin is present in covalently bound complexes with small peptides and carbohydrates as a somewhat ill-defined complex known as 'niacytin' (Mason *et al.*, 1973). The esterified niacin in these complexes is normally not available for absorption. However, the bioavailability of niacin can be increased by treatment with alkali to hydrolyse the esters. This process has long been a part of the preparation of tortillas in central America where corn is soaked in lime water prior to use. The rarity of the deficiency disease, pellagra, in this part of the world is almost certainly due to this process making niacin available. In maize (corn) at least 70% of the niacin is unavailable (Carter and Carpenter, 1982). In some other foods, for example coffee, niacin is present as a methylated derivative called trigonelline (1-methylnicotinic acid) which functions as a plant hormone. Nicotinic acid is liberated in coffee by roasting (Baessler *et al.*, 1992) and this makes coffee a useful source of niacin. Gut flora may synthesize niacin but its subsequent absorption is uncertain. Human milk contains a higher concentration of preformed niacin than cow's milk (Table 4.1) and a lactating woman

Table 4.1. Niacin, tryptophan and niacin equivalents in foods (values expressed as mg/1000 kcal unless otherwise stated).

	Preformed niacin	Tryptophan	Niacin equivalents	Niacin (mg/100 kcal)
Beef	2.47	1280	23.80	4.6
Pork	1.15	61	2.17	0.8–5.6
Wheat flour	2.48	297	7.43	3.4–6.5
Maize (corn) flour	4.97	106	6.74	1.4–2.9
Rice	4.52	290	9.35	1.6[1], 4.7[2]
Eggs (whole)	0.60	1150	19.80	0.1
Cow's milk	1.21	673	12.40	0.2
Human milk	2.46	443	9.84	–
Beans	–	–	–	0.5–2.4
Brussels sprouts	–	–	–	0.9
Carrots	–	–	–	0.6
Peas	–	–	–	0.9–25.0
Potatoes	–	–	–	1.5
Apples	–	–	–	0.6
Bananas	–	–	–	0.7
Peaches	–	–	–	1.0
Peanuts	–	–	–	17.2

Adapted from Combs (1992).
[1]Polished rice.
[2]Unpolished rice.

typically secretes 1.0–1.3 mg of preformed niacin daily in 750 ml of milk (National Research Council, 1989).

Absorption

The preformed niacin in foods of animal origin, NADH and NADPH (Fig. 4.3), is released by hydrolysis in the intestinal lumen by pyrophosphatase to nicotinamide ribonucleotide and riboside. There is no deamination to nicotinic acid (Henderson, 1983). Nicotinamide is rapidly absorbed in the stomach and small intestine. At low concentrations niacin is absorbed by a sodium-dependent facilitated diffusion process, whereas at higher concentrations absorption is mainly by passive diffusion (Sadoogh-Abasian and Evered, 1980). If high doses of niacin are taken, e.g. 3 g, 85% is excreted in the urine (Friedrich, 1987). Nicotinic acid is more slowly eliminated from the blood plasma if it is taken as nicotinic acid esters. As with some other vitamins, for instance ascorbic acid, slow release forms of niacin have been tested.

Transport

Niacin circulates in the plasma as unbound forms of both the acid and the amide. Each enters peripheral tissues by passive diffusion. However, some tissues are apparently able to take up niacin by facilitated diffusion. Erythrocytes take up nicotinic acid by their anion-transport system and a Na-dependent saturable transport system has been demonstrated in the renal tubules of the rabbit. Energy-dependent transport systems have been demonstrated in the brain (Combs, 1992). The site of the blood–cerebrospinal fluid barrier, the choroid plexus, appears to have separate systems for the uptake/release of both nicotinic acid and nicotinamide. Brain cells, themselves, have a high affinity transport

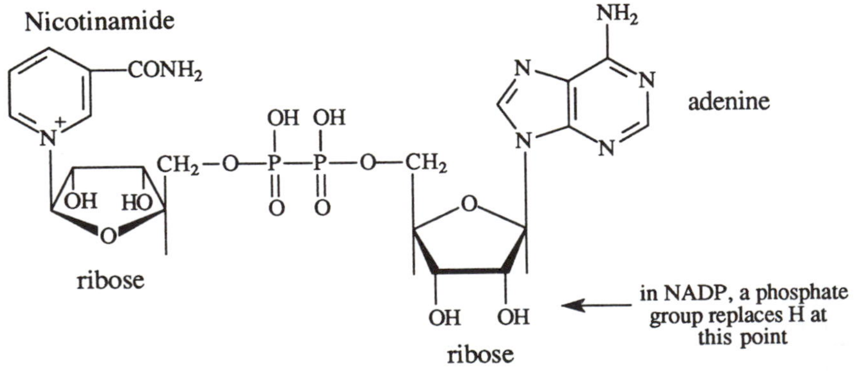

Fig. 4.3. Nicotinamide coenzymes, NAD and NADP. Source: Bender, 1980.

system for nicotinamide. Thus, the homeostasis of brain nicotinamide, but not nicotinic acid, is controlled by these two processes and nicotinamide is able to enter readily.

Tissue Storage

Niacin is trapped in the tissues by conversion to the pyridine nucleotide NAD(H) and NADP(H). Most is found as NAD(H) and in the oxidized form (NAD) (Table 4.2). In contrast, most of the NADP is found in the reduced form (NADPH).

Metabolism

All animal species, to varying degree, are able to synthesize the metabolically active forms of niacin, NADH and NADPH, from the indispensable amino acid tryptophan (Fig. 4.2). Key steps in this biosynthesis are the oxidative cleavage of the tryptophan pyrrole ring by the enzyme tryptophan pyrrolase to form *N*-formyl kynurenine. This compound is ring hydroxylated by a FAD-dependent enzyme, kynurenine-3-hydroxylase, to yield 3-hydroxykynurenine. This product can be deaminated to yield xanthurenic acid which is excreted in the urine. Alternatively, its side-chain can be cleaved of the amino acid, alanine, to yield 3-hydroxyanthranilic acid (3-OH-AA). This step is catalysed by the pyridoxine-dependent enzyme, kynureninase. The ring of 3-OH-AA is oxidatively opened by an Fe^{2+}-dioxygenase, 3-hydroxyanthranilic acid oxygenase to form a semi-stable compound, α-amino-β-carbonyl-muconic-ε-semialdehyde. This is a critical point in the pathway as the semi-stable compound can be converted by a number of steps to acetyl CoA. Alternatively, the semi-aldehyde can spontaneously cyclize to form quinolinic acid which, by the action of quinolate phosphoribosyl-transferase, yields nicotinic acid mononucleotide. The final conversion to NAD is catalysed by NAD synthetase.

Table 4.2. Pyridine nucleotide contents (mg kg^{-1}) of various rat tissues.

	NAD^+	NADH	$NADP^+$	NADPH
Blood	55	36	5	3
Testes	80	71	2	6
Pancreas	80	78	2	12
Lung	108	52	9	18
Thymus	116	35	2	12
Brain	133	88	2	8
Kidney	223	212	3	54
Heart	299	184	4	33
Liver	370	204	6	205

Adapted from Combs (1992).

All the various tissues of the body are apparently able to synthesize their own pyridine nucleotides, but an exchange between the tissues occurs at the level of nicotinamide which is rapidly transported between tissues. Thus, in the rat, nicotinic acid is the most important source of the coenzymes in the liver, kidneys, brain and erythrocytes, whilst nicotinamide appears to be a better precursor in the testes and ovaries.

We have seen earlier that the conversion of tryptophan to NAD is a rather inefficient process with approximately 60 mg of tryptophan being required to synthesize 1 mg of niacin. Even so, it has been suggested (Bender, 1992) that normal intakes are sufficient for the synthesis of the body's requirement of niacin.

Vitamin B_6 plays an important part in the conversion of tryptophan to niacin (Fig. 4.2). In fact, pyridoxal phosphate-dependent enzymes are involved at four points in the pathway. There are two transaminases: one catalyses the conversion of kynurenine to kynurenic acid and of 3-hydroxykynurenine to xanthurenic acid; and kynureninase which catalyses the conversion of kynurenine to anthranilic acid and of 3-hydroxykynurenine to 3-hydroxyanthranilic acid. Thus, pyridoxine deficiency reduces the efficiency of conversion of tryptophan to niacin. In spite of this, however, it is only the last mentioned conversion which is affected by deficiency of pyridoxal phosphate. The requirement of the transaminase for pyridoxal phosphate is smaller by a factor of 10^{-5} M (10^{-8} M compared with 10^{-3} M for kynureninase). Thus deficiency of pyridoxal phosphate reduces the overall conversion of tryptophan to niacin by reducing the production of 3-hydroxyanthranilic acid. The urinary excretion of kynurenic acid and xanthurenic acid is not blocked and this forms the basis of the method of assessing pyridoxal status by measuring the excretion of xanthurenic acid after a tryptophan load (also see Chapter 5). Since zinc is an essential cofactor of pyridoxal phosphate, deficiency of this mineral may also impair niacin production from tryptophan by reducing the production of pyridoxal phosphate.

The end result of the catabolism of pyridine nucleotides is 1-methlynicotinamide which can be oxidized to a variety of products that are excreted in the urine. The nucleotides are first cleaved hydrolytically at their two β-glycosidic bonds by NADP glycohydrolase to release nicotinamide. The amide can be deaminated to form nicotinic acid and re-converted into NAD or it can be methylated (mainly in the liver) by nicotinamide methylase to yield the 1-methyl product. It is of interest that nicotinic acid appears not to be methylated by animals. Trigonelline (1-methyl-nicotinic acid) is found, however, in the urine of coffee drinkers due to its presence in coffee.

A variety of water-soluble metabolites of niacin are excreted in the urine. At normal levels of niacin intake the major metabolites are 1-methylnicotinamide and its oxidation product, 1-methyl-6-pyridone 3-carboxamide. Much smaller amounts of nicotinic acid and nicotinamide are excreted. At higher rates of niacin intake, 65–85% of the total is excreted unchanged. It is worthy of note that at all levels of niacin intake the amide tends to be excreted more extensively as its metabolites than as the acid. Furthermore, since the biological turnover of the vitamers is determined primarily by rate of accretion, turnover of nicotinamide is faster than that of nicotinic acid.

Metabolic Functions

Niacin is the essential component of the enzyme cofactors NAD(H) and NADP(H). These are central electron carriers in cells and function in about 200 reactions in the metabolism of carbohydrates, fatty acids and amino acids in which there is a transfer of H ions. The H-ion transport by these nucleotides are two electron transfers in which the hydride ion serves as a carrier for both electrons. The transfer is stereo-specific and involves C-4 of the pyridine ring. Both NAD and NADP are involved in the same type of metabolic reactions, functioning as hydrogen acceptors in oxidation reactions and in the reduced form acting as hydrogen donors in reductive reactions.

In these functions it is the nicotinamide ring of the cofactor that is important. In the oxidized form this has an overall positive charge. The process of reduction of the coenzyme involves transfer from the substrate of two hydrogen atoms (Fig. 4.4). One of these can be visualized as separating into a proton (H ion) and an electron. The electron is transferred to the nicotinamide ring where it neutralizes the positive charge of the oxidized form, while the proton remains associated with, but not chemically bound to, the cofactor. The other hydrogen atom from the substrate is added to the unsaturated bond system at the 4-position of the ring, leading to formation of the reduced forms of NAD and NADP. Because of these changes in charge, it is conventional to write the oxidized form of NAD as NAD^+ and the reduced form as $NADH+H^+$. The same notation is used for NADP, so that the oxidized form is written as $NADP^+$ and the reduced form

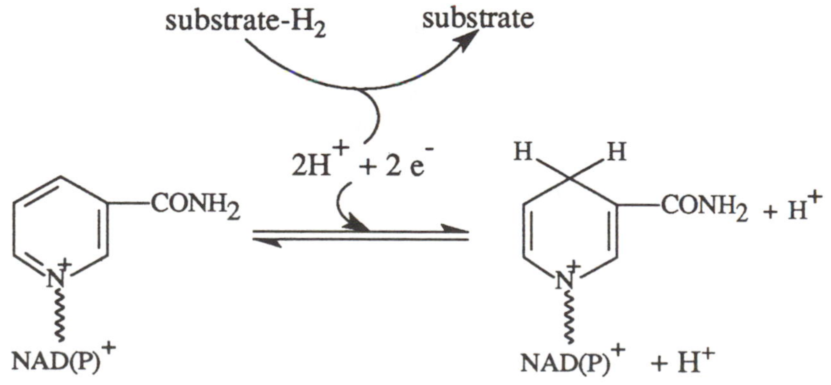

Fig. 4.4. Oxidation and reduction of NAD(P).

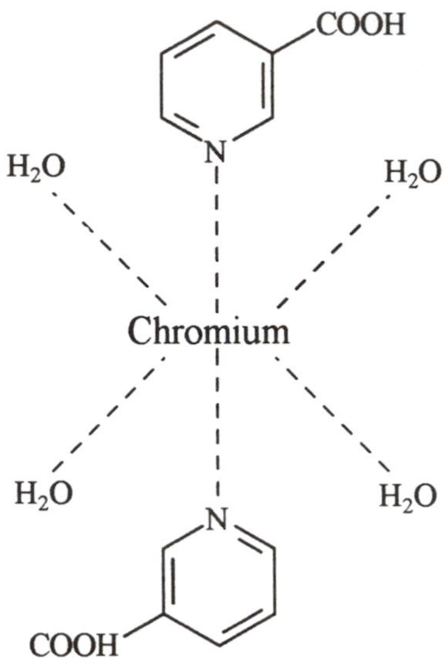

Fig. 4.5. A nicotinic acid–chromium complex as glucose tolerance factor.

as NADPH+H^+. Sometimes it is more convenient, but less correct, to write $NADH_2$ for the reduced form of NAD and $NADPH_2$ for the reduced form of NADP.

Although NAD and NADP are frequently referred to as coenzymes this is not strictly correct. The term 'cofactor' is used to signify any non-protein compound involved in an enzyme reaction apart from the obvious substrates. A coenzyme is generally considered to be covalently bound to the enzyme protein relatively tightly. The term 'prosthetic' group is reserved for compounds that are always covalently bound to the enzyme protein, as are the flavin coenzymes (Bender, 1980). Most correctly, NAD and NADP should be considered as cosubstrates of the enzymes with which they are associated.

NAD(H) and NADP(H) are similar in their mechanism of action and in their structure, but despite this, they have quite different metabolic roles and most dehydrogenases are specific for one or the other. Some of the important pyridine nucleotide-dependent enzymes are shown in Table 4.3.

Niacin is part of the chromium-containing 'glucose tolerance factor' of yeast (Fig. 4.5), which enhances the response to insulin. Its role in that factor is not at present clear. Free niacin does not have a similar effect.

Deficiency

The disease 'pellagra', as well as niacin-deficiency diseases in animals, is properly viewed as the result of a multi-factorial dietary deficiency rather than being due specifically to a deficiency of niacin. In addition to tryptophan and

Table 4.3. Some important pyridine nucleotide-dependent enzymes of animals.

Enzyme	Pyridine nucleotide
Carbohydrate metabolism	
3-Phosphoglyceraldehyde dehydrogenase	NADH
Glucose-6-phosphate dehydrogenase	NADPH
6-Phosphogluconate dehydrogenase	NADPH
Lactate dehydrogenase	NADH
Alcohol dehydrogenase	NADH
Lipid metabolism	
Glycerophosphate dehydrogenase	NADH
Hydroxyacyl CoA dehydrogenase	NADH
3-Ketoacyl ACP reductase	NADPH
Enoyl-ACP reductase	NADPH
3-Hydroxy-3-methylglutaryl-CoA reductase	NADPH
Amino acid metabolism	
Glutamate dehydrogenase	NADH/NADPH
Other	
Glutathione reductase	NADPH
Dihydrofolate reductase	NADPH
Thioredoxin-NADP reductase	NADPH
4-Hydroxybenzoate hydroxylase	NADPH
NADH dehydrogenase/NADH-ubiquinone reductase complex	NADH
NADPH-cytochrome P450 reductase	NADPH

Adapted from Combs (1992).

pyridoxine supplies being determinants of niacin status, there is evidence that excess intakes of the branched chain amino acid, leucine, may be involved in some areas (Gopalan and Jay Rao, 1975). In healthy volunteers, oral leucine has been shown to cause a disturbance in tryptophan metabolism (Belavady *et al.*, 1963). Some studies in rats have yielded results that support the view that excess leucine impairs the synthesis of NAD from tryptophan. Other results have been negative with regard to this relationship. It therefore seems prudent to consider that at the present time the role of excess leucine in the aetiology of pellagra is not clear.

In humans niacin deficiency results in changes in the skin, gastrointestinal tract and nervous system. The most prominent changes are, in fact, those that occur in the skin. The changes are most pronounced in the parts of the skin exposed to sunlight – the face, neck, backs of the hands, and the feet. The neck changes are referred to as 'Casal's collar'. The lesions in some patients resemble those occurring in early sunburn: cracking, desquamation, hyperkeratosis and hyperpigmentation. These changes in the skin constitute what is described as the 'dermatitis' of pellagra. Lesions in the gastrointestinal tract include angular stomatitis, cheilosis, glossitis as well as alterations in different parts of the tract from the buccal cavity to the intestine. The changes in the intestine are responsible for the 'diarrhoea' of pellagra. The third characteristic sign of pellagra, 'dementia' or 'delirium', is due to the neurological changes caused by the disorder. Anxiety, depression, apathy, headache, irritability and tremors are all features which may be manifest. Studies aimed at elucidating the biochemical basis for the neurological changes have focused on tryptophan metabolism and particularly on the neurotransmitter, serotonin, which is formed from tryptophan in the brain. Platelets have some of the properties of brain nerve endings, or synaptosomes, and have been used for the study of biogenic amines (Stahl, 1977). Platelet serotonin levels have been found to be lower in Hyderabad (India) pellagrins with mental depression compared with normal subject pellagrins whose mental state was apparently normal (Krishnaswamy and Ramana Murthy, 1970). Pellagra in Hyderabad is thought to be caused or exacerbated by the high leucine content of the diet and the Indian workers reported that adding 30 g leucine kg^{-1} to the diet of rats greatly reduced brain serotonin concentration (Krishnaswamy and Roghuram, 1972). This effect was obliterated by the addition of 3 g kg^{-1} diet of isoleucine or by small amounts (20 mg) of nicotinic acid or nicotinamide.

Requirements

The mean amount of niacin required to prevent or cure pellagra in experimental subjects maintained on niacin-deficient diets and in energy balance has been found to be 5.5 mg/1000 kcal (Horwitt *et al.*, 1956). As indicated above, and under normal conditions for an adult in nitrogen balance, the amount of tryptophan present in dietary protein provides adequate niacin without the need for any preformed vitamin. In the UK, median protein intake has been estimated to be 84.0 g day^{-1} by men and 61.8 g day^{-1} by women (Gregory *et al.*, 1990). At an equivalence of 60 mg tryptophan : 1 mg of niacin, this alone is equivalent to 17.6 mg day^{-1} of niacin for men and 13.0 mg day^{-1} for women. In the UK, the estimated average requirement for individuals of all ages is 5.5 mg NE/1000 kcal (Department of Health, 1991). The recommendation for Canadians and Americans is set at approximately 7 mg NE per 1000 kcal day^{-1}, and that the intake should not be lower than that based on 2000 kcal day^{-1} (i.e. 14 mg NE).

Measurement of Niacin Status

The most convenient method of assessing niacin status in humans is by measurement of the excretion of the metabolite, *N*-methylnicotinamide (NMN) in the urine (De Lange and Joubert, 1964). Alternatively, the ratio of NMN to

Table 4.4. An interpretive guideline for biochemical assessment of niacin.

Urinary metabolite(s)	Inadequate	Adequate
N'-Methylnicotinamide:		
mg/6 hr	< 0.2	> 0.6
mg g^{-1} creatinine	< 0.5	> 1.6
N'-methyl-2-pyridone-5-carboxylamide/ *N'*-methylnicotinamide ratio:	< 1.0	> 1.0

creatinine may be determined in urine collected between 10:00 a.m. and 12:00 a.m. Collection over this period is said to yield values that are similar to those obtained in 24-hour urine samples (Goldsmith and Miller, 1967). In addition to *N'*-methylnicotinamide, *N'*-methyl-2-pyridine-5-carboxylamide is a major end-product of niacin metabolism in humans. In adults, the former metabolite accounts for 20–30% and the latter accounts for 40–60% of niacin excreted in urine, when the vitamin is taken in its physiological level (De Lange and Joubert, 1964). The ratio of *N'*-methyl pyridone-5-carboxylamide and N'-methyl-nicotinamide in urine is another preferred method of assessing niacin status. An interpretive guideline for various methods of niacin assessment is given in Table 4.4.

References

Baessler, K.H., Gruehn, E., Loew, D. and Pietrzik, K. (1992) *Vitamin-lexikon.* Gustav Fischer Verlag, Stuttgart, pp. 154–173.

Belavady, B., Srikantia, S.G. and Gopalan, C. (1963) The effect of oral administration of leucine on the metabolism of tryptophan. *Biochemical Journal* 87, 652–655.

Bender, D.A. (1980) Niacin. In: Barker, B.M. and Bender, D.A. (eds) *Vitamins in Medicine,* 4th edn, Vol. 1. Heinemann Medical, London, pp. 315–347.

Bender, D.A. (1992) Factors affecting the equivalence of dietary tryptophan and niacin. In: *Advances in Tryptophan and Niacin.* Fugita Health University Press, Toyoake.

Carter, E.G.A. and Carpenter, K.J. (1982) The bioavailability for humans of bound niacin from wheat bran. *American Journal of Clinical Nutrition* 36, 855–861.

Combs, G.F. (1992) In: *The Vitamins.* Academic Press, New York, pp 289–309.

De Lange, D.J. and Joubert, C.P. (1964) Assessment of nicotinic acid status in population groups. *American Journal of Clinical Nutrition* 15, 169–174.

Department of Health (1991) *Dietary reference values for food energy and nutrients for the United Kingdom.* Report on Health and Social Subjects no. 41. H.M. Stationery Office, London.

Dickerson, J.W.T. and Williams, C.M. (1990) Vitamin-related disorders. In: Cohen, R.D., Lewis, B., Alberti, K.G.M.M. and Denman, A.M. (eds) *The Metabolic and Molecular Basis of Acquired Disease,* Vol. 1. London, Baillière Tindall, pp. 634–669.

Friedrich, W. (1987) *Handbook der Vitamine.* Urban & Schwarzenberg, Munchen.

Goldsmith, G.A. and Miller, O.N. (1967) Niacin. In: Gyorgy, P. and Pearson, W.N. (eds) *The vitamins*, 2nd edn., Vol. VII. Academic Press, New York and London, Chapter 5.

Gopalan, C. and Jay Rao, K.S. (1975) Pellagra and amino acid imbalance. *Vitamins and Hormones* 33, 505–528.

Gregory, J., Foster, K., Tyler, H. and Wiseman, M. (1990) *The Dietary and Nutritional Survey of British Adults.* Office of Population Censuses and Surveys. H.M. Stationery Office, London.

Henderson, L.M. (1983) Niacin. *Annual Review of Nutrition* 3, 289–307.

Horwitt, M.K., Harvey, C.C., Rothwell, W.S., Cutler, J.L. and Haffron, D. (1956) Tryptophan-niacin relationships in man. *Journal of Nutrition* 60, Suppl. 1, 1–43.

Krishnaswamy, K. and Ramany Murthy, D.–S.–V. (1970) Mental changes and platelet serotonin in pellagrins. *Clinica Chimica Acta* 27, 301–304.

Krishnaswamy, K. and Roghuram, T.–C. (1972) Effect of leucine and isoleucine on brain serotonin concentration in rats. *Life Sciences* 11, 1191–1197.

Mason, J.B., Gibson, N. and Kodicek, E. (1973) The chemical nature of the bound nicotinic acid of wheat bran: studies of nicotinic acid-containing macromolecules. *British Journal of Nutrition* 30, 297–311.

National Research Council (1989) In: *Recommended dietary allowances*, 10th edn. National Academy Press, Washington D.C., pp. 137–142.

Sadoogh-Abasian, F. and Evered, D.F. (1980) Absorption of nicotinic acid and nicotinamide from rat small intestine *in vitro. Biochimica Biophysica Acta* 598, 385–391.

Stahl, S.–M. (1977) The human platelet. A diagnostic and research tool for the study of biogenic amines in psychiatric and neurologic disorders. *Archives of General Psychiatry* 34, 509–516.

Pyridoxine (Vitamin B_6) 5

Vitamin B_6 was first recognized as an antidermatitis factor in rats by Gyorgy in 1934. Subsequently, the vitamin was isolated from rice and yeast in pure form by a number of separate groups of workers in 1938 and synthesized in the United States and Germany in 1939. The first metabolic role was given to pyridoxal phosphate by Gunsalus and his associates (1944) when they reported that this compound was the cofactor in the enzymic non-oxidative decarboxylation of amino acids.

Chemistry

Vitamin B_6 is the generic descriptor for 3-hydroxy-2-methyl-pyridine derivatives (Fig. 5.1) having the biological activity of pyridoxine. In nature, it occurs in three forms (or vitamers) which include pyridoxine (pyridoxol), pyridoxal and pyridoxamine, in which alcohol, aldehyde and amine groups, respectively, are located at the 4-position of the pyridine ring. The main excretory form of the vitamin is 4-pyridoxic acid which cannot be converted to a metabolically active form *in vivo* and therefore is not a vitamer. All three vitamers are interchangeable and therefore are comparably active. The metabolic conversion of the

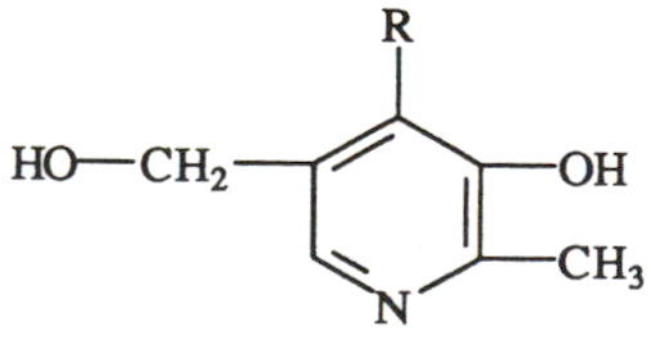

R=	
CH_2OH	pyridoxine (pyridoxol)
H-C=O	pyridoxal
H_2-C-NH_2	pyridoxamine
COOH	pyridoxic acid

Fig. 5.1. Structures of vitamin B_6.

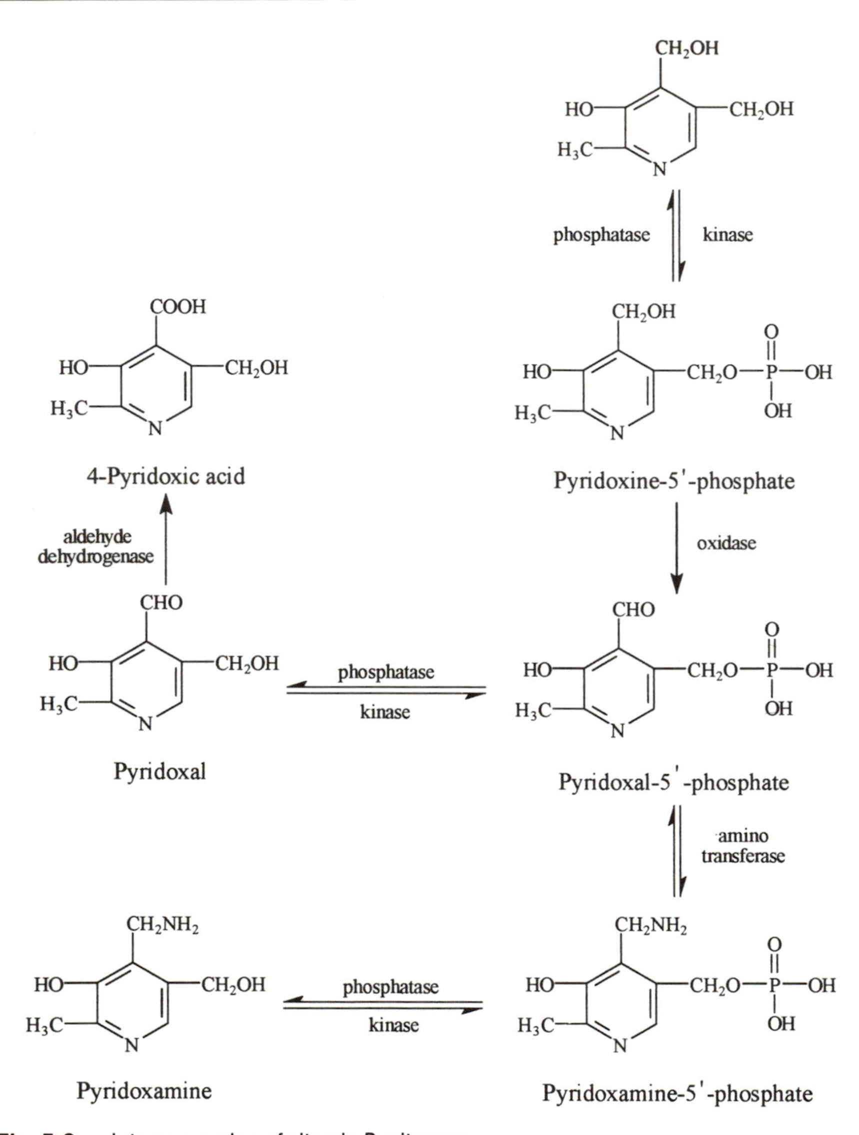

Fig. 5.2. Interconversion of vitamin B_6 vitamers.

vitamers is shown in Fig. 5.2. Pyridoxal-5′-phosphate and sometimes pyridoxamine 5′-phosphate are the coenzyme forms of vitamin B_6. Pyridoxine hydrochloride is the commonly available synthetic form. It has a molecular weight of 205.6, and is readily soluble in water. Pyridoxine is stable in acid solutions, but rapidly destroyed by light. Overall, it is far more stable than either pyridoxal or pyridoxamine. Hence, vitamin B_6 in plants which contain essentially pyridoxine is lost less than the vitamin in animal foods which contain mostly pyridoxal and pyridoxamine.

Sources

The vitamin is widely distributed in foods with no food being a particularly rich source. The richest sources are meats and cereals; fruits are generally poor sources and the form of vitamin B_6 in some vegetables is present as unavailable glycosides. In many foods the vitamin is bound to protein. Some examples of food contents of vitamin B_6 are shown in Table 5.1. In cereal grains the vitamin is concentrated in the germ and aleuronic layer; thus during the refining process of flour production, the removal of these layers results in a considerable reduction in their vitamin content. Plant tissues contain vitamin B_6 mostly as pyridoxine while animal tissues contain mostly pyridoxal and pyridoxamine. Vitamin B_6 is also synthesized by intestinal flora but there seems to be a question as to whether much of this is absorbed.

The vitamin B_6 in foods is stable under acidic conditions but unstable under neutral or alkaline conditions especially when exposed to heat or light. Because pyridoxine is more stable than pyridoxal or pyridoxamine, cooking losses of vitamin B_6 from plants are generally less than those from meats. However, since with the roasting and grilling of meats there is considerable loss of moisture, the concentration of the vitamin in meats cooked in these ways may be little different from those present in the raw meats (see Table 5.1). If foods are stored for a long time, that is for a year, losses of vitamin B_6 may be substantial (25–50%).

Bioavailability

The losses of vitamin B_6 that occur during the heating of food may be due to decomposition of the vitamin or loss of availability due to the formation of reaction products. Thus, the more reactive vitamers, pyridoxal and pyridoxamine and their phosphates, can bind to amino or sulphydryl groups of amino acid residues from proteins to form aldimines or Schiff bases. These in turn may be stabilized by chelating in the presence of metal ions. Under reducing conditions a stable aldimine or ketomine is formed, for example, pyridoxyl-ε-aminolysine.

Protein-bound forms of vitamin B_6 have a low availability in man. In rats, pyridoxyl-ε-aminolysine was found to possess approximately half the activity of the free form (Gregory and Kirk, 1977). Cysteine-bound pyridoxal has been identified in heat stabilized milk. Partial destruction of vitamin B_6 during sterilization and low availability of the remaining bound vitamin has resulted in symptoms of clinical deficiency in infants fed a commercial sterilized formula

Table 5.1. Concentrations (mg per 100 g) of vitamin B_6 in some common foods[1].

Cereals	
Bran (wheat)	1.38
Flour	
wholemeal	0.50
brown (85%)	0.30
white (72%)	0.15
Rice	
unpolished	0.55
polished	0.17
Bread	
Wholemeal	0.14
White	0.04
Dairy products	
Cow's milk (whole)	0.04
Human milk	0.01
Eggs	
Whole, raw	0.11
Meats	
Beef, lean average raw	0.32
Steak, raw lean + fat	0.27
Lamb, average raw	0.25
leg roast lean	0.22
Chicken, raw meat only	0.42
roast meat only	0.26
Pork, lean average raw	0.45
leg roast lean	0.41
Vegetables	
Green	0.16–0.18
Root	0.06–0.15
Potatoes new boiled	0.20
Fruits	
Various	0.02–0.04

[1]Sources: Paul and Southgate (1978); Combs (1992).

(Davies *et al.*, 1959). Pyridoxamine can react with carbonyl groups of reducing sugars and in the presence of ascorbic acid the inactive 6-hydroxy-pyridoxine can be formed (Tadera *et al.*, 1988).

Table 5.2. Concentrations (μg per 100 g) and proportions of vitamin B6 derivatives in selected foods[1].

	Total B6	%P	% Free	%PNG
Vegetables				
Carrots	206	18	29	54
Fruits				
Apple raw	104	33	52	15
Apple juice	87	26	65	9
Orange	83	39	33	28
Orange juice	55	42	36	23
Cereals				
Whole wheat bread	79	39	33	30
White bread	16	42	51	8
Nuts/seeds				
Almonds	137	5	95	0
Sunflower seeds	605	14	34	52
Soyabean	267	34	49	18

[1] Sources: Reynolds, 1988; Bitsh and Schramm, 1992.
%P = % phosphorylated vitamers; % Free = % non-phosphorylated vitamers; % PNG = % PN glucosides.

Vitamin B6 can react with equimolar amounts of glucose to form 5-*O*-β-glucopyranosyl-pyridoxine, a compound first identified in rice (Gregory and Ink, 1987). In rats, this compound was relatively well absorbed, with about 80% of the glucoside rapidly excreted in the urine. Relative to pyridoxine, the glucoside had a bioavailability of only 20–30% (Gregory *et al.*, 1991; Gilbert and Gregory, 1992). In man, orally administered glucoside was 58% utilized whereas given intravenously the glucoside was only 30% available. This difference suggests that the β-glucosidases of the intestinal mucosa, micro-flora or both contributed to the increased availability of the orally administered glucoside.

Glucosidic vitamin B6 accounts for 8–50% of the total in plant material. Generally, plant components with storage functions such as root vegetables, e.g. carrots, have the highest concentration of glucosidic B6 (Table 5.2). In some plant seeds, two glucosidic forms have been found. Thus, rice bran contains higher conjugated glucosides. Almonds seem to be unusual amongst nuts and seed in not containing any glucosidic B6 and this is due to their high natural glucosidase activity.

It appears that 70–80% of the vitamin B6 in the average American diet is available (Reynolds, 1988). In spite of the fact that glucosidic B6 is not found in animals, small amounts have been found in the breast milk of lactating vegetarian women; there was a correlation between the glucoside content of food and breast milk samples (Reynolds, 1988).

Absorption and Transport

All the forms of vitamin B_6 are freely absorbed by passive diffusion in the jejunum and ileum. There is some doubt whether the vitamin synthesized by microflora in the colon is absorbed since it is formed distal to the main absorption sites. This does not apply, of course, to coprophagous animals. In ruminants, microflora produce vitamin B_6 in the rumen in adequate amounts to meet their needs. The small intestine is able to absorb amounts of vitamin B_6 greatly in excess of physiological need. During the process of absorption phosphorylation and protein-binding occur in the mucosal cells and in the blood. The removal of the phosphate of pyridoxal and pyridoxamine phosphates is catalysed by the membrane-bound alkaline phosphatases during absorption. The resulting free vitamers, together with any free vitamers absorbed from the intestine, are phosphorylated in the jejunal mucosa by pyridoxal kinase. The phosphorylated forms of pyridoxine and pyridoxamine are then oxidized to pyridoxal phosphate.

Vitamin B_6 is transported in the blood predominantly as pyridoxal phosphate, which is bound to albumin in the plasma and haemoglobin in the erythrocytes, by Schiff-base linkages. Since pyridoxal crosses cell membranes more readily than pyridoxal phosphate, it is the form taken up by the tissues. This suggests that phosphatases are important for tissue uptake.

The predominant form of vitamin B_6 in the cells is pyridoxal phosphate produced by phosphorylation of pyridoxal by pyridoxal kinase. This is the main form in which the vitamin is stored in the body, though there is also some pyridoxamine phosphate. The total body store of vitamin B_6 has been estimated to be 40–150 mg, which is enough to meet 20–75 days' needs (Combs, 1992). In the tissues, the greatest levels are found bound to various proteins in liver, brain, kidney, spleen and muscle. In muscle, pyridoxal phosphate is bound mostly to glycogen phosphorylase. The plasma levels of different vitaminers of vitamin B_6 vary considerably in different species (Table 5.3).

Metabolism

The different vitamers of pyridoxine are interconvertible (Ink and Henderson, 1984), by reactions involving phosphorylation/dephosphorylation, oxidation/reduction and amination/deamination (Fig. 5.2). As mentioned above, because the non-phosphorylated vitamers cross cell membranes more readily, phosphorylation seems to be a way of retaining the vitamin within the cells. The limiting enzyme in vitamin B_6 metabolism appears to be pyridoxine phosphate

Table 5.3. Concentrations (nM) of vitamers of B_6 in the plasma of several species[1].

Species	Pal	Palp	Pine	Pam	PamP	Pac
Human	13	62	33	6	3	40
Pig	139	29	167	–	–	139
Calf	96	308	50	–	9	91
Sheep	57	626	43	–	466	318
Dog	268	417	66	–	65	109
Cat	139	2443	93	44	271	17

[1]Adapted from Combs (1992).
Pal = pyridoxal; Palp = pyridoxyl PO_4; Pine = pyridoxine; PamP = pyridoxamine PO_4; Pac = pyridoxic acid

oxidase which requires flavin mononucleotide (FMN). It follows that deprivation of riboflavin may reduce the conversion of pyridoxine and pyridoxamine to the active coenzyme, pyridoxal phosphate.

The metabolism of vitamin B_6 takes place mainly in the liver which contains all the necessary enzymes. In the liver, the major forms, pyridoxal phosphate and pyridoxamine phosphate, constitute an endogenous pool which is not readily accessible to newly formed molecules. The latter constitute a second pool which is readily mobilized, interconverted and passed into the blood.

The binding of pyridoxal phosphate to albumin protects it from degradation in the bloodstream. In the liver it is dephosphorylated and oxidized, probably by the FAD-dependent aldehyde oxidase and the NAD-dependent aldehyde dehydrogenase to yield 4-pyridoxic acid. This derivative is not biologically active and is excreted in the urine quantitatively after parenteral administration of the vitamin.

Pyridoxic acid is, then, the major excretory product of vitamin B_6 metabolism and is not detected in the urine of vitamin B_6-deficient individuals. The excretion of 4-pyridoxic acid can be used as a measure of vitamin B_6 status. Adequately nourished men and women excrete 0.5–1.2 mg day^{-1} and 0.40–1.1 mg day^{-1} respectively, in this form. Excretion of less than 0.5 mg (men) and 0.4 mg (women) is considered evidence of an inadequate intake (Combs, 1992).

Metabolic Functions

Over 60 vitamin B_6-dependent enzymes are known, and all of them act on amino acids in the reactions which include transamination, decarboxylation, dehydration, desulphydration, racemization, cleavage and synthesis. Examples of some of these enzymes are listed in Table 5.4. All the enzymes use pyridoxal phosphate (PLP) as the coenzyme. PLP also serves as a co-factor for the phosphorylases and acts as a modulator of protein structure. The mechanism of action

Table 5.4. Some important pyridoxal phosphate-dependent enzymes[1].

Type of reaction	Enzyme
Decarboxylations	Aspartate 1-decarboxylase
	Glutamate decarboxylase
	Ornithine decarboxylase
	Aromatic amino acid decarboxylase
	Histidine decarboxylase
R-group interconversions	δ-Aminolevulinic acid synthase
Transaminations	Aspartate aminotransferase
	Alanine aminotransferase
	γ-Aminobutyrate aminotransferase
	Cysteine aminotransferase
	Tyrosine aminotransferase
	Leucine aminotransferase
	Ornithine aminotransferase
	Glutamine aminotransferase
	Branched chain amino acid aminotransferase
	Serine-pyruvate aminotransferase
	Aromatic amino acid transferase
	Histidine aminotransferase
Racemization	Cystathionine β-synthase
α-β-Elimination	Serine dehydratase
γ-Elimination	Cystathionine γ-lyase
	Kynureninase

of PLP with its various apoenzymes always involves the formation of a Schiff base between the keto-C of the coenzyme and the ε-amino group of a specific lysol residue of the apoenzyme (Combs, 1992). Some of the reactions catalysed by the vitamin B_6-dependent enzymes are discussed below.

Transaminases (aminotransferases)

These enzymes are responsible for the catalysing transamination reactions in which α-amino acids are reversibly converted to α-keto acids with the concurrent transfer of the amino moieties to different α-keto acids, forming different amino acids. Most amino acids undergo these reversible transamination reactions, and these reactions are responsible for the synthesis of the non-essential amino acids from keto acids. The most common keto acids which accept amino (NH_2) groups to form amino acids include oxaloacetate (forming aspartate),

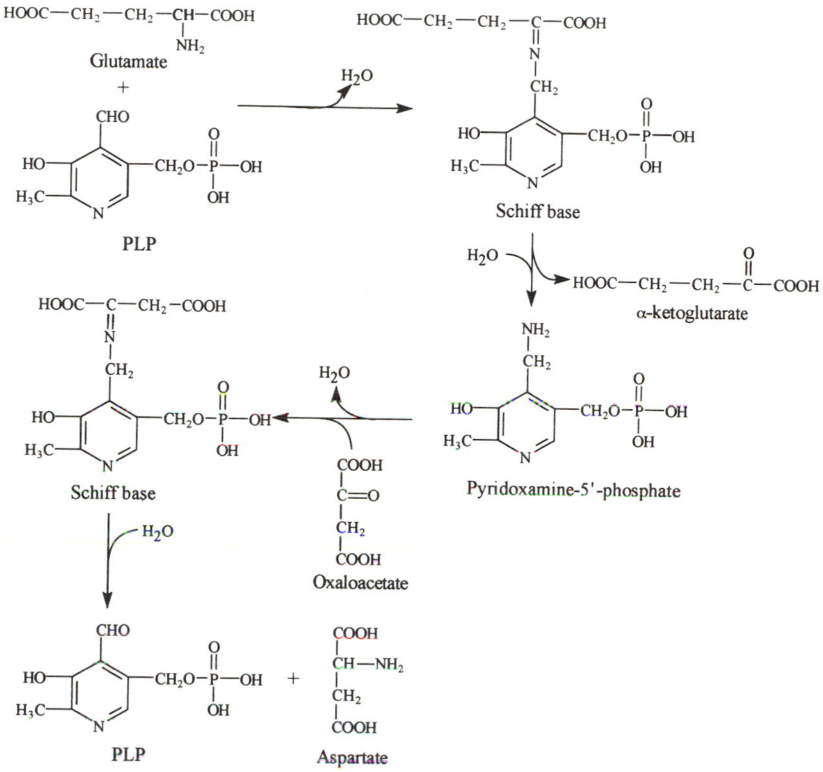

Fig. 5.3. Mechanism of action of vitamin B_6 in the synthesis of aspartate from keto-acids by transamination.

pyruvate (forming alanine), α-ketoglutarate (forming glutamate), and glyoxylate (forming glycine).

PLP forms an essential part of the active site of transaminases; all PLP-dependent reactions of amino acids stem from the condensation of an amino acid and the aldehyde of PLP forming a Schiff base (—C=N—CH= ⇌ —C—N=CH—). This intermediate can be rearranged releasing a keto acid forming enzyme-bound pyridoxamine phosphate. During transamination, the enzyme-bound pyridoxamine phosphate serves as an intermediate carrier of amino groups. In a subsequent step, the enzyme-bound pyridoxamine reacts with another keto acid to form again a Schiff base. This base is then hydrolysed to regenerate PLP and to form a non-essential amino acid corresponding to the second keto acid. The overall reaction to transamination in the process of aspartic acid synthesis from keto acids, glutamate and oxaloacetate, is shown in Fig. 5.3.

Table 5.5. Some neuroactive amines derived from amino acids or their derivatives under the action of vitamin B_6-dependent decarboxylases in the body.

Amino acid/derivative	Amine
Histidine	Histamine
5-Hydroxytryptophan	5-hydroxytryptamine (serotonin)
Aspartic acid	β-Alanine
Glutamic acid	γ-Aminobutyric acid (GABA)
Tyrosine	Noradrenaline/adrenaline Tyramine/dopamine
Cysteic	Taurine

Decarboxylases

These enzymes are PLP dependent and responsible for catalysing decarboxylation of amino acids. They attack amino acid at its carboxyl group and catalyse the decarboxylation, yielding CO_2 and an amino (Eq. 5.1). A large number of amino acid decarboxylases utilizing PLP are known. Some important neuroactive amines derived from amino acids or their derivatives are listed in Table 5.5. Examples of reactions involving vitamin B_6 are the biosynthesis of the neurotransmitters, serotonin (tryptophan decarboxylase) and γ-aminobutyric acid (glutamate decarboxylase), and adrenaline and noradrenaline (tyrosine decarboxylase). Hence, vitamin B_6 plays an important role in neurological functions. The mechanism of action of PLP is the decarboxylation of amino acids which also involves the formation of a Schiff base (Fig. 5.4).

$$\text{R-CH}_2\text{-NH}_2\text{-COOH} \xrightarrow[\text{Decarboxylases}]{\text{PLP}} \text{R-CH}_2\text{-NH}_2 + \text{CO}_2 \qquad \text{(Eq. 5.1)}$$

Enzymes involved in cysteine synthesis

The sulphydryl group of cysteine is derived from methionine via the homocysteine portion of the molecule (Fig. 5.5). The carbon skeleton of cysteine comes from serine. The two amino acids (e.g. homocysteine and serine) condense to form cystathionine under the influence of cystathionine synthase for which PLP is the coenzyme. Hydrolytic cleavage of cystathionine forms homoserine and cysteine. The cleavage involves the PLP-dependent cystathionase. In subsequent steps, cysteine can be degraded either to pyruvate by action of a desulphydrase or through oxidation to cysteic acid, which is finally

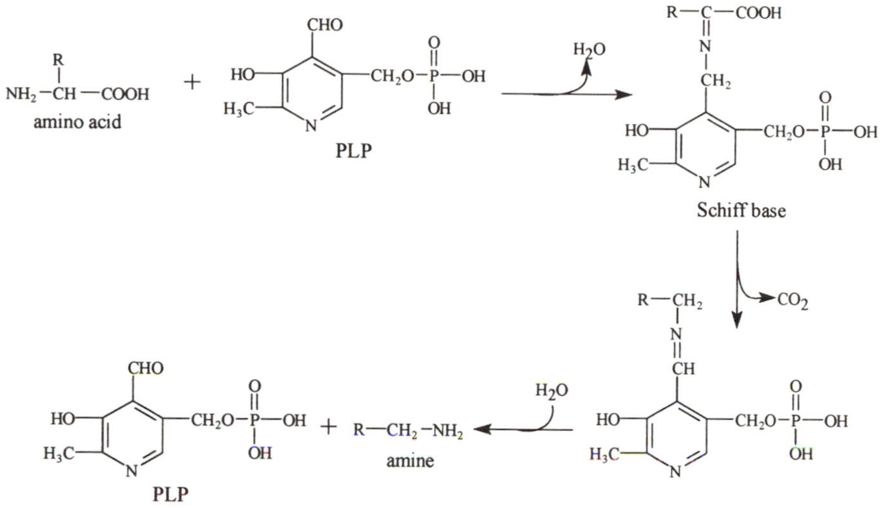

Fig. 5.4. Mechanism of action of pyridoxal-5′-phosphate (PLP) in the synthesis of biogenic amines by decarboxylation reaction.

decarboxylated to taurine by the PLP-dependent enzyme decarboxylase. Both the desulphydrase and decarboxylase enzymes are also dependent on PLP for their activities. Through pyruvate, the carbon skeletons of cysteic can enter the Krebs cycle and be used for energy or, through phosphoenolpyruvate, can be converted to glucose via the gluconeogenic process.

Kynureninase

This is another important enzyme for which PLP acts as a cofactor through formation of the Schiff base. This enzyme is involved in the degradation of tryptophan. Kynureninase catalyses the cleavage of 3-hydroxyanthranilate, which through stages forms the pyridine ring of nicotinamide (Fig. 5.6). This is an important pathway by which nicotinamide is synthesized in the body. Thus, a patient's symptoms may indicate a niacin deficiency when in fact the dietary deficiency is that of vitamin B6 (see Chapter 4).

Other vitamin B_6-dependent enzymes

Over half of the vitamin B6 in the body is accounted for by its involvement as a coenzyme of glycogen phosphorylase in the utilization of glycogen in muscle. The essential role of the vitamin in this connection is clear, but what is not clear is its mechanism of action; it may be directly involved in catalysis or in a

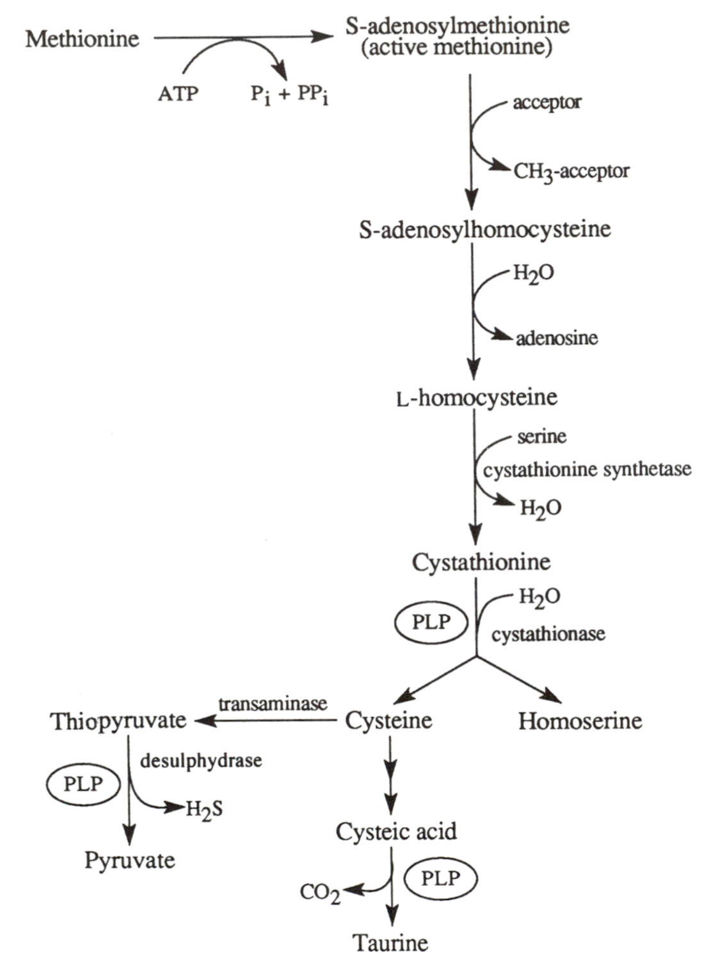

Fig. 5.5. Metabolism of sulphur amino acids involving pyridoxal-5′-phosphate (PLP).

structural role, e.g. as an allosteric effector of catalysis. Pyridoxal phosphate may also modulate steroid hormone receptors. It inhibits the induction of hepatic tyrosine aminotransferase by glucocorticoids, probably by forming Schiff-base linkages to the DNA-binding site of the receptor-steroid complex, to inhibit the binding to DNA and displace the complex from the nucleus (Combs, 1992).

Pyridoxal and its phosphate bind to haemoglobin and enhance the oxygen-binding capacity of the protein. This binding inhibits sickling in sickle cell haemoglobin. The binding of PLP to haemoglobin occurs at two sites on the β-chains – the *N*-terminal valine and lysine 82 residues; whereas pyridoxal binds at the *N*-terminal valine residues of the α-chains (Combs, 1992). Vitamin B_6 is also involved in the synthesis of haem, which is synthesized from glycine

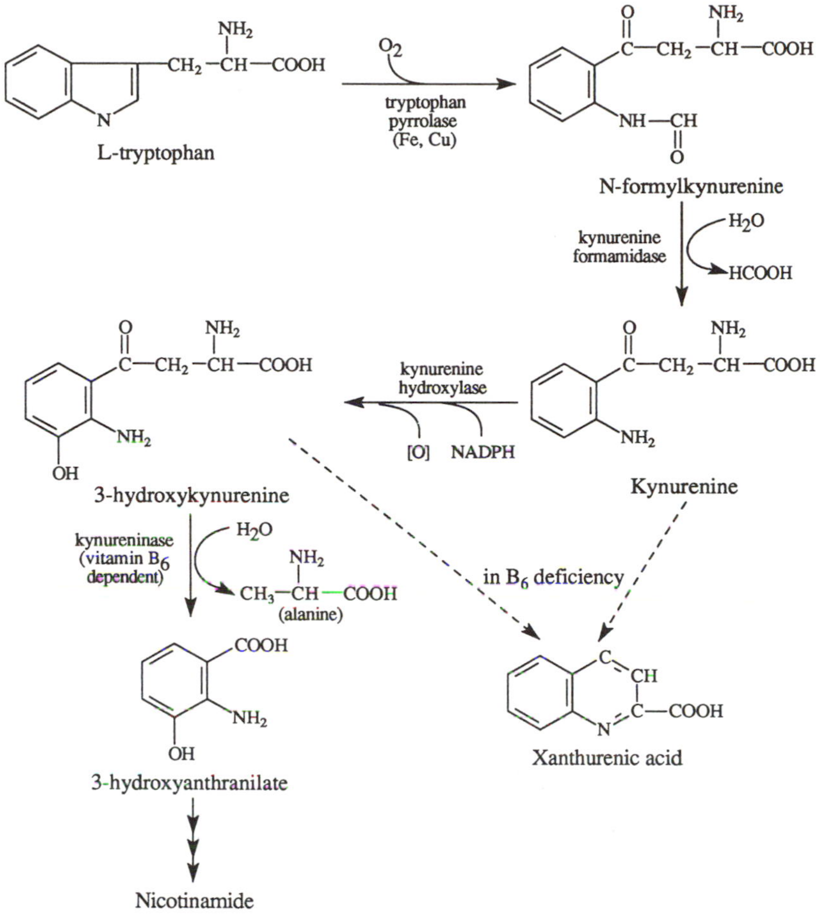

Fig. 5.6. Role of vitamin B_6 on the degradation of tryptophan.

and succinyl-CoA. Glycine provides all the nitrogens and succinyl-CoA provides most of the carbons in the porphyrin rings. Both the substances condense in the mitochondria to form δ-aminolevulinic acid, catalysed by δ-aminolevulinic acid synthetase which is PLP-dependent (Eq. 5.2). It is thought that the PLP is necessary in this reaction to activate glycine forming a Schiff base, whereby the α-carbon of glycine can be combined with the carbonyl carbon of succinate. The product of the condensation reaction between succinate and glycine is α-amino-β-aminolevulinic acid.

$$\text{Glycine} + \text{Succinyl-CoA} \xrightarrow[\text{PLP}]{\delta\text{-aminolevulinate synthetase}} \delta\text{-aminolevulinate} \rightarrow \text{Haem} \quad \text{(Eq. 5.2)}$$

Vitamin B_6 Deficiency

Substantial stores, possible availability of vitamin synthesized by intestinal flora and wide distribution in both plant and animal foods may contribute to the fact that deficiency of vitamin B_6 is rare in man. The first definite cases were reported in children in 1939 who were given a dried milk preparation in which the vitamin had been destroyed by overheating. These children showed an ill-defined syndrome which included weakness, irritability, nervousness, susceptibility to noise, insomnia, difficulty in walking and weight loss. Although some 300 children were involved and responded to vitamin B_6, only comparatively few of the children who received the food (about three per thousand) developed symptoms. Some of the children required 2.25–2.40 mg of the vitamin per day for recovery, well in excess of the recommended amount of 0.2–0.4 mg, suggesting variation in requirement. Human volunteers on a vitamin B_6-deficient diet become irritable, depressed and lose their sense of responsibility. Filiform hypertrophy of the lingual papillae, aphthous stomatitis, nasolabial seborrhoea and an acneiform rash of the forehead occurred. Abnormal electroencephalograms were also observed (Sauberlich and Canham, 1974).

Abnormalities of the nervous system are probably tc be explained in terms of the involvement of vitamin B_6 in the synthesis of neurotransmitters. The levels of brain GABA are reduced (Roberts *et al.*, 1964) and so too is the activity of glutamic acid decarboxylase, although the total amount of the enzyme protein is greatly increased (Bayoumi and Smith, 1972). However, the levels of other neurotransmitters (serotonin, dopamine, adrenaline and noradrenaline) are not reduced in spite of the fact that vitamin B_6 is a co-factor for the relevant enzymes. The explanation (Barker and Bender, 1980) appears to lie in the higher affinity of the aromatic amino acid decarboxylase apoenzyme compared with glutamic acid decarboxylase.

The vitamin B_6 responsive hypochromic microcytic anaemia occasionally found in humans, and in deficient animals, is probably due to depressed synthesis of δ-aminolevulinic acid (Eq. 5.2). Synthesis of this intermediate in the formation of haem and porphyrins is pyridoxal phosphate dependent. The changes in the skin and those in lipid metabolism that occur in vitamin B_6 deficiency cannot at present be explained in terms of the human metabolism of the vitamin.

In addition to the deficiency syndrome, low blood levels of vitamin B_6 occur in a number of clinical conditions (Wilson and Davis, 1983). Two kinds of deficiency can be distinguished: 'conditional deficiency', which occurs when the dietary intake is adequate with respect to recommended values; and 'relative deficiency', which occurs when the primary intake is inadequate in relation to increased demands due to metabolic activity, infection, pregnancy and infancy (Dickerson and Williams, 1990).

Conditional deficiency may be due to: (i) defective intestinal absorption because of coeliac disease, gastroenteritis, Crohn's disease, jejunal-ileal bypass and kwashiorkor; (ii) defective cellular and intercellular transport; and (iii) impaired oxidation or phosphorylation mechanisms in vitamin B_6 metabolism. Some examples of the latter include primary acquired deficiency in which there may be a block in the conversion of pyridoxine to pyridoxine phosphate, possibly due to riboflavin deficiency; toxaemia of pregnancy in which the toxaemic placenta may not be able to convert pyridoxal phosphate due to a low level of pyridoxal phosphate kinase; and low levels of this enzyme in the serum of newborn preterm infants may be one factor increasing their risk of vitamin B_6 depletion.

Requirements

Calculations from the half-life of 33 days derived from the estimated body pool size give an average daily requirement of 0.6–3.8 mg day^{-1}. Estimates based on changes in tryptophan and methionine metabolism, and blood concentrations of vitamin B_6 during depletion and repletion of adults maintained on controlled diets, have yielded rather more useful values. The requirement has been shown to be greater and depletion to develop faster on high protein intakes (80–160 g day^{-1}) than on low protein intakes (30–50 g day^{-1}). Based on these studies, the 'average estimated requirement' was estimated at 13 μg g^{-1} protein for individuals of one year of age and over (Department of Health, 1991). The corresponding 'reference nutrient intake' or 'recommended daily allowance' for the UK is 15 μg g^{-1} protein. The estimated allowance of vitamin B_6 per day is 1.4 mg for males aged 19 years and over and 1.2 mg for females aged 15 years and over. The value for infants during the first six months is 0.2 mg day^{-1}. The values for adults in the US are 2.0 mg day^{-1} for men and 1.5 mg day^{-1} for women (National Research Council, 1989).

Assessment of Vitamin B_6 Status

Status of vitamin B_6 may be assessed by measuring serum levels of pyridoxal phosphate and the urinary excretion of 4-pyridoxic acid. Alternatively, the activity of dependent enzymes and particularly their degree of stimulation or activation may be measured. An enzyme often used is red cell aspartate transaminase (AST) (Marsh *et al.*, 1955). As discussed earlier, the reaction by which kynurenine and hydroxykynurenine are converted to hydroxyanthranilic acid is catalysed by the enzyme kynureninase, which requires PLP as coenzyme. A deficiency of vitamin B_6 results in some degree of failure to catabolize these kynurenine derivatives, which thus reach various extrahepatic tissues where they are converted to xanthurenic acid (Fig. 5.6). This abnormal metabolite has been identified in the urine when dietary intake of vitamin B_6 is inadequate. The excretion of xanthurenic acid can be markedly increased by giving a loading dose (100 mg kg^{-1} body weight) of tryptophan. A high excretion of xanthurenic acid (> 25 mg per 6 hours) following a loading dose of tryptophan, is indicative of vitamin B_6 deficiency. A description of these and other methods of assessing vitamin B_6 status is given by Gibson (1990).

References

Barker, B.M. and Bender, D.A. (1980) Vitamin B_6. In: Barker, B.M. and Bender, D.A. (eds) *Vitamins in Medicine*, 4th edn, Vol. 1. Heinemann Medical, London, pp. 348–380.

Bayoumi, R.A. and Smith W.R.D. (1972) Some effects of dietary vitamin B_6 on γ-amino-butyric acid metabolism in developing rat brain. *Journal of Neurochemistry* 19, 1883–1897.

Bitsh, R. and Schramm, W. (1992) Free and bound vitamin B_6 derivatives in plant foods. *Chemical Reactions in Foods II*, Fecs Event No. 174, 285–290.

Combs, G.V. (1992) Vitamin B_6. In: *The Vitamins*. Academic Press, New York, pp. 313–328.

Davies, M.K., Gregory, M.E. and Henry, K.M. (1959) The effect of heat on the vitamin B_6 of milk. II. A comparison of biological and micro-biological tests of evaporated milk. *Journal of Dairy Research* 26, 215–220.

Department of Health (1991) Dietary references values for food energy and nutrients for the United Kingdom. *Report on Health and Social Subjects*, No. 41, H.M. Stationery Office, London.

Dickerson, J.W.T. and Williams, C.M. (1990) Vitamin-related disorders. In: Cohen, R.D., Lewin, B., Alberti, K.G.M.M., and Denman, A.M. (eds) *The Metabolic and Molecular Basis of Acquired Disease*. Baillière Tindall, London, pp. 634–669.

Gibson, R.S. (1990) In: *Principles of Nutritional Assessment*. Oxford University Press, New York, pp. 391–397.

Gilbert, J.A. and Gregory, J.F. (1992) Pyridoxine-5-β-D-glucoside affects the metabolic utilization of pyridoxine in rats. *Journal of Nutrition* 122, 1029–1035.

Gregory, I.F. and Ink, S.L. (1987) Identification and quantification of pyridoxine-β-glucoside as a major form of vitamin B_6 in plant derived foods. *Journal of Agricultural and Food Chemistry* 35, 76–82.

Gregory, J.F. and Kirk, J.R. (1977) Effect of ε-pyridoxyllysine bound to dietary protein on the vitamin B_6 status of rats. *Journal of Nutrition* 110, 995–1005.

Gregory, J.F., Trumbo, P.R., Bailey, L.B., Baumgartner, T.G. and Cerda, J.J. (1991) Bioavailability of pyridoxine-5-β-D-glucoside determined in humans by stable-isotope methods. *Journal of Nutrition* 121, 177–186.

Gunsalus, I.C., Bellamy, W.S. and Umbreit, W.W. (1944) A phosphorylated derivative of pyridoxal as the coenzyme of tyrosine decarboxylase. *Journal of Biological Chemistry* 155, 685–689.

Gyorgy, P. (1934) Vitamin B_2 and pellagra-like dermatitis in rats. *Nature* 133, 498–499

Ink, S.L. and Henderson, L.M. (1984) Vitamin B_6 metabolism. *Annual Review of Nutrition* 4, 455–470.

Marsh, M.E., Greenberg, L.D. and Rhinehart, J.F. (1955) The relationship between B_6 investigation and transaminase activity. *Journal of Nutrition* 56, 115–127.

National Research Council (1989) In: *Recommended Dietary Allowances*. 10th edn. National Academy Press, Washington, D.C.

Paul, A.A. and Southgate, D.A.T. (1978) McCance and Widdowson's *'The Composition of Foods'*, Fourth Revised Edition. Her Majesty's Stationery Office, London.

Reynolds, R.D. (1988) Bioavailability of vitamin B_6 from plant food. *American Journal of Clinical Nutrition* 48, 863–867.

Roberts, E., Vein, J. and Simonsen, D.G. (1964) γ-Aminobutyric acid and neuronal function, a speculative synthesis. *Vitamins and Hormones* 22, 503–559.

Sauberlich, H.E. and Canham, J.E. (1974) Vitamin B_6. In: Goodhart, R.S. and Shils, M.E. (eds) Modern Nutrition in Health and Disease, 5th edn. Lea and Febiger, Philadelphia, pp. 203–209.

Tadera, K., Kaneko, T., and Yagi, F. (1988) Isolation and structural elucidation of three new pyridoxine glucosides in rice bran. *Journal of Nutritional Science and Vitaminology* 34, 167–177.

Wilson, R.G. and David, R.E. (1983) Clinical chemistry of vitamin B_6. *Advances in Clinical Chemistry* 23, 1–68.

Biotin 6

Chemistry

Biotin is a sulphur containing bicyclic compound in which tetrahydrothiophene and imidazolidone rings are fused, and there is a valeric acid as side chain (Fig. 6.1). The molecular structure of biotin contains three asymmetric carbon atoms. Eight different stereoisomers are therefore possible, of which only the dextro-rotatory, so-called D-biotin (*cis*-form, β-isomer: mol. wt 244.3), is ordinarily found in nature. This is the only isomer that is biologically active as a coenzyme. Biotin is sensitive to heat, especially under conditions that support simultaneous lipid peroxidation. Therefore, solvent extraction, heat curing, and canning of foods can result in losses of the vitamin in appreciable amounts.

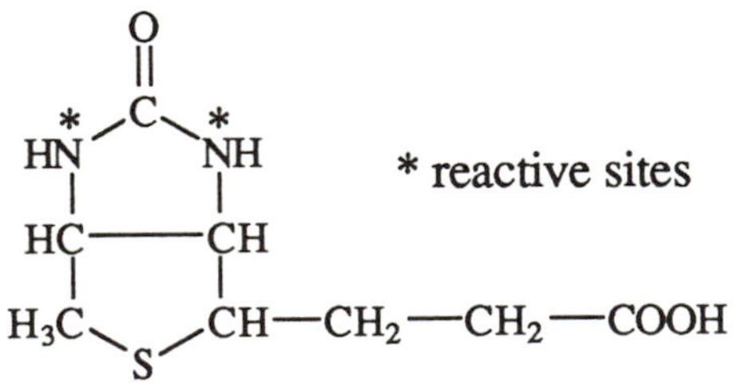

O
C
HN* NH*
HC CH
H_3C S $CH-CH_2-CH_2-COOH$

* reactive sites

Fig. 6.1. Structure of biotin.

Sources

Biotin is widely distributed in nature and is present in almost all foods. The best dietary sources of biotin are organ meats, egg yolk and yeast, followed by some vegetables such as cauliflower, legumes, mushrooms and nuts. All other meats, dairy products and cereals are generally considered relatively poor sources.

Biotin is synthesized by many different microorganisms and fungi. Careful balance studies in man have shown that the combined urinary and faecal excretion of this vitamin exceeds its dietary intake. Furthermore, biotin deficiency can be induced more readily in animals that have been given sulphonamide drugs which reduce intestinal bacteria to a minimum. This evidence suggests that a large proportion of the biotin requirement is probably supplied by the action of intestinal bacteria (McCormick, 1975; Bonjour, 1977). Indeed, a variety of microorganisms including the bacterium *Azotobacter agilis*, the yeast *Sporobolomyces* and the mould *Aspergillus oryzae* have been found to be able to biosynthesize biotin. Microorganisms can synthesize biotin, but the extent to which this biotin is available to humans remains uncertain. It is, therefore, difficult to arrive at a quantitative requirement of this vitamin. Usual western diets contain between 150 and 300 μg day^{-1} which is probably supplemented by intestinal synthesized biotin.

Absorption and Transport

Much of the biotin in food occurs bound to protein. This protein-bound biotin is first liberated as biotinyl peptides or biocytin (see below) by the action of gastrointestinal proteases. Subsequently, free biotin is released from the ε-amide bond with lysine by the intestinal enzyme, biotinidase (Dakshinamurti *et al.*, 1987). Little is known about the mechanism of absorption of biotin in humans. Animal studies have indicated that it is absorbed by facilitated diffusion at low concentrations and by simple diffusion at high concentrations. It is thought that biotin is transported in the circulation by a biotin-binding glycoprotein. The vitamin is generally cleared from the circulation more rapidly in deficient than in normal rats, suggesting that the tissue uptake of biotin may be related to its need.

Activation of Biotin as a Coenzyme

Within a cell D-biotin is covalently bound by an amide linkage to an ε-amino group of a lysine moiety in the enzyme in which biotin functions as a cofactor in carboxylation reactions. The combination of biotin–enzyme (biotinyllysine) is termed 'biocytin' (commonly referred to as a non-avidin binding protein) in which the amide linkage is facilitated by ATP (Fig. 6.2). Biotin is first cleared from biocytin by an enzyme, called biotinidase. It then functions in carboxylation reactions, whereby bicarbonate is covalently attached to biotin, forming carboxybiocytin. Charging the biotin enzyme with CO_2 is an endergonic process requiring ATP and Mg^{2+}. The CO_2 molecule attached to the nitrogen of the biotin molecule is the active form of CO_2 which is transferred to one of several

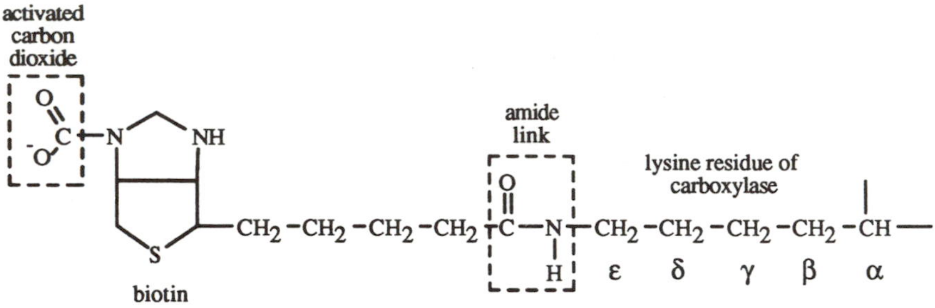

Fig. 6.2. Carboxybiotin.

substrates, and thereby participates in many important carboxylation reactions.

Metabolic Role

A number of carboxylation reactions involving biocytin permit the conversion of odd-numbered hydrocarbons into even-numbered hydrocarbons which are metabolized in the major metabolic pathways involving carbohydrates and lipids. Four of these reactions are known to be important in mammalian metabolism (Dakshinamurti and Chauhan, 1988). The carboxylase enzymes involved in the reactions are pyruvate carboxylase, acetyl-CoA carboxylase, propionyl-CoA carboxylase, and 3-methylcrotonyl-CoA carboxylase. Biotin serves as the prosthetic group of these enzymes, and thereby functions as a mobile carboxyl carrier. These biotin-dependent enzymes have important roles in gluconeogenesis and lipogenesis, fatty acid biosynthesis, proprionate metabolism, and in the catabolism of leucine.

Pyruvate carboxylase

This biotin-dependent enzyme is present in the mitochondrial matrix. It not only provides the four carbon units for the TCA cycle but also catalyses the formation of oxaloacetate from pyruvate, an important step for gluconeogenesis (Fig. 6.3). Pyruvic acid is used for gluconeogenesis if there is a surplus of ATP in the cell. When there is a deficiency of ATP, the oxaloacetate enters the TCA cycle upon condensation with acetyl CoA.

Pyruvic carboxylase is generally activated when there is a high level of acetyl CoA. The rise in acetyl CoA may be due to a high level of ATP so that the acetyl CoA is not needed for the production of energy through the TCA cycle, or

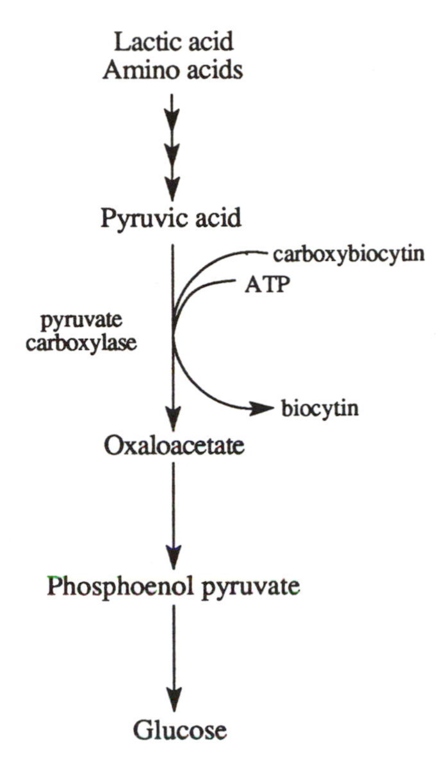

Fig. 6.3. Role of biotin in gluconeogenesis.

the acetyl CoA entry into the TCA cycle may be slowed down by a lack of carbohydrate intermediate, principally oxaloacetate. In either case, the stimulation of the pyruvate carboxylase reaction will provide oxaloacetate that can either lead through phosphoenolpyruvate to the synthesis of glucose, or combine with acetyl CoA to form citrate, bringing acetyl CoA into the TCA cycle for the production of energy.

Acetyl CoA carboxylase

This is another biotin carboxylase, which is a key regulatory enzyme in the extramitochondrial biosynthetic pathway for fatty acids. The first reaction of fatty acid synthesis is the carboxylation of acetyl CoA (a two-carbon compound) to yield malonyl CoA (a three-carbon compound) taking place in the cytosol (Fig. 6.4). Although acetyl CoA is the building block of long-chain fatty acids, their biosynthesis in animal tissues is not a direct reversal of fatty acid oxidation.

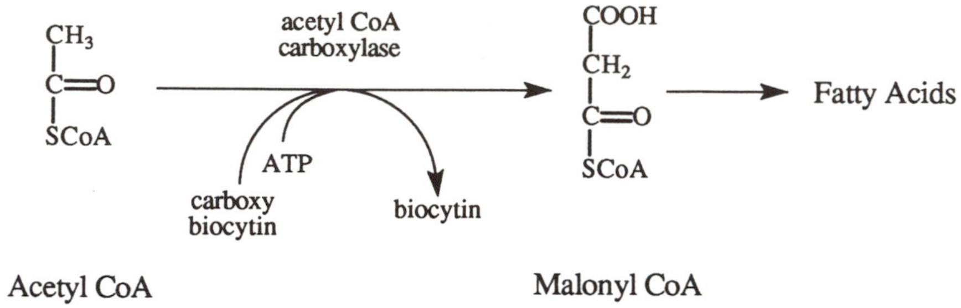

CH_3–C(=O)–SCoA

acetyl CoA carboxylase

ATP

carboxy biocytin → biocytin

HOOC–CH_2–C(=O)–SCoA → Fatty Acids

Acetyl CoA Malonyl CoA

Fig. 6.4. Role of biotin in the biosynthesis of fatty acids.

Instead, acetyl CoA is carboxylated to form malonyl CoA in the presence of the enzyme acetyl CoA carboxylase in a reaction dependent on ATP and biotin (Fig. 6.4). The positive modulator of acetyl CoA carboxylase is citric acid. This acid is not consumed in fatty acid synthesis, but it participates by carrying acetyl CoA out of the mitochondria, where it was formed from pyruvate, and into the cytosol, where fatty acid synthesis takes place.

Propionyl CoA carboxylase

Propionic acid is an oxidation product of fatty acids containing an odd number of carbon atoms. It may also arise as a catabolic product of branched chain amino acids, or as a fermentation product of microorganisms in the gastrointestinal tract. The biotin-dependent enzyme, propionyl CoA carboxylase, is known to add an activated CO_2 to propionyl CoA and forms the D-isomer of methylmalonyl CoA in the degradative path of odd-carbon fatty acids (Fig. 6.5). Methylmalonyl CoA is then converted to succinyl CoA (also see Chapter 9), an intermediate in the TCA cycle.

β-Methylcrotonyl CoA carboxylase

Biotin is also involved in the activation of β-methylcrotonyl CoA carboxylase, which plays a role in the catabolism of the ketogenic amino acid, leucine. This biotin-dependent enzyme, with mechanisms similar to the ones already discussed, converts β-methylcrotonyl CoA to β-methylglutaconyl CoA (Fig. 6.6). This reaction is an intermediate step in the oxidative degradation of leucine to acetoacetate and acetyl CoA.

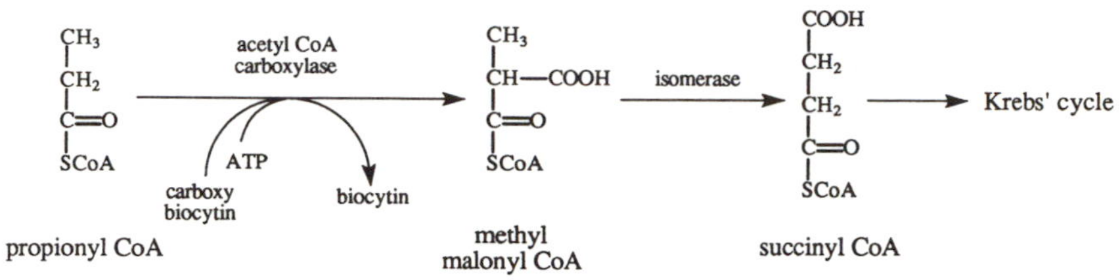

Fig. 6.5. Role of biotin in the metabolism of odd carbon fatty acids.

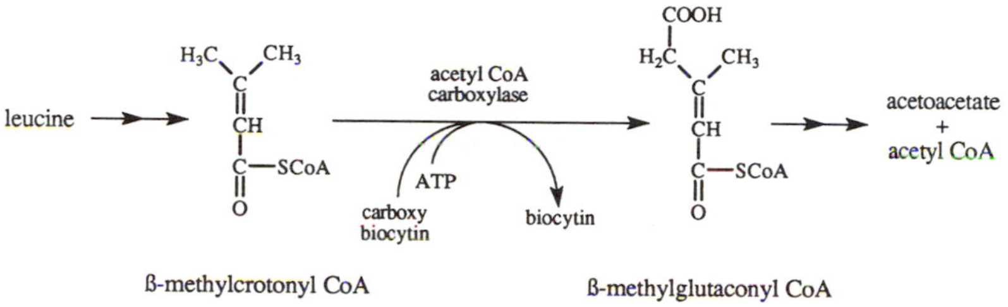

Fig. 6.6. Role of biotin in the catabolism of leucine.

Desaturases

In addition to the above metabolic reactions involving biotin-dependent carboxylases, recent studies have shown that biotin can also be an essential factor for the metabolism of essential fatty acids (Marshall, 1987). An impairment in chain elongation of linoleic acid, and hence reduced biosynthesis of prostaglandins, has thus been demonstrated in biotin-deficient rats and chicks. Since desaturase enzymes are involved in the chain elongation of linoleic acid (Fig. 6.7), biotin is thought to be involved in activating these enzymes.

Deficiency

In spite of the key carboxylation reactions in crucial metabolic pathways performed by biotin-requiring enzymes, no naturally occurring biotin-deficiency symptoms have been reported in adults. This is essentially because biotin is widespread in food. The amount obtained from an average western diet plus any benefit derived from bacterial synthesis appear to be enough to satisfy the daily requirement of the vitamin. Although biotin deficiency is rarely seen in humans, its low levels in blood and urine have been reported in infants with seborrhoeic dermatitis, alcoholics and pregnant women (Bonjour, 1977). However, most reports found in the literature describing biotin deficiency are based on the

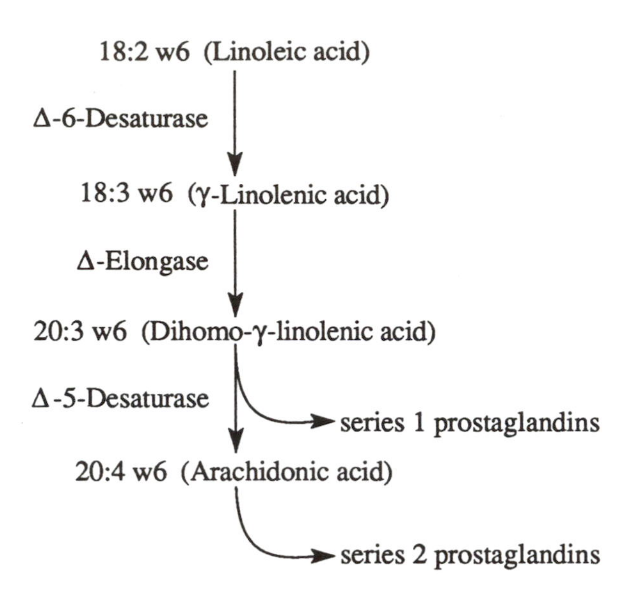

Fig. 6.7. Biosynthesis of prostaglandins (desaturases are possibly biotin dependent).

antagonistic effect of raw eggs and the genetical defects of biotin-dependent enzymes.

Avidin-induced deficiency

Raw egg white contains avidin, a glycoprotein (mol. wt 60,000), which has a high affinity for biotin. It binds biotin in a complex that the body neither can break to release biotin nor can absorb the whole complex molecule. The complex is undissociated over a wide range of pH, but it is broken down by heating or irradiation. Experimentally it has been shown that the diet has to provide as much as 30% of its calories from raw egg white in order to induce a biotin deficiency. Isolated cases have been reported of biotin deficiency induced by excessive consumption of raw eggs (Mistry, 1980). Seborrhoeic dermatitis is the major deficiency symptom in man. Loss of appetite, alopecia together with organic aciduria and a propensity toward ketosis have been observed in experimentally produced deficiency. It should be pointed out that the occasional ingestion of raw egg white is not going to contribute toward precipitating a deficiency state. Furthermore, avidin is heat labile; thus the cooking of eggs destroys the biotin-binding property of raw egg albumin.

Deficiency of carboxylases

There appear to be two clinical forms of biotin deficiency: a neonatal form and a juvenile form. Both conditions are referred to as multiple carboxylase deficiency (MCD), in which deficiencies of more than one biotin-dependent carboxylase are seen (Dakshinamurti and Chauhan, 1988). The neonatal form manifests itself in the first few weeks of life with metabolic acidosis and ketosis accompanied by increased urinary excretions of 3-methyl-crotonylglycine and 3-hydroxypropionic acid. The excessive levels of the urinary metabolites are believed to be the reflections of deficiencies of propionyl-CoA and β-methylcrotonyl-CoA carboxylases. The juvenile form, appearing after 2–3 months, presents with alopecia, erythematous rash, keratoconjunctivitis, ataxia, and lactic acidosis (Thoene *et al.*, 1981). This condition has been related to the deficiency of biotinidase (Wolf *et al.*, 1983).

The underlying causes of both neonatal and juvenile forms of biotin deficiency appear to be diverse. They may be caused by a defect in carboxylase apoenzymes as well as by a failure in intestinal absorption. Some of these cases have been reported to respond to oral or parenteral administration of 10 mg or more of biotin per day with normalization of the clinical and biochemical abnormalities (Mercer and Baugh, 1978). The therapeutic effectiveness of oral biotin, 5 mg twice daily to a child with the deficiency of carboxylases, is shown in Fig. 6.8.

Other contributory factors to biotin deficiency

Since a proportion of biotin is believed to be produced by endogenous bacteria, treatment with antibiotics for a long time can potentially precipitate biotin deficiency. This is especially true if raw egg whites are taken while on antibiotics (Bonjour, 1977). Biotin deficiency has also been reported in persons with Laennec's cirrhosis taking raw eggs. Until recently, the solutions for total parenteral nutrition (TPN) did not contain biotin. As a result, long term TPN resulted in a number of cases of biotin deficiency, associated with clinical signs, such as skin rash, alopecia and neurological symptoms (Mock *et al.*, 1985). Addition of biotin in amounts of 100–1000 mg day^{-1} to the parenteral nutrition preparation reverses or prevents symptomatic biotin deficiency. Currently available commercial solutions for TPN do, however, contain biotin.

Sudden infant death syndrome (SIDS)

SIDS or 'cot death' or 'crib death' as more popularly called in North America is defined as an unexpected death of an apparently well infant. It is one of the most common causes of death in infancy with a peak incidence at around 2–4 months

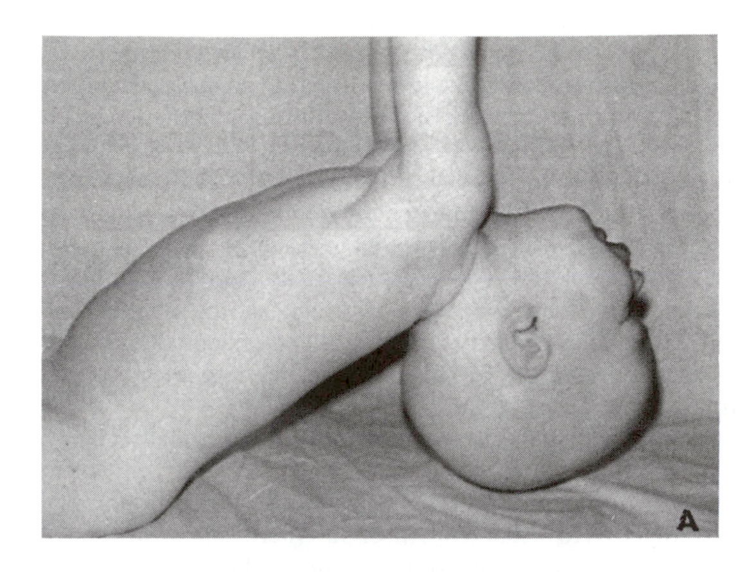

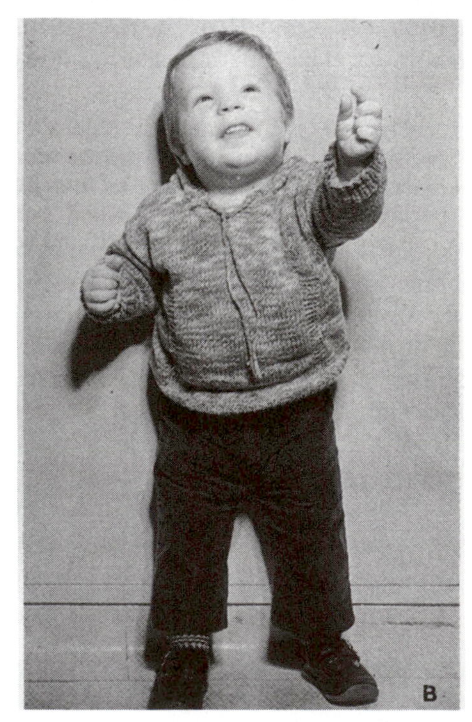

Fig. 6.8. (A) A child (14 months old) showing alopecia and lack of head control. (B) The same child after 4 months of biotin therapy (5 mg twice daily) showing normal posture and hair growth. Source: McLaren, 1981 (with permission).

Table 6.1. Biotin values in normal human biological fluids.

Biological fluids	Mean	Range
Blood (pg ml^{-1})	485	215–750
Serum (pg ml^{-1})	400	200–700
Urine (μg 24 h^{-1})	29	6–50

Adapted from Baker (1985).

of age. Its occurrence is world wide. The cause of death is cardiorespiratory arrest although the aetiology is still unknown.

Reports have suggested a link between SIDS and biotin deficiency (Johnson and Hood, 1980). Such a deficiency would result in a lowering of blood sugar levels, perhaps due to a deficiency of pyruvate carboxylase. Normally this would be of little significance but under stress the body's extra demand for sugar cannot be met and fatal hypoglycaemia may result. Typical examples of mild stress are infection, a missed meal, excessive heat or cold or a changed environment. Liver samples of infants who died of unexplained causes were found to have low levels of biotin. Deaths from SIDS are more common among bottle-fed babies. It may be significant that some infant formulas are deficient in biotin. In North America, however, infant formulas are now required to contain 1.5 mg biotin per 100 kcal.

Requirements

Because of the uncertain contribution of gut microbial synthesis to total biotin availability, a recommended dietary allowance for the vitamin cannot be established. In North America and Western Europe the daily dietary intake of biotin for an adult is estimated to be between 50 and 300 μg. Since there is no evidence of spontaneous biotin deficiency in human adults, the amount usually ingested, in addition to vitamins available from intestinal synthesis, is thought to be well in excess of minimal biotin requirements under normal conditions. A range of 30–100 μg day^{-1} is provisionally recommended to be safe and adequate for adults (National Research Council, 1989). In Canada the suggested intake for biotin is expressed as 1.5 μg kg^{-1} of body weight per day for all age groups (Health and Welfare Canada, 1990).

Assessment of Biotin Status

Biochemical assessment of biotin is normally carried out by measuring its concentration in blood, serum and urine. Urinary biotin level is thought to be a better reflection of dietary intake than its blood or serum level. Normal biotin values in human biological fluids are shown in Table 6.1.

References

Baker, H. (1985) Assessment of biotin status: clinical implications. *Annals of the New York Academy of Sciences* 447, 129–132.

Bonjour, J.P. (1977) Biotin in man's nutrition and therapy – a review. *International Journal of Vitamin and Nutrition Research* 47, 107–118.

Dakshinamurti, K. and Chauhan, J. (1988) Regulation of biotin enzymes. *Annual Review of Nutrition* 8, 211–233.

Dakshinamurti, K., Chauhan, J. and Ebrahim, H. (1987) Intestinal absorption of biotin and biocytin in the rat. *Bioscience Report* 7, 667–673.

Health and Welfare Canada (1990) *The Report of the Scientific Review Committee*, Canadian Government Publishing Centre, Ottawa.

Johnson, A.R. and Hood, R.L. (1980) Biotin and the sudden infant death syndrome. *Nature* 285, 159–160.

McCormick, D.B. (1975) Biotin. *Nutrition Review* 33, 97–102.

McLaren, D.S. (1981) *A Colour Atlas of Nutritional Disorders*. Wolfe Medical Publications, London.

Marshall, M.W. (1987) The nutritional importance of biotin – an update. *Nutrition Today* 22, 26–30.

Mercer, L. and Baugh, C.M. (1978) Effect of specific nutrient deficiencies in man: Biotin. In: Rechcigl, M. (ed.) *CRC Handbook Series in Nutrition and Food* CRC Press, Florida.

Mistry, S.P. (1980) Biotin. In: Barker, B.M. and Bender, D.A. (eds) *Vitamins in Medicine*, 4th edn, Vol. 1. William Heinemann, London.

Mock, D.M., Baswell, D.L. and Baker, H. (1985) Biotin deficiency complicating parenteral alimentation: Diagnosis, metabolic repercussions and treatment. *Annals of the New York Academy of Sciences* 447, 314–333.

National Research Council (1989) *Recommended Dietary Allowances*, 10th edn., National Academy Press, Washington, D.C.

Thoene, J., Baker, H., Yoshino, M. and Sweetman, L. (1981) Biotin-responsive carboxylase deficiency associated with sub-normal plasma and urinary biotin. *New England Journal of Medicine* 304, 817–820.

Wolf, B., Grier, R.E., Allen, R.J., Goodman, S.I. and Kien, C.L. (1983) Biotinidase deficiency: the enzymatic defect in late-onset multiple carboxylase deficiency. *Clinica Chimica Acta* 131, 273–281.

Pantothenic Acid 7

Unlike the other B-vitamins, pantothenic acid does not function independently as a coenzyme, rather it forms a portion of a much larger coenzyme, coenzyme A (CoA or CoASH). Our understanding of the metabolic role of pantothenic acid began in 1945 with the discovery of CoA and the subsequent identification of pantothenic acid as one of the constituents of CoA.

Chemistry

Chemically, pantothenic acid consists of pentoic acid linked by a peptide bond to β-alanine (Fig. 7.1). There is no enzyme system in mammals either for forming the amide bond between β-alanine and pentoic acid or for hydrolysing it. Its chemical name is (+) α,γ-dihydroxy-β-β-dimethylbutyryl-β-alanine (mol. wt 219). It is hydrolysed in acid or alkaline media to pantoic acid and alanine.

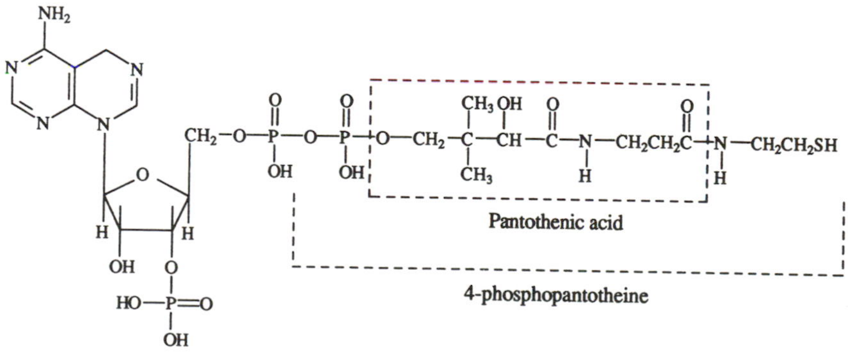

Pantothenic acid

4-phosphopantotheine

Fig. 7.1. Coenzyme A showing pantothenic acid as one of its components.

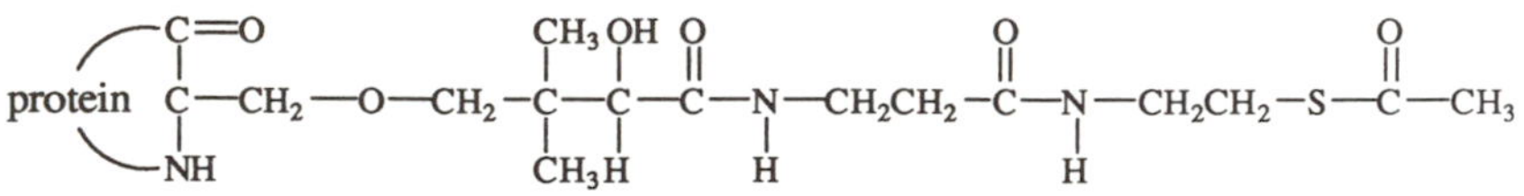

$$\text{protein}\ \overset{\overset{\displaystyle C{=}O}{|}}{\underset{\underset{\displaystyle NH}{|}}{C}}{-}CH_2{-}O{-}CH_2{-}\overset{\overset{\displaystyle CH_3}{|}}{\underset{\underset{\displaystyle CH_3}{|}}{C}}{-}\overset{\overset{\displaystyle OH}{|}}{\underset{\underset{\displaystyle H}{|}}{C}}{-}\overset{\overset{\displaystyle O}{||}}{C}{-}\underset{\underset{\displaystyle H}{|}}{N}{-}CH_2CH_2{-}\overset{\overset{\displaystyle O}{||}}{C}{-}\underset{\underset{\displaystyle H}{|}}{N}{-}CH_2CH_2{-}S{-}\overset{\overset{\displaystyle O}{||}}{C}{-}CH_3$$

Fig. 7.2. Acyl carrier protein containing pantothenic acid.

The naturally occurring form of pantothenic acid is the D-isomer. It is optically active; only the dextro-rotatory form is effective as a vitamin. The L-isomer has no biological activity but in rodents, it competitively inhibits the D-isomer.

Free pantothenic acid is an unstable, extremely hygroscopic oil. It is used mainly in the form of calcium and sodium salts, which are readily soluble in water. D-Panthenol, the analogue of D-pantothenic acid, has also assumed great importance, possessing full pantothenic acid activity.

Coenzyme A consists of pantothenic acid, β-mercaptoethylamine and adenosine diphosphate (Fig. 7.1). Pantothenic acid as CoA does not pass through membranes readily; most tissues seem to make enough to meet their requirements (Novelli, 1953). It is the terminal sulphydryl group of cysteine which is the metabolically active part of the coenzyme (Fig. 7.1). In addition to the presence of the pantothenic acid molecule as a part of CoA, it is also part of the acyl carrier protein (ACP) involved in fatty acid synthesis (Fig. 7.2). The 4′-phosphopantethenyl residue in the protein is probably derived from CoA and is attached to a seryl residue of the protein by a phospho-diester bond.

The exact pathway of synthesis of pantothenic acid is not completely understood but it seems likely that organisms which synthesize the vitamin do so by condensing β-alanine with di-hydroxybutyric acid or its lactone. Microorganisms synthesize it in the rumen of cattle and sheep (Peterson and Peterson, 1945).

Sources

Pantothenic acid is widely distributed in plant and animal tissues. In fact the name of this vitamin is derived from the Greek word 'pentos', meaning everywhere. In foods, pantothenic acid occurs mainly in bound forms (CoA and acyl carrier protein); it may also occur in its free form. It is particularly abundant in animal tissues, whole grain cereals, and legumes. It also occurs in lesser amounts in vegetables and fruits. Pantothenic acid is also possibly synthesized by the gut flora. However, the amounts produced and the availability from this source are unknown. The vitamin is also added to breakfast cereals, dietetic foods, and infant formulas. Pantothenic acid contents in some of the principal sources in nature are given in Table 7.1.

Table 7.1. Principal sources of pantothenic acid.

Food	mg 100 g^{-1}
Pork liver	7.0
Calf kidney	3.9
Calf heart	2.5
Eggs	2.9
Wheat bran	2.9
Soybeans	1.7
Lentils	1.4
Broccoli	1.2

Absorption

As we have seen, pantothenic acid occurs in many foods as a constituent of CoA and as a component of acyl carrier protein (ACP). For absorption the free vitamin must be released from the complexes by hydrolytic digestion. In the lumen of the intestine both CoA and ACP are degraded to release the vitamin as 4′-phosphopantotheine. This compound is then dephosphorylated to yield pantatheine which is converted by intestinal pantetheinase to pantothenic acid.

In the rat, mouse and chick an intestinal pantothenic acid transporter has been demonstrated, which is saturable, Na-dependent and energy-requiring (Combs, 1992). At high levels the vitamin is also absorbed by simple diffusion throughout the small intestine. The alcohol, panthenol, appears to be absorbed somewhat faster than the acid form.

The vitamin is transported in the free acid form in the plasma at a concentration of about 1 μg ml^{-1} (Fox, 1984). Levels in erythrocytes are higher than in plasma. A mechanism of transmembrane transport of pantothenic acid has been identified in the adipocytes of rats (Sugarman and Munro, 1980). In tissues, most of the pantothenic acid is found present as CoA, followed by lesser amounts of ACP, with only small amounts of free pantothenic acid. Levels of CoA and pantothenic acid vary with the tissue (250–4500 μg/100 g tissue) and with the nutritional status of the subject. Distribution of the vitamin is under metabolic and hormonal control. In rats the adrenal gland contains a high concentration and this suggests a close relationship with adrenal function.

Excretion

No degradation products of pantothenic acid are known; 1–7 mg of intact pantothenic acid are excreted in the urine and this is dependent on dose but not age. Excretion is reduced during fasting and in non-insulin-dependent diabetes;

in both cases the plasma concentration of pantothenic acid is increased. Excretion is controlled by tubular reabsorption and this may be controlled by insulin.

Metabolism

As already indicated, all tissues have the ability to synthesize CoA from dietary pantothenic acid. The steps in this pathway are shown in Fig. 7.3. In each tissue the rate-limiting step is the phosphorylation of pantothenic acid to 4′-phosphopantothenic acid by the enzyme, pantothenate kinase. This is one of the four steps that require ATP. Thus 4 moles of ATP are required for the synthesis of 1 mole of CoA from 1 mole of pantothenic acid. The synthesis of CoA is inhibited strongly by acetyl, malonyl, and propionyl CoA and more weakly by CoA and long-chain acyl CoAs. Acetaldehyde, a metabolite of ethanol, also inhibits the conversion of pantothenic acid to CoA. In this connection, it is of interest that alcoholics excrete in the urine a large percentage of the pantothenic

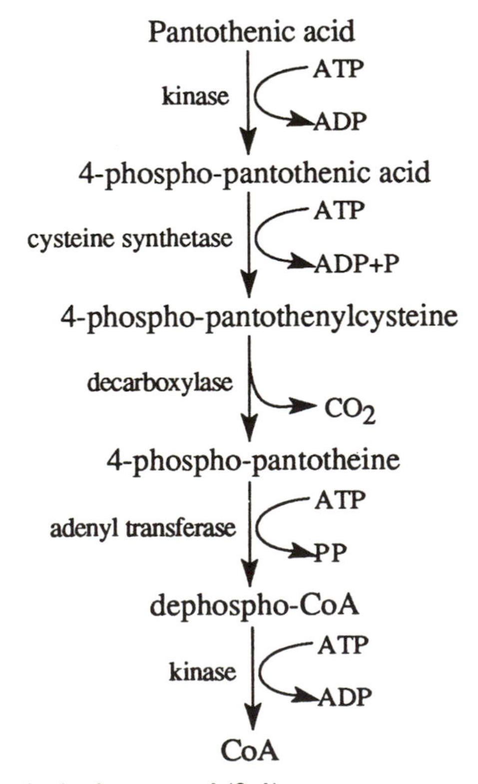

Fig. 7.3. The biosynthesis of coenzyme A (CoA).

acid they consume and that this condition is corrected by the withdrawal of alcohol (Combs, 1992).

Metabolic Functions

As a component of CoA and ACP, pantothenic acid is involved in fatty acid synthesis (Jeffrey, 1982). It is thus involved in lipid metabolism, the synthesis of cholesterol and steroids. It is also involved in oxidation of pyruvate and α-ketoglutarate, and hence in the generation of energy. As a component of acetyl CoA it is involved in many acetylations, including the conversion of choline to acetylcholine, the acetylation of drugs such as sulphonamides, the conversion of aminosugars to acetylhexosamines and the production of *N*-acetylated proteins. Other important thioesters of CoA that participate in the intermediary metabolism include succinyl CoA, propionyl CoA and malonyl CoA. Pantothenic acid as CoA is also involved in the detoxification of benzoic acid (Eq. 7.1) and the formation of porphyrin for haem synthesis (Eq. 7.2).

$$\text{Benzoic acid} + \text{ATP} \longrightarrow \text{Benzoyl AMP} \xrightarrow{\text{CoA}} \text{Benzoyl CoA} \xrightarrow{\text{Glycine}} \text{Benzoylglycine (hippuric acid)} \qquad (7.1)$$

$$\alpha\text{-ketoglutarate} + \text{CoA} + \text{NAD} \longrightarrow \text{succinyl CoA} \xrightarrow{\text{Glycine}} \delta\text{-aminolevulenic acid} \longrightarrow \text{porphyrins} \longrightarrow \text{haem} \qquad (7.2)$$

In summary, pantothenic acid as CoA and ACP plays a fundamental role in metabolism and particularly as a link between carbohydrates, amino acids and lipids and their final metabolic products. Thus, acetyl CoA or active acetate is formed in the oxidative decarboxylation of pyruvic acid, β-oxidation of fatty acids, and degradation of some amino acids. It can then be transferred to various acceptors in the synthesis of the various substances or oxidized completely in the citric acid cycle.

Deficiency

The widespread distribution of pantothenic acid in foodstuffs is one factor contributing to the rare occurrence of deficiency in man except as part of general protein–energy malnutrition (Sauberlich, 1974). During World War II prisoners of war in the Philippines, Japan and Burma developed severe malnutrition and B-vitamin deficiencies. Among the manifestations of nutritional deprivation in these prisoners was a condition called 'burning feet' (Jeffrey, 1982). This condition took 3–4 months to develop. Gradual onset of numbness and tingling in the toes was followed by burning pains and then shooting pains. This symptom was associated with other neurological and mental symptoms and clinical evidence of B-vitamin deficiency. In India it was observed that whilst thiamin and niacin relieved most of the symptoms of B-vitamin deficiency, pantothenic acid was required for the relief of the 'burning foot' syndrome. These observations have led to the presumption that the 'burning foot' condition was due to pantothenic acid deficiency. Incidentally, it was found that subjects with this condition had a reduced ability to acetylate para-aminobenzoic acid.

There have been a number of reports that patients with clinical evidence of diseases due to deficiencies of B-group vitamins, e.g. pellagra, beriberi and riboflavin deficiency, had low plasma levels and low urinary excretion of pantothenic acid. Anorectic hospital patients have also been observed to have similar evidence of low pantothenic acid status.

Elevated serum copper levels seen in some malnourished Bantu with pellagra have been reduced by injecting large doses (1–1.5 g) of pantothenic acid intramuscularly (Findlay and Venter, 1958). Copper levels are not lowered in healthy individuals by this treatment.

Human volunteers on a deficient diet have been found to develop vomiting, malaise, abdominal distress and burning cramps. Later the subjects developed tenderness in the heels, fatigue and insomnia. When the deficiency was exacerbated by giving a pantothenic acid antagonist, ω-methyl-pantothenic acid, the subjects developed abdominal pain, nausea, personality changes, insomnia, weakness and cramps in the legs and paraesthesiae in hands and feet. These latter symptoms are reminiscent of those described above which occurred in malnourished prisoners of war. Other signs of deficiency that have been described are skin changes (dermatitis, achromotrichia, alopecia), gastrointestinal ulcers, hepatic steatosis, thymic necrosis, adrenal hypertrophy, ataxia and paralysis (Combs, 1992).

Requirements

Though blood concentrations and urinary excretion of pantothenic acid have been measured and a range of values has been reported for individuals of different ages (Jeffrey, 1982), it is difficult to interpret them in terms of dietary need (Fox, 1984). No recommended intake is set for pantothenic acid. Intakes in the UK as estimated from National Food Survey records were 5.1 mg in 1979 and 6.07 mg in 1986 (Lewis and Buss, 1988). The median intake for adult males was 6.1 mg day^{-1} and for females 4.4 mg day^{-1}. British mothers have been reported to have mean intakes between 3.4 and 5.3 mg day^{-1} during pregnancy and lactation, though it has been suggested that these values are up to 0.6 mg below the true values (Black *et al.*, 1986). In the UK it has been suggested (Department of Health, 1991) that intakes between 3 and 7 mg day^{-1} must be adequate, even during pregnancy and lactation. Infant formula diets should contain at least 2.0 mg pantothenic acid per litre, providing 1.7 mg day^{-1} (DHSS, 1980). In Canada and the United States, 2.3 mg/1000 kcal or 5–7 mg day^{-1} is considered adequate. The formula applies to children, pregnant and lactating women.

References

Black, A.E., Wiles, S.J. and Paul, A.A. (1986) The nutrient intakes of pregnant and lactating mothers of good socioeconomic status in Cambridge, United Kingdom: some implications for recommended daily allowances of minor nutrients. *British Journal of Nutrition* 56, 59–72.

Combs, G.F. (1992) Pantothenic acid. In: *The Vitamins*. Academic Press, New York, pp. 345–356.

Department of Health (1991) *Dietary Reference Values for Food Energy and Nutrients for the United Kingdom*. Report on Health and Social Subjects 41: Her Majesty's Stationery Office, London.

Department of Health and Social Security (1980) *Artificial Feeds for the Young Infant*. Report on Health and Social Subjects 18. Her Majesty's Stationery Office, London.

Findlay, G.H. and Venter, I.J. (1958) Preliminary and short report. An effect of pantothenic acid on serum copper values in human pellagra. *Journal of Investigative Dermatology* 31, 11.

Fox, H.M. (1984) Pantothenic acid. In: Machlin, L. (ed.) *Handbook of Vitamins: Nutritional Biochemical and Chemical Aspects*. Marcel Dekker, New York, pp. 437–457.

Jeffrey, D.M. (1982) Pantothenic acid. In: Barker, B.M. and Bender, D.A. (eds) *Vitamins in Medicine*, 4th edn., Vol. 2. Heinemann Medical, London, pp. 69–91.

Lewis, M. and Buss, D.H. (1988) Trace elements. 5. Minerals and vitamins in the British household food supply. *British Journal of Nutrition* 60, 413–424.

Novelli, G.D. (1953) Metabolic function of pantothenic acid. *Physiological Reviews* 33, 525–543.

Peterson, W.H. and Peterson, M.S. (1945) Relation of bacteria to vitamins and other growth factors. *Bacteriological Reviews* 9, 49–109.

Sauberlich, H.E. (1974) Pantothenic acid. In: Goodhart, R.S. and Shils, M.E. (eds.) *Modern Nutrition in Health and Disease*, 5th edn. Lea & Febiger, Philadelphia, pp. 203–209.

Sugarman, B.S. and Munro, H.N. (1980) 14 c-pantothenate accumulation by isolated adipocytes from adult rats of different ages. *Journal of Nutrition* 110, 2297–2301.

Pteroylglutamic Acid (Folic Acid, Folacin) 8

The history of folic acid began in the early 1930s when Dr Lucy Wills in India observed that pregnant women with macrocytic anaemia responded to Marmite, a preparation of autolysed yeast (Wills, 1932). A decade later, a substance was isolated from spinach, alfalfa, and yeast, which had a growth promoting factor for the microorganism, *Lactobacillus casei*. In 1941, Mitchell and his associates coined the term 'folic acid' for this substance. The name was derived from the Latin word 'folium' meaning 'leaf'. The isolated substance was eventually purified and identified by a large team of American industrial chemists (Angier *et al.*, 1945).

Chemistry

The folic acid molecule consists of a pteridine nucleus, *p*-aminobenzoic acid (PABA), L-glutamic acid, and single carbon substituent groups (such as formyl, methyl and methylene). The pteridine ring and PABA are components of pteroic acid and this could be regarded as the parent compound to which L-glutamic acid is then conjugated (Fig. 8.1). Because of these structural components, folic acid is also called pteroylglutamic acid (mol. wt 441). The pteridine nucleus is composed of two rings, which include pyrimidine and pyrazine. The pyrazine ring is the portion of the pteridine nucleus that undergoes reductions forming dihydro- and tetrahydro-pteroylglutamic acid (folic acid) (Fig. 8.2). The substituent groups are attached to either N^5, N^{10} or to both N^5 and N^{10} positions of the pyrazine ring (Fig. 8.3).

Much of the naturally occurring folic acid is conjugated with polyglutamate molecules (Fig. 8.1); the number of glutamate molecules can be as many as seven. The glutamate residues are bound by peptide linkages to the gamma carboxylic group of the glutamate. It is estimated that over 150 different forms of folic acid could exist in nature (Baugh and Krumdieck, 1971). These include folic acid, unsubstituted or substituted with one of several possible single carbon

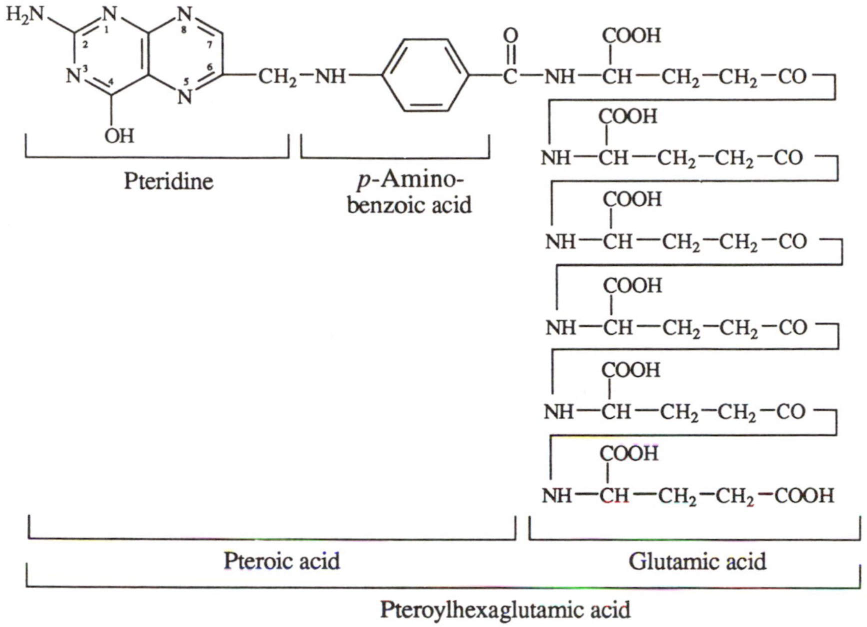

Fig. 8.1. Structure of folic acid.

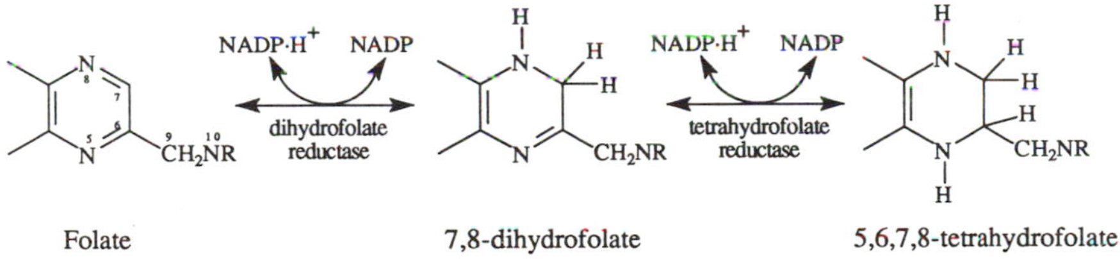

Fig. 8.2. Reduction of folate (R= *p*-aminobenzoyl glutamic acid).

substituents, its state of oxidation and length of side chain with glutamic acid residues.

Tetrahydrofolate (FH_4), unsubstituted or substituted with single carbon units (Fig. 8.3) and bearing polyglutamic acid residues, is the active coenzyme form of folic acid. The IUPAC-IUB used the collective term folacin(s) to refer to all biologically active forms of folic acid (IUPAC-IUB Commission on Biochemical Nomenclature, 1966). Since the synonym, folate, is widely used, both terms will be used in this chapter.

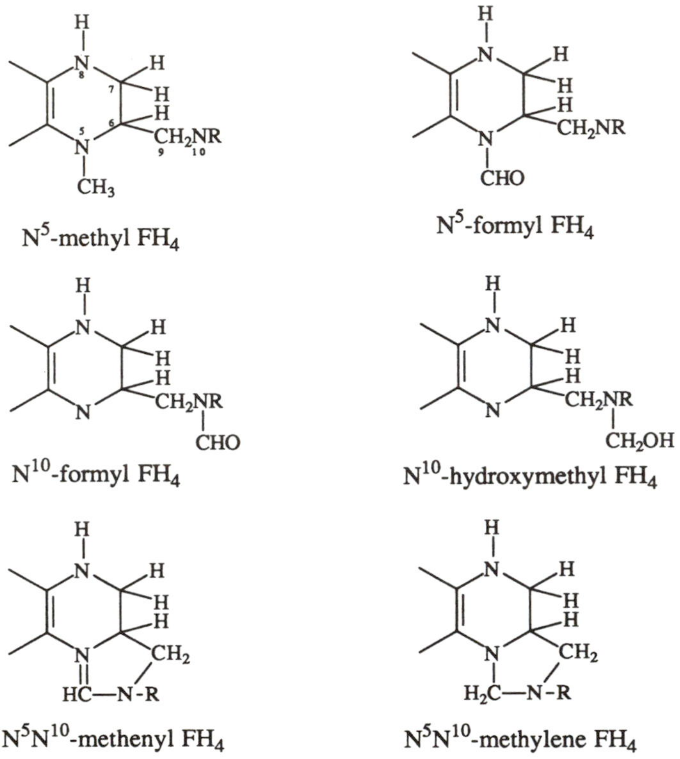

N^5-methyl FH_4

N^5-formyl FH_4

N^{10}-formyl FH_4

N^{10}-hydroxymethyl FH_4

N^5N^{10}-methenyl FH_4

N^5N^{10}-methylene FH_4

Fig. 8.3. Tetrahydrofolic acid (FH_4) carrying one-carbon units of varying oxidation states.

Sources

Folic acid is synthesized in nature by microorganisms and plants. A step in the synthesis of folate involves coupling the pteridine molecule to p-aminobenzoic acid (PABA) to form pteroic acid (Fig. 8.4). This coupling to PABA to form pteroic acid is lost in mammals, which can synthesize large quantities of pteridines. Mammals are thus unable to synthesize folic acid, making them auxotrophic for this vitamin.

Folic acids synthesized by plants and microorganisms are passed on to higher animals, and thus are present in both vegetables and animal foods. The principal sources include liver, dark green leafy vegetables, dry beans and peas, wheat germ, and yeast. Other significant sources include egg yolk, broccoli, whole grain products, peanuts and almonds. Most fruits and vegetables contain small amounts of folic acid. Meat is generally a poor source with the exception of organ meat, such as liver and kidney.

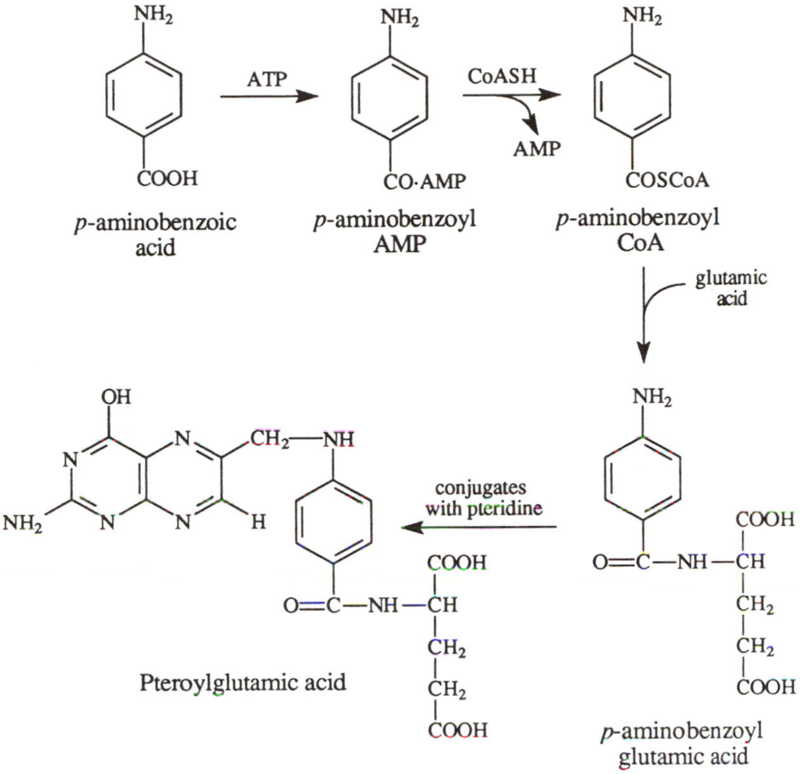

Fig. 8.4. Biosynthesis of folic acid in plants and microorganisms.

Both monoglutamate and polyglutamate forms of folic acid are present in foods; the most part, however, is in the form of polyglutamate. Most of the folic acids are in their reduced forms, making them labile and easily oxidizable. Ascorbic acid, due to its reducing property, helps to prevent oxidative destruction of reduced folates. Folates in food are generally sensitive to changes in pH, presence of oxidizing agents, and exposure to heat and light. The vitamin is unstable unless refrigerated. Fresh leafy vegetables may lose up to 70% of the vitamin within three days unless they are stored in the refrigerator; up to 95% may be lost in cooking water.

Absorption and Transport

As discussed above, much of the folic acid in food is conjugated with polyglutamate molecules. Prior to its absorption, the excess glutamates are deconjugated by a hydrolytic enzyme located in the membrane of the intestinal mucosa (Reisenauer *et al.*, 1977). The enzyme is called γ-glutamylcarboxypeptidase. In much of the literature this enzyme is referred to by the trivial name conjugase. The hydrolytic enzyme is present primarily in the lumen and also in the brush border of the intestine (Reisenauer *et al.*, 1985). Folic acid in food, not already in the reduced form, must also be converted to FH_4 in the process of absorption. This reduction occurs in two steps involving NADPH-dependent dihydrofolate reductase and tetrahydrofolate reductase (Fig. 8.2). The reduced monoglutamate form of folic acid is then absorbed by an energy-dependent active transport mechanism from the proximal third of the small intestine. Absorption increases in a slightly acidic medium (pH 6.3) but decreases at an alkaline or markedly acidic pH. The availability of folate from a typical North American diet has been suggested to be about 50–75% (Herbert, 1987b).

Transport of folate is complex and poorly understood. It is assumed that dietary folates, after hydrolysis, reduction and absorption from the intestine, are transported in plasma to the liver and other tissues as reduced monoglutamate derivatives bound to albumin and folate binding proteins (FBPs). The latter have been found to be present in the liver, brush border of the small intestine as well as in extracellular fluids, such as milk, plasma, and cerebrospinal fluid, and in cellular membranes and tissues (Wagner, 1982). The FBPs tend to bind reduced folate with much greater affinity (*c.* 100-fold) than the oxidized form.

The predominant form of circulatory folic acid is 5-methyltetrahydrofolate. It is thought that various monoglutamate forms of folic acid are converted into 5-methyl FH_4 in enterocytes. Most of this folate derivative is quickly secreted into the bile after reaching the liver. Subsequently, it becomes available for distribution to peripheral tissues via enterohepatic circulation (Fig. 8.5), which may account for 50% of the total folate that reaches peripheral tissues (Steinberg, 1984). The enterohepatic circulation from the intestine to the liver is thus important to the maintenance of plasma folate levels.

The monoglutamate forms of folic acid, predominantly as 5-methyl FH_4, are taken up by cells and are metabolized intracellularly to polyglutamate forms, which, due to their greater molecular size, are retained by tissues better than monoglutamate forms. The liver is the primary storage site for folic acid (50% of the body content), and thereby becomes a concentrated source of the vitamin.

Metabolic Role

As discussed earlier the biologically active form of folic acid is the reduced form, FH_4 (Fig. 8.2). The reactive part of this coenzyme molecule occurs in the pteridine portion at nitrogens 5 and 10 (Fig. 8.1). These nitrogens can receive and transfer one-carbon units of varying oxidation states (Fig. 8.3). The least oxidized one-carbon unit carried by FH_4 is the methyl (— CH_3) group, the next

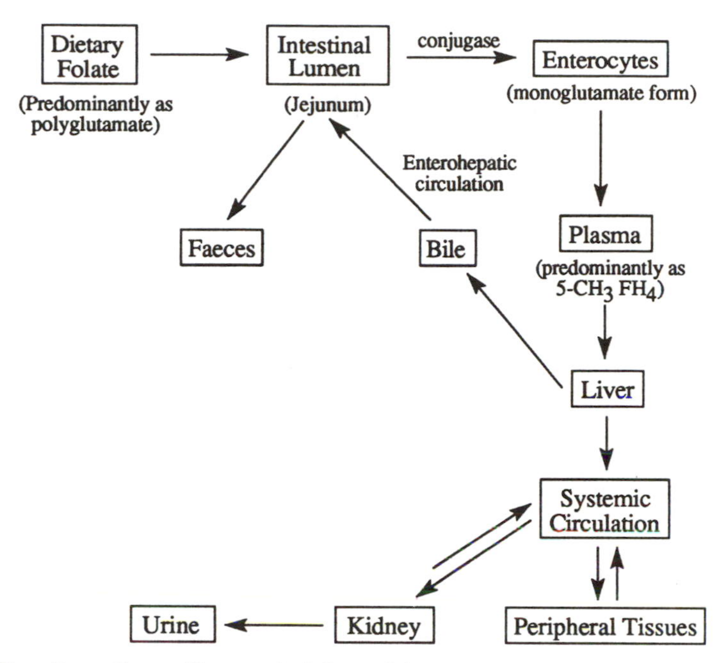

Fig. 8.5. Absorption and transport of dietary folate.

most oxidized unit is the methylene (— CH_2—) group, and the most oxidized one-carbon units carried by FH_4 include the formimino (— CH=NH), formyl (—CH= O), and methenyl (— CH =) groups (the most highly oxidized form of one-carbon unit is CO_2, which is transferred by biotin; see Chapter 6). Tetrahydrofolate due to its ability to accept and release different oxidative states of single-carbon units plays a vital role in many metabolic reactions. Folate derivatives carrying various single-carbon units are interconvertible except that the conversion of 5, 10-methylene FH_4 to N^5-methyl FH_4 is not reversible. The interconversions of the coenzyme forms of folate and some of the reactions in which they participate are shown in Fig. 8.6. The single carbon units transferred by FH_4 appear to be important in certain methylation reactions, in the biosynthesis of nucleic acid bases, and in some amino acid interconversions.

Methionine synthesis

The majority of naturally occurring folates, when ingested, are converted to 5-methyl FH_4 and stored in the liver. As shown in Fig. 8.6, the 5-methyl FH_4 is converted to FH_4 by the vitamin B_{12}- dependent methionine synthetase reaction before entering the cellular pool of active folates. Through this reaction,

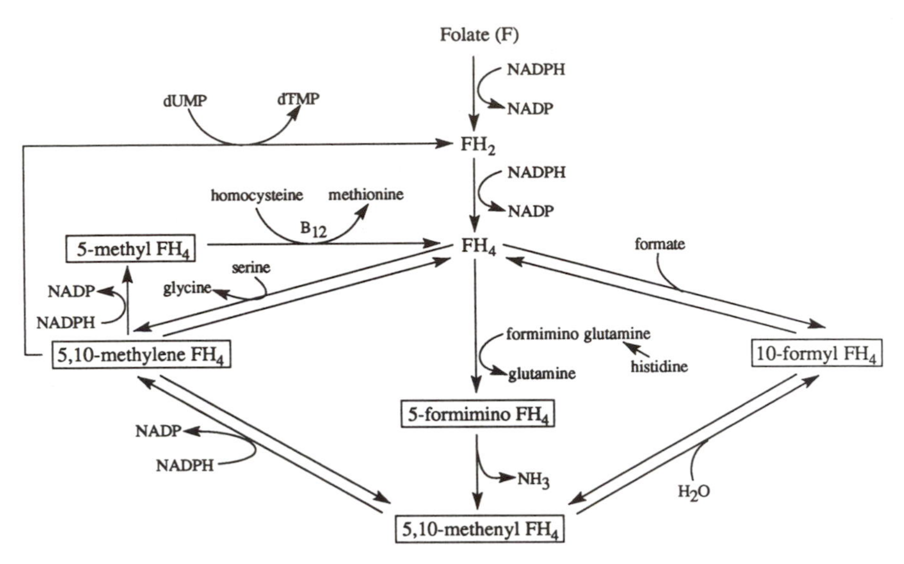

Fig. 8.6. Interconversions of folate coenzymes.

the methyl group of methyl FH_4 is transferred to homocysteine, generating methionine and FH_4. Hence, vitamin B_{12} deficiency leads to disturbed metabolism of both folate and methionine. Folate is trapped as methyl FH_4 leading to folate deficiency when there is vitamin B_{12} deficiency (for details, see Chapter 9).

Biosynthesis of nucleic acids

The purine bases, such as adenine and guanine, and pyrimidine bases such as thymine, are important constituents of the nucleic acids. The active folate coenzymes are involved in the synthesis of these nucleic acid bases, and therefore in DNA synthesis.

Active folate coenzymes, acting as formyl carriers, are required for the incorporation of the one-carbon units into positions 2 and 8 of the purine ring. 10-formyl FH_4 and 5,10-methenyl FH_4 are the specific formyl donors for carbons 2 and 8 of the purine molecule, respectively (Fig. 8.7). In reaction 1, glycinamide-ribosyl-P is formylated to form formyl glycinamide ribosyl-5-P. This reaction requires 5,10-methenyl FH_4 and the enzyme transformylase to transfer the one-carbon moiety which becomes position 8 of the purine nucleus. In a subsequent step (reaction 2), an amination occurs at carbon 4 of the formylated glycinamide, where glutamine acts as the amino donor. The added nitrogen will be in position 3 in the purine. Ring closure forms an amino-

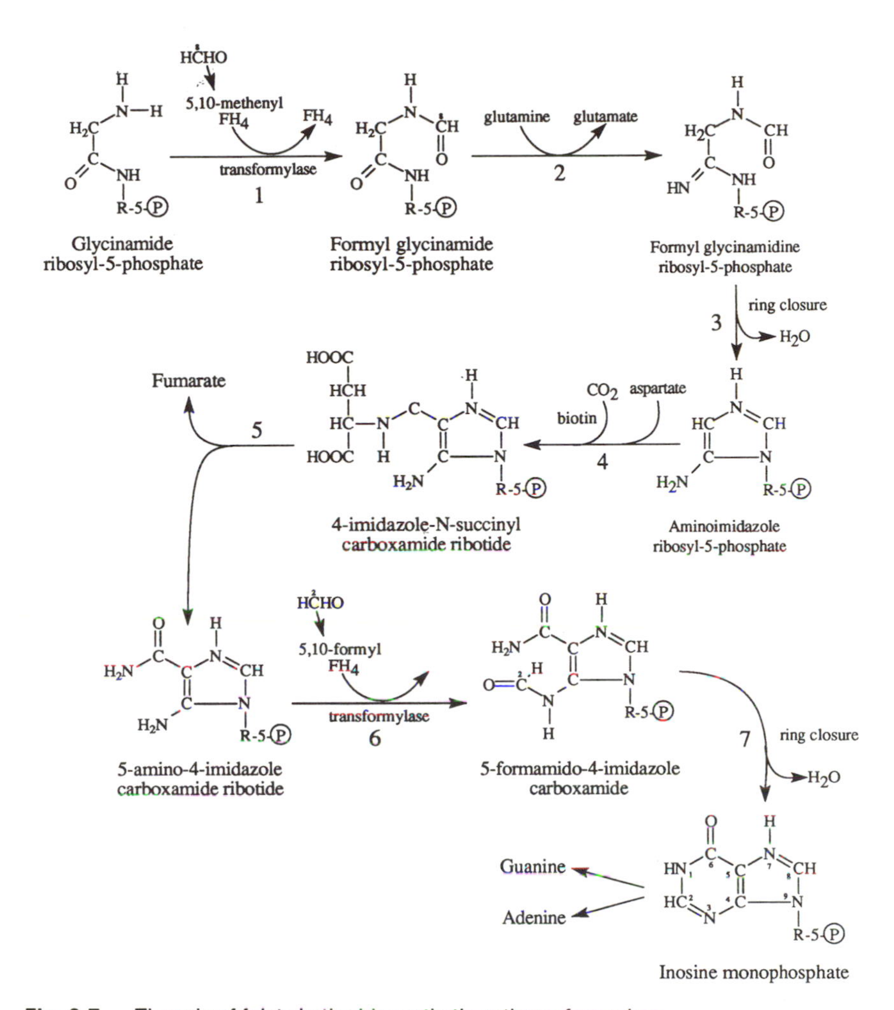

Fig. 8.7. The role of folate in the biosynthetic pathway for purines.

imidazole ribosyl-5-P (reaction 3), which progresses to 5-amino-4-imidazole-N-succinyl carboxamide ribotide (reaction 4). This is formed by addition of a carbamoyl group to the precursor compound. The source of the carbon is CO_2, and the source of the nitrogen is the amino nitrogen of aspartic acid. The utilization of CO_2 is dependent upon biotin (see Chapter 6). In the next step (reaction 5), fumaric acid is split off, leaving 5-amino-4-imidazole-carboxamide ribotide. The latter compound is then formylated to 5-formamido-4-imidazole-carboxamide ribotide (reaction 6). This reaction requires 10-formyl FH_4 and the enzyme transformylase to transfer the one-carbon moiety which will become carbon 2 of the purine nucleus. Ring closure occurs in a subsequent step

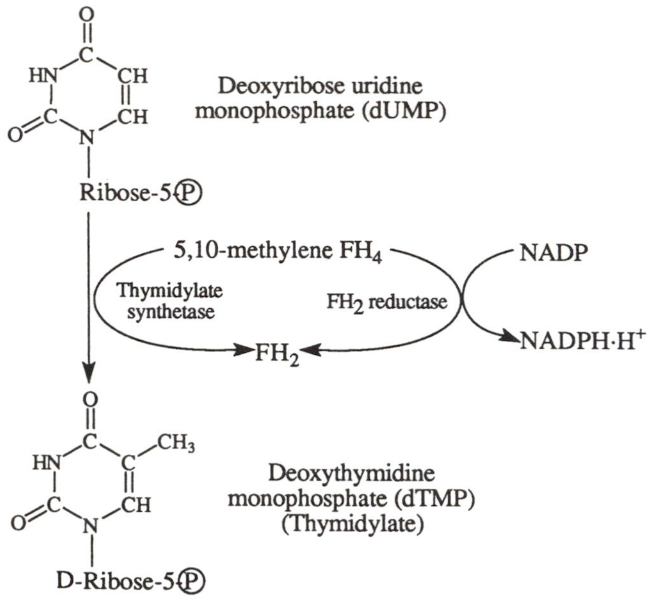

Deoxyribose uridine monophosphate (dUMP)

Ribose-5-Ⓟ

5,10-methylene FH_4

Thymidylate synthetase

FH_2 reductase

NADP

$NADPH{\cdot}H^+$

FH_2

Deoxythymidine monophosphate (dTMP) (Thymidylate)

D-Ribose-5-Ⓟ

Fig. 8.8. The role of folate in the biosynthesis of thymidylate.

(reaction 7) forming inosine monophosphate from which guanine and adenine nucleotides are derived.

In addition to purine synthesis, folic acid coenzymes are involved in the synthesis of thymidylate, and hence pyrimidines. Thymidylate is synthesized from deoxyuridylate by the action of thymidylate synthetase (Fig. 8.8). Active folate is responsible for the donation of the methyl group of thymidylate. The single-carbon unit transferred to deoxythymidylate comes from 5,10-methylene FH_4. In this biosynthetic reaction, folate is not only the carrier of a one-carbon unit but also the two hydrogens involved in reducing the methylene carbon to methyl.

Amino acid interconversions

Besides the synthesis of methionine from homocysteine, active folate coenzymes are involved in the interconversions of other amino acids. The interconversion of serine to glycine is one example. Serine hydroxymethyl transferase, a pyridoxal phosphate (PLP)-dependent enzyme, catalyses the reversible transfer of formaldehyde from serine to FH_4 to generate 5,10-methylene FH_4 and glycine (Eq. 8.1). The generated 5,10-methylene FH_4 can act as a coenzyme in a number of reactions, such as synthesis of pyrimidine and methionine. The FH_4 derivative can be oxidized by NADP to 5,10-methenyl FH_4 for the synthesis of purines.

$$\text{Serine} + \text{FH}_4 \xrightarrow[\text{Hydroxymethyl transferase}]{\text{PLP}} \text{Glycine} + 5,10\text{-methylene FH}_4 \quad (8.1)$$

Folate is also involved in the metabolism of formiminoglutamate. This histidine metabolic product is catalysed by formimino transferase, which transfers the formimino group to FH_4 to generate glutamate and 5-formimino FH_4 (Eq. 8.2). The latter is further metabolized by deamination to 5,10-methenyl FH_4 (Fig. 8.6).

$$\text{Histidine} \longrightarrow \text{Formiminoglut} \xrightarrow[\text{Formimino transferase}]{\text{FH}_4} \text{Glut} + 5\text{-formimino FH}_4 \quad (8.2)$$

Signs of Folate Deficiency

The most significant role of folic acid appears to be its involvement in the syntheses of purine and pyrimidine, and hence in DNA synthesis. Limitation of the vitamin interferes with DNA replication and the number of cell divisions. Consequently, one of the most noticeable effects of folate deficiency is the occurrence of abnormalities in those cells with the most rapid rates of multiplication, which include the intestinal mucosa, regenerating liver and the bone marrow. The turnover rates of the villus epithelial cells of the small intestine and the red blood cells are rapid; they need to be regenerated at every 30 hours and 120 days, respectively.

Signs and symptoms of folate deficiency related to the gastrointestinal tract include soreness of the tongue which is often accompanied by shiny redness and ulcers, cheilosis, loss of appetite, abdominal pain, and occasional diarrhoea. Folate is also essential for the formation and maturation of both red and white blood cells in the bone marrow. The most consistent and prominent manifestations of folate deficiency are the changes in these blood cells. Abnormally large red blood cells (RBCs) form when the newly formed immature red blood cells fail to mature and lose their nuclei. In these conditions, the RBCs grow larger, their number decreases but the amount of haemoglobin they contain does not decrease (Fig. 8.9). This is a condition, called macrocytic or megaloblastic anaemia, which is essentially a consequence of decreased DNA synthesis and failure of the cells to divide properly, coupled with the continued formation of RNA. In contrast to folate deficiency, iron deficiency leads to smaller size of the RBCs (a microcytic anaemia), which is caused by interfering with the attainment of the optimal amount of haemoglobin (hypochromic).

In folate deficiency, macrocytic anaemia is often accompanied by other clinical signs such as tiredness, breathlessness, pallor of the skin and the mucous

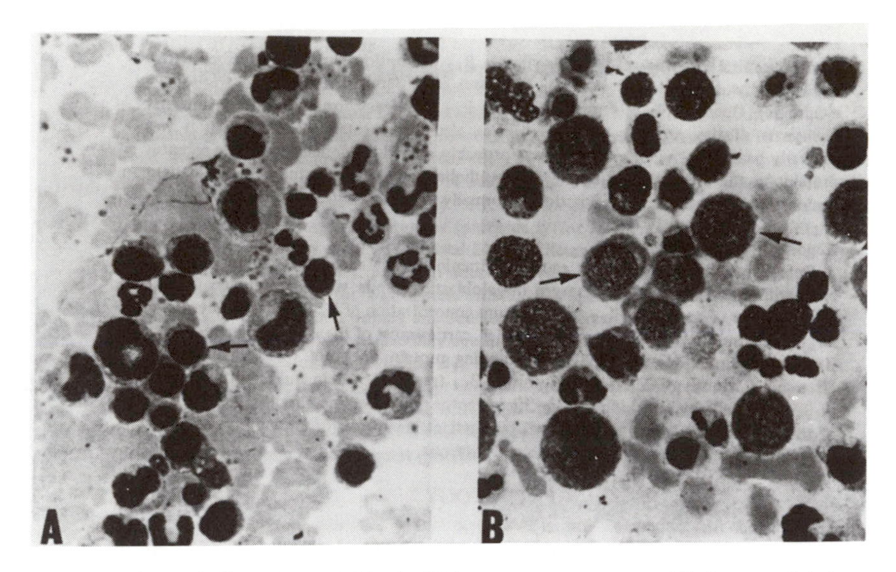

Fig. 8.9. Normal (A) and megaloblastic (B) bone marrow smears. (A) Arrow points to condensed nuclear material of a normal orthochromatic erythroblast; (B) Arrow points to an orthochromatic megaloblastic erythroblast which contains lacy nuclear chromatin material. Source: Rothenberg and Cotter, 1978 (with permission).

membranes, and systolic murmurs. It is important to note that in the early stage of folate deficiency, macrocytosis may occur before anaemia develops. Also in early folate deficiency, there may be an increase in neutrophil hypersegmentation (shift to predominance of five-lobed over three-lobed neutrophils) without abnormality in other haematological indices (Bills and Spatz, 1977).

It should be noted that the signs and symptoms of folate deficiency are indistinguishable from the haematological manifestations of vitamin B_{12} deficiency (see Chapter 9). The two vitamin deficiencies differ in their clinical presentation in that only vitamin B_{12} deficiency results in damage to myelin leading to neurological abnormalities.

Factors Affecting Folate Status

Dietary factors

It is clear from the previous sections that the bio-availability of folic acid depends not only on its presence in the diet but also on the presence of other nutrients, such as ascorbic acid, niacin, and vitamin B_{12}. Niacin helps to activate folates through promoting their reduction, ascorbic acid prevents the oxidation of already reduced folates, and vitamin B_{12} makes folates metabolically available in the body for their biological roles. Therefore, an adequate intake of folate alone may not be enough to maintain a positive balance of the vitamin. Another unique characteristic of this vitamin is that the majority of folates in foods are present as reduced forms, which are susceptible to oxidation. Methods of cooking and storage may result in considerable loss of the vitamin; steaming or frying in an open pan may cause over 70% loss (Chanarin, 1980).

There are certain foods that are grossly deficient in folate: for example, goat's milk contains only one-tenth of that present in human milk. The incidence of megaloblastic anaemia has been reported frequently in infants reared on goat's milk (Becroft and Holland, 1966). Poor diet and inappropriate methods of cooking are usually the major causes of folate deficiency in many parts of the world, especially in developing countries.

Malabsorption

As indicated above, folate polyglutamate forms are the major type of food folate. These polyglutamate forms must be hydrolysed to the monoglutamate form before dietary folates can be absorbed from food. This action is performed by conjugases present in the lumen and brush border of the intestine. Malabsorption syndromes, such as coeliac disease, tropical sprue and Crohn's disease, adversely affect the absorption of folate because of reduced levels of the hydrolysing enzymes.

There are foods, such as yeast and pulses, which contain intestinal conjugase inhibitors. Presence of these factors in diets may affect considerably the absorption of dietary folates (Colman, 1977). It is also thought that the brush-border folate hydrolytic conjugase enzyme requires zinc to be activated (Halsted *et al.*, 1986). Thus, the food folate absorption decreases in the presence of zinc deficiency.

Increased requirement

Pregnancy

There is an increased demand for folate in pregnancy on account of markedly accelerated cell multiplication with fetal and placental growth, uterine enlargement and expansion of blood volume (Rodriguez, 1978). There appears to be an increasing folate clearance from the blood following parenteral administration as pregnancy advances. The maternal serum and red blood cell folate levels also fall progressively as pregnancy continues. Although iron deficiency anaemia is most often associated with pregnancy, megaloblastosis due to folate deficiency is frequently seen and is more common in twin pregnancy. The incidence of folate-responsive megaloblastic anaemia runs at about 2.5–5.0% in pregnant women in developed countries and is considerably higher in the developing countries (Knipsheer, 1975). Daily maternal requirements for dietary folic acid are often accepted as being twice that of the non-pregnant state. The increased requirement for folate during pregnancy is in part caused by accelerated folate breakdown (McPartlin *et al.*, 1993).

Women who have low serum and erythrocyte folic acid concentration are at risk of producing babies with neural tube defect (NTD), particularly spina bifida (Schorah and Smithells, 1991). Giving a multivitamin preparation (including folate) peri-conceptually was found to reduce the number of expected NTDs in women who had already given birth to at least one child with spina bifida. A subsequent multicentre placebo-controlled study (MRC Vitamin Study Research Group, 1991) confirmed that a folic acid supplement taken peri-conceptually reduced the risk of NTD. The group suggested that public health measures should be taken to ensure that the diet of all women who may bear children contains an adequate amount of folic acid. The neural tube closes late in the 4th week after conception and it is important that adequate amounts of folate are available at this time.

Infancy and Childhood

Blood folate levels appear to fall during the first few months of life, indicating a high demand for the vitamin during this fast growing period (Shojania, 1984). This fall generally is more marked in premature than in full-term infants. Many factors may account for folate deficiency at this time of life, especially in pre-term babies. Premature birth does not allow the passage across the placenta of the same amount of folate that a full-term infant normally receives. The destruction of folate caused by the sterilization of milk by boiling has a relatively greater affect in premature babies with their already low reserves of the vitamin (Ek and Magnus, 1980).

Adolescents

Accelerated growth and increased lean body mass are the characteristics of adolescents. Rapidly dividing tissues during the adolescent growth spurt increases requirements of most nutrients and especially for folate. Because of this increased need, folate status appears to be of concern during the age of this rapid growth, especially in low-income families, whose diets are likely to be poor in quality (Sauberlich, 1990).

Elderly population

Low folate status has been reported to be fairly common in the elderly of Europe and North America as determined by dietary and biochemical analyses (Bates *et al.*, 1980; Rosenberg *et al.*, 1982; Bailey *et al.*, 1984; Marcus and Freedman, 1985). The folate status of elderly people may be affected by a variety of factors including chronic disease, polypharmacy (see below), cigarette smoking, alcohol consumption, impaired bio-availability, and poor dietary habits.

Haemolytic anaemia

Haemolytic anaemia is associated with a hyperactive bone marrow; this results in an increased DNA synthesis and hence increased requirements for folate. A marginally folate-deficient subject who develops a haemolytic anaemia may produce a megaloblastic bone marrow. The increased cell turnover will naturally require more folate in these circumstances.

Disturbances of folate metabolism

Drug-induced disturbances

An overt deficiency of folic acid is common among chronic alcoholics (Hoyumpa, 1986). Poor dietary intake, impaired absorption, and altered metabolism of folate can all contribute to poor folate status. In addition to alcohol, there are a number of drugs which act as folic acid antagonists (for details see Chapter 18). They act at different points of its metabolic pathway. Some of these drugs are similar in structure to folate, and they are usually involved in inhibition of dihydrofolate reductase, thus preventing the normal formation of FH_4 and its subsequent derivatives. The consequences are that all the effects of folate deficiency ensue. Examples of drugs in this category are methotrexate, which is a chemotherapeutic agent, pyrimethamine which is used as a malarial therapeutic agent, triamterene which is a diuretic, and the antibacterial agent, trimethoprim. In addition, the chronic administration of drugs, including anticonvulsants such as diphenylhydantoin, primidone and barbiturates, that induce the mixed function oxygenase (MFO) system, is associated with an increased requirement for folic acid due to its involvement in the synthesis of cytochrome P450 (Labadarios *et al.*, 1978). Patients receiving these drugs often show evidence of interference with folate function and megaloblastic changes in the bone marrow.

Genetical disturbances

A number of congenital disturbances of folate metabolism are known to exist (Erbe, 1975). These rare conditions are associated with defective absorption and utilization of the vitamin due to deficiencies of enzymes involved in its metabolism. The enzymes found to be affected include folate reductase, methyltransferase, and formiminotransferase. The result is a megaloblastic anaemia together with mental retardation, convulsions, and a neurological motor disorder.

Requirements

Folacin is synthesized in large quantities by microorganisms in the colon, making faecal excretion an unreliable indicator of dietary folacin absorption

(Herbert *et al.*, 1984). Furthermore, the body pool of folacin is conserved by enterohepatic circulation (Fig. 8.5). These factors have caused difficulty in deriving reliable turnover rates of the vitamin. However, normally nourished individuals have been found to excrete 5–40 μg of folate in the urine each day (Herbert, 1968). Using ^{14}C-pteroylglutamic acid in a healthy subject, the biological half-life of the vitamin has been reported to be 101 days (Krumdieck *et al.*, 1978). The total body pool of folacin in a normal adult male is estimated to be 5–10 mg (average 7.5 mg) (Herbert, 1971). Assuming that half of this is lost every 100 days, the daily turnover of tissue folacin would amount to $7.5 \text{ mg} \times 1/100 \times 0.5 = 37.5 \text{ μg day}^{-1}$ (Food and Nutrition Board, 1980).

On the basis of a maximum body pool, minimum availability from diets (50%), body turnover rate and average dietary intakes by populations in different countries (Table 8.1), the daily allowance of folate for adults in North America is set at around 3.0 μg kg^{-1} day^{-1} (Food and Nutrition Board, 1989; Health and Welfare Canada, 1990); this amount is thought to be sufficient to maintain normal body stores. The allowances, however, are set at higher levels during the earlier part of life and during pregnancy, reflecting the increased needs for rapid growth and development involved during these life cycles. The daily allowances of folate at different stages of life in different countries are given in Table 8.2.

Table 8.1. Daily folate intake in different countries.

Country	Average intake (μg day^{-1})	Reference
UK	212	Poh Tan *et al.* (1984)
USA	227	Life Sciences Research Office (1984)
Canada		
male	205	Health and Welfare Canada (1977)
female	149	

Assessment of Folate Status

Serum folate

Serum folate concentration is a sensitive indicator of short-term folate balance and recent dietary intake. It increases rapidly following the ingestion of folate-containing foods, and decreases abruptly on a folate-deficient diet. Although the serum folate level reflects dietary folate intake and hence recent dietary folate status, it provides no information on the size of the folate tissue stores (Herbert, 1987a).

Table 8.2. Daily recommended intake of folate in different countries.

	UK[1] (μg)	USA[2] (μg kg^{-1} body wt)	Canada[3] (μg kg^{-1} body wt)
0–1 year	50	3.6	4.0
1–12 years	50–200	3.3	3.5
12 years –adult	200	3.0	3.1
Pragnancy	200 + 100	3.0 (+ 200 μg)	3.1 (+ 200 μg)
Lactation	200 + 60	3.0 (+ 100 μg)	3.1 (+ 100 μg)

[1]Department of Health (1991).
[2]Food and Nutrition Board, National Research Council (1989).
[3]Health and Welfare Canada (1990).

RBC folate

The folate tissue stores can be best determined by measuring erythrocyte folate concentrations (Herbert, 1990). A reduction in liver folate stores is paralleled by a reduction in erythrocyte folate levels. Since only the youngest red cells in the bone marrow require folate to synthesize DNA, the red cell concentration of folate is a measure of the status of this vitamin at the time the erythrocyte is synthesized. In other words, the life span of a red cell (120 days) corresponds to the period of time that normal folate stores can be maintained in the liver on a folate deficient diet. Thus, in contrast to the immediate fluctuations reflecting previous folate intake indicated by serum folate concentrations, erythrocyte levels of folate reflect tissue status over a longer period of time, making the latter measurements more reliable for indicating risk of development of folate deficiency.

A subnormal value for RBC folate is generally suggestive of folate deficiency of some standing and severity. However, such a low value is by itself is inadequate to diagnose folate deficiency, because it also occurs in vitamin B_{12} deficiency. Because of the close relationship between folate and vitamin B_{12} in the development of megaloblastic anaemia, it is important that a differential diagnosis is made; this aspect is dealt with in Chapter 9.

FIGLU test

An analysis of urine for the presence of formiminoglutamic acid (FIGLU) can also be useful in evaluating folate status of an individual. This effect is accentuated when a loading dose of histidine is fed (histidine load test). The amino acid is converted to FIGLU, which is then removed from the tissues by reacting with FH_4 (Fig. 8.10). In the presence of folate deficiency, FIGLU cannot be

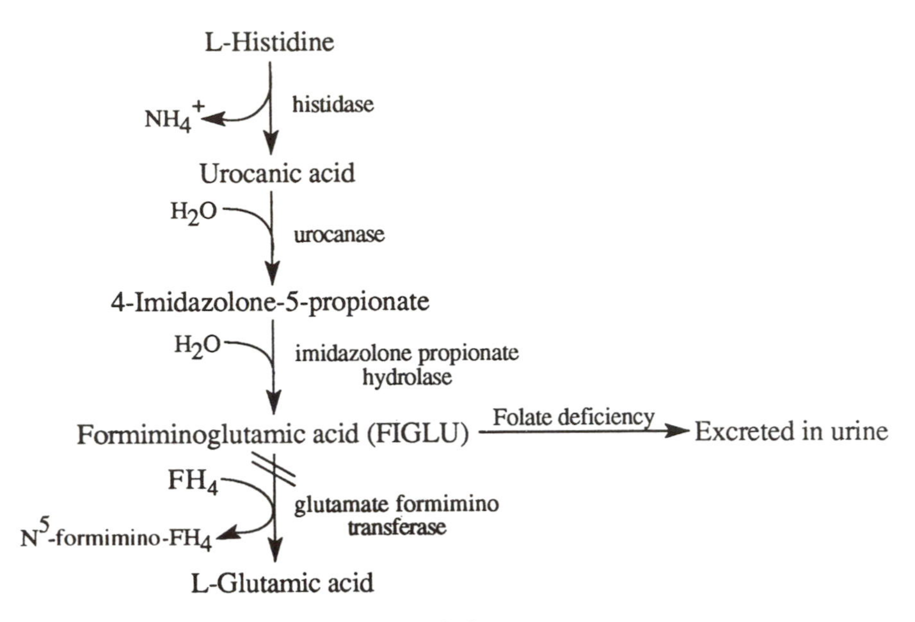

Fig. 8.10. Folate-dependent histidine metabolism.

Table 8.3. Interpretive guideline for biochemical assessment of folate status.

Measurement	Deficient	Marginal	Adequate
Serum (ng ml^{-1})	< 3	3.0–6.0	> 6
RBC (ng ml^{-1})	< 140	140–160	> 160
FIGLU excretion following histidine load (mol/24 h urine)	> 200	–	< 200

Adapted from Wagner (1984).
Conversion factor to SI units (nmol l^{-1}) = × 2.266.

further metabolized; consequently, a large quantity of this metabolite is excreted in the urine. A FIGLU excretion of > 200 μmol 24 h^{-1} (or > 100 μmol 8 h^{-1}), following a histidine loading dose (5 g), is observed in patients with folate deficiency (Lascelles and Donaldson, 1989). An interpretive guideline for all biochemical assessments for folate status is shown in Table 8.3.

In summary, folate deficiency occurs in stages. The first stage of folate deficiency is associated with depressed serum folate levels. The cut-off point used for low serum folate concentrations is < 3 ng ml^{-1} (<6.8 nmol l^{-1}). Serum folate will decrease into the range of deficiency from the normal range within 1–2 weeks of low folate intake. Tissue depletion is the next stage where RBC folate level is decreased below 140 ng ml^{-1} (317 nmol l^{-1}). The fall of RBC folate to its deficiency level usually takes place over 120 days as a new folate-deficient population of erythrocytes enters the circulation. Functional changes characterized by megaloblastic anaemia develop as the final stage of tissue folate depletion. Megaloblastic anaemia can be diagnosed by identifying abnormally large cells in a peripheral blood smear. An elevated mean cell volume and mean cell haemoglobin are frequently accompanied by large red cells. These functional changes do not generally appear until 3–4 months after beginning a folate-deficient diet (Herbert, 1962).

References

Angier, R.B., Bothe, J.H., Hutchings, B.L., Mowat, J.H., Semb, J., Stockstad, E.L.R., SubbaRow, Y., Waller, C.W., Cosulich, D.B., Fahrenbach, M.J., Hultquist, M.E., Kuh, E., Northey, E.H., Seeger, D.R., Sickles, J.P. and Smith, J.M. (1945) Synthesis of a compound identical with the *L. casei* factor isolated from liver. *Science* 102, 227–228.

Bailey, L.B., Cerda, J.J., Block, B.A., Burby, M.J., Vargas, L., Chandler, C.J. and Halsted, C.H. (1984) Effects of age on poly- and monoglutamyl folacin absorption in human subjects. *Journal of Nutrition* 114, 1770–1776.

Bates, C.J., Fleming, M. and Paul, A.A. (1980) Folate status and its relation to vitamin C in healthy elderly men and women. *Age and Ageing* 9, 241–248.

Baugh, C.M. and Krumdieck, C.L. (1971) Naturally occurring folates. *Annals of the New York Academy of Sciences* 186, 7–28.

Becroft, D.M. and Holland, J.T. (1966) Goat's milk and megoblastic anaemia of infancy. *New Zealand Medical Journal* 65, 303–305.

Bills, T. and Spatz, L. (1977) Neutrophilic hypersegmentation as an indicator of incipient folic acid deficiency. *American Journal of Clinical Pathology* 68, 263–267.

Chanarin, I. (1980) The folates. In: Barker, B.M. and Bender, D.A. (eds.) *Vitamins in Medicine*, 4th edn, Vol. 1. William Heinemann, London, pp. 247–314.

Colman, N. (1977) Folate deficiency in humans. *Advances in Nutrition Research* 1, 77–124.

Department of Health (1991) *Dietary Reference Values for Food Energy and Nutrients for the United Kingdom*. Her Majesty's Stationery Office, London.

Ek, J. and Magnus, E. (1980) Plasma and red cell folacin in cow's milk-fed infants and children during the first 2 years of life: The significance of boiling pasteurized milk. *American Journal of Clinical Nutrition* 33, 1220–1224.

Erbe, R.W. (1975) Inborn errors of folate metabolism. *New England Journal of Medicine* 293, 753–757.

Food and Nutrition Board, National Research Council (1980) Folacin. In: *Recommended Dietary Allowances*. National Academy of Sciences, Washington, D.C., pp. 106–113.

Food and Nutrition Board, National Research Council (1989) *Recommended Dietary Allowances.* National Academy Press, Washington, D.C.

Halsted, C.H., Beer, W.H., Chandler, C.J., Ross, K., Wolfe, B.M., Bailey, L. and Cerda, J.J. (1986) Clinical studies of intestinal folate conjugases. *Journal of Laboratory Clinical Medicine* 107, 228–232.

Health and Welfare Canada (1977) *Food Consumption Patterns Report.* Bureau of Nutritional Sciences, Ottawa, Canada.

Health and Welfare Canada (1990) *Nutrition Recommendations: The Report of the Scientific Review Committee.* Ministry of Supply and Services, Ottawa.

Herbert, V. (1962) Experimental nutritional folate deficiency in man. *Transactions of the Assocation of American Physicians* 75, 307–320.

Herbert, V. (1968) Nutritional requirements for vitamin B_{12} and folic acid. *American Journal of Clinical Nutrition* 21, 743–752.

Herbert, V. (1971) Predicting nutrient deficiency by formula. *New England Journal of Medicine* 284, 976–977.

Herbert, V. (1987a) Making sense of laboratory tests of folate status: folate requirements to sustain normality. *American Journal of Hematology* 26, 199–207.

Herbert, V. (1987b) Recommended dietary intakes (RDI) of folate in humans. *American Journal of Clinical Nutrition* 45, 661–670.

Herbert, V. (1990) Development of human folate deficiency. In: Picciano, M.F., Stokstad, E.L.R. and Gregory, J.F. (eds.) *Evaluation of Folic Acid Metabolism in Nutrition and Disease,* Alan R. Liss, New York, pp. 195–210.

Herbert, V., Drivas, G., Manusselis, C., Mackler, B., Eng, J. and Schwartz, E. (1984) Are colon bacteria a major source of cobalamin analogues in human tissues? Twenty-four hour human stool contains only about 5 μg of cobalamin but about 100 μg of apparent analogue (and 200 μg of folate) *Transactions of the Association of American Physicians* 97, 161–171.

Hoyumpa, A.M. (1986) Mechanisms of vitamin deficiencies in alcoholism. *Alcoholism: Clinical and Experimental Research* 10, 573–581.

IUPAC-IUB Commission on Biochemical Nomenclature (1966) Tentative rules: Nomenclature and symbols of folic acid and related compounds. *Journal of Biological Chemistry* 241, 2991–2992.

Knipsheer, R.J.U.L. (1975) Megaloblastic anaemia in pregnancy and folate. In: Eskes, T.K.A.B. (ed.) *Aspects of Obstetrics Today,* Excerpta Medica/American Elsevier, New York.

Krumdieck, C.L., Fukushima, K., Fukushima, T., Shiota, T. and Butterworth, Jr., C.F. (1978) A long-term study of the excretion of folate and pterins in a human subject after ingestion of 14 folic acid, with observations of the effect of diphenylhydantoin administration. *American Journal of Clinical Nutrition* 31, 88–93.

Labadarios, D., Dickerson, J.W.T., Parke, D.V., Lucas, E.G. and Obuwa, G.H. (1978) The effects of chronic drug administration on hepatic enzyme induction and folate metabolism. *British Journal of Clinical Pharmacology* 5, 167–173.

Lascelles, P.T. and Donaldson, D. (1989) *Diagnostic Function Tests in Chemical Pathology.* Kluwer Academic Publishers, United Kingdom, pp. 59–60.

Life Sciences Research Office (1984) Assessment of the folate nutritional status of the US population based on data collected in the Second National Health and Nutrition Examination Survey, 1976–1980. Federation of American Societies for Experimental Biology. Bethesda, MD.

McPartlin, J., Halligan, A., Scott, J.M., Darling, M. and Weir, D.G. (1993) Accelerated folate breakdown in pregnancy. *Lancet* 341, 148–149.

Marcus, D.L. and Freedman, M.L. (1985) Folic acid deficiency in the elderly. *American Geriatric Society* 33, 552–558.

Mitchell, H.K., Snell, E.E. and Williams, R.J. (1941) The concentration of folic acid. *Journal of American Chemistry Society* 63, 2284.

MRC Vitamin Study Research Group (1991) Prevention of neural tube defects: Results of the Medical Research Council Vitamin Group. *Lancet* 338, 13–37.

Poh Tan, S., Wenlock, R.W. and Buss, D.H. (1984) Folic acid content of the diet in various types of British household. *Human Nutrition: Applied Nutrition* 38A, 17–22.

Reisenauer, A.M., Krumdieck, C.L. and Halsted, C.H. (1977) Folate conjugase: two separate activities in human jejunum. *Science* 198, 196–197.

Reisenauer, A.M., Halsted, C.H., Jacobs, L.R. and Wolfe, B.M. (1985) Human intestinal folate conjugase: adaptation after jejunoileal bypass. *American Journal of Clinical Nutrition* 42, 660–665.

Rodriguez, M.S. (1978) A conspectus of research on folacin requirements of man. *Journal of Nutrition* 108, 1983–2103.

Rosenberg, I.H., Bowman, B.B. and Cooper, B.A. (1982) Folate nutrition in the elderly. *American Journal of Clinical Nutrition* 36, 1060–1066.

Rothenberg, S.P. and Cotter, R. (1978) Nutrient deficiencies in man: Vitamin B_{12}. In: *CRC Handbook Series in Nutrition and Food: Nutritional Disorder*, Volume III. CRC Press, Florida, pp. 69–88.

Sauberlich, H.E. (1990) Evaluation of folate nutrition in population groups. In: Picciano, M.F., Stokstad, E.L.R. and Gregory, J.F. (eds) *Folic Acid Metabolism in Health and Disease* Wiley-Liss Inc., New York, pp.212–235.

Schorah, C.J. and Smithells, R.W. (1991) Maternal vitamin nutrition and malformations of the neural tube. *Nutrition Research Review* 4, 33–49.

Shojania, A.M. (1984) Folic acid and vitamin B_{12} deficiency in pregnancy and in the neonatal period. *Clinical Perinatology* 11, 433–456.

Steinberg, S.E. (1984) Mechanisms of folate homeostasis. *American Journal of Physiology* 246, G319–324.

Wagner, C. (1982) Cellular folate binding proteins: Functions and significance. *Annual Review of Nutrition* 2, 229–248.

Wagner, C. (1984) Folic acid. In: *Present Knowledge of Nutrition*, 5th edn. Nutrition Foundation. Washington, D.C., pp. 332–346.

Wills, L. (1933) The nature of the hemopoietic factor in marmite. *Lancet* 224, 1283–1286.

Vitamin B_{12} (Cobalamins)

9

Structure

In 1948, vitamin B_{12} was first isolated from liver as a red crystalline substance (Rickes *et al.*, 1948). It was the last of the vitamins to be isolated. Vitamin B_{12} is very different from all other vitamins in that it contains a mineral, cobalt. The cobalt is chelated by a 'corrin' ring system (Fig. 9.1), which is very similar to that of the porphyrins except that two of the four pyrrole rings (rings 1 and 4) are joined directly rather than through a single methylidyne carbon. Below the corrin ring system there is a 5,6-dimethyl-benzimidazole riboside that is linked with one end to the central cobalt atom and with the other end from the ribose moiety through phosphate and aminopropanol to a side chain on ring 4 of the tetrapyrrole nucleus. A cyanide group may be coordinately bound to the cobalt atom of the vitamin B_{12} molecule (Fig. 9.1); this form is called 'cyanocobalamin' (mol. wt 1355). The cyanide group may be removed and substituted by a hydroxy, aqua, nitro, methyl, or a 5′-adenosyl group, called 'hydroxycobalamin', 'aquacobalamin', 'nitrocobalamin', 'methylcobalamin' and '5′-adenosylcobalamin', respectively. Of these, methylcobalamin and adenosylcobalamin are the two coenzymes that are involved in most vitamin B_{12} metabolic reactions.

The cobalt atom in the vitamin B_{12} molecule may be uni-, di-, or trivalent having 8, 7, or 6 electrons, respectively. There may also be either 2, 1, or 0 axial ligands present in the vitamin B_{12} molecule so that it is a 6, 5, or 4 coordinate. Cyanocobalamin is a 6 coordinate, and the compound is unchanged in neutral solution. The charge on the cobalt atom, cob(III)alamin, is neutralized by negative charges on the corrin ring, the phosphate of the nucleotide and the cyanide. The reduction of cob(III)alamin gives rise to cob(II)alamin and cob(I)alamin.

Crystalline vitamin B_{12} is stable to heating at 100°C, and aqueous solutions at pH 4–7, and can be autoclaved with very little loss. However, the cobalamins are very sensitive to light and alkaline, and in the presence of reducing agents.

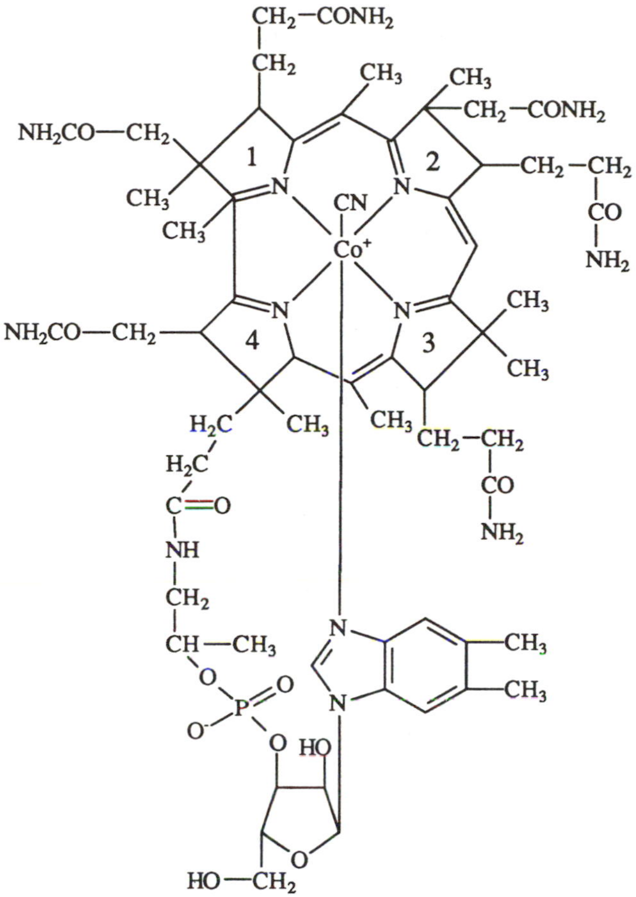

Fig. 9.1. Structure of vitamin B_{12}.

Photolysis results in the conversion of cob(III)alamin to cob(II)alamin and cob(I)alamin which are stable only under anaerobic conditions. Of all cobamides, cyanocobalamin is the most stable form and therefore it is in this form that the vitamin is produced commercially from bacterial fermentation. It must, however, be converted in the body to one of the naturally occurring forms, such as adenosyl- and methylcobalamins before it can be an active coenzyme. Adenosyl- and methylcobalamins are particularly sensitive to light.

Sources

The ultimate sources of vitamin B_{12} are microorganisms, which have the ability to synthesize it. There is no evidence for its synthesis by the tissues of higher plants or animals. The activity of microorganisms in synthesizing vitamin B_{12} extends to the bacteria of the intestine. The microbial flora of the rumen in ruminant animals are able to make the vitamin, accounting for the richer vitamin B_{12} content of livers from ruminant animals compared with other animals such as the pig.

Commercially, vitamin B_{12} is produced as a fermentation product of microorganisms such as *Streptomyces olivaceous, Streptomyces griseus*, or *Bacillus megatherium*. Microorganisms in the intestinal tract of humans are able to synthesize the vitamin, but in the colon it is not absorbed, making faeces concentrated in vitamin B_{12}. Hence faeces become a significant source of the vitamin for those animals such as rats, rabbits and fowl, that practise coprophagy.

Vitamin B_{12} is generally passed on to higher animals via a chain of reactions. Contamination of food by soil and bacterial activity associated with food storage provide some of the dietary sources of the vitamin. Clams, crabs and oysters scavenge microorganisms from their surrounding water, and hence concentrate vitamin B_{12}, which is in turn passed on to whatever creature devours them.

It should be noted that plant foods contain no detectable vitamin B_{12}. The only plant sources found to contain appreciable amounts of vitamin B_{12} are algae, particularly the nori (*Porphyria tenera* and the spirulina); the bioavailability or biological activity of the vitamin from these sources, however, is questionable (Dagnelie *et al.*, 1991). The exclusive reliable sources of this vitamin for humans are therefore animal foods except when plant foods are contaminated with microorganisms. Legumes and root vegetables, due to their contaminations with microorganisms, become the only sources of vitamin B_{12} for vegans (strict vegetarians). Of the animal foods, organ meats, lamb and fish are rich sources of the vitamin (Table 9.1).

Absorption and Transport

The vitamin B_{12} present in food is bound in coenzyme form to protein; it is released by cooking or through the action of the gastric acidification and/or proteolytic enzyme pepsin. The absorption of vitamin B_{12} is a unique process (Rose *et al.*, 1984). The released dietary vitamin is exposed to both intrinsic factor (IF), a glycoprotein secreted by the gastric parietal cells which also produce hydrochloric acid, and the R proteins, which are secreted by the salivary glands and the stomach (Grasbeck, 1984; Seetharam and Alpers, 1985). In the stomach, the free vitamin B_{12} binds more readily to the R protein than to the IF (Fig. 9.2). It is thought that R proteins provide protection to the vitamin against its intestinal bacterial utilization. Upon entrance of the R–vitamin B_{12} complex into the intestinal lumen, R proteins are destroyed by the pancreatic proteases and the vitamin binds to the IF, which travels to the intestine from the stomach.

Table 9.1. Vitamin B_{12} content of some common foods of animal origin.

Sources	Vitamin B_{12} (µg per 100 g net weight)
High:	> 10
Liver	
Lamb	
Kidney	
Heart	
Clams and oysters	
Medium:	3–10
Dried fat-free milk	
Crabs	
Salmon	
Sardines	
Egg yolk	
Moderate:	1–3
Mussels	
Lobster	
Scallops	
Flounder	
Tuna	
Fermenting cheese	
Low:	< 1
Cream	
Whole milk	
Cottage cheese	

Source: Rothenberg and Cotter (1978).

In the intestinal lumen the IF binds with vitamin B_{12} resulting in the formation of a complex containing two moles of the vitamin per mole of IF dimer. This combination of vitamin B_{12} with IF serves to transport the vitamin to specific receptors on the ileal brush border without danger of its degradation by intestinal enzymes. The IF–vitamin B_{12} complex becomes attached at the distal ileum to the receptor sites; the process of attachment requires the presence of calcium ions, and an optimum pH in excess of 5.6. The mechanism of entry of vitamin B_{12} into the cells is not clearly understood. Possibly the IF–vitamin B_{12} complex actually enters the ileal cell as a whole, and within the cell the vitamin B_{12} splits from IF. It is also claimed that there is a splitting factor in the intestinal mucosa. Absorption of the IF–vitamin B_{12} complex to the distal ileum receptor sites does not require energy, but the actual uptake of the vitamin into the cell is oxygen and glucose dependent. Absorption may also occur by passive diffusion bypassing the IF-mediated mechanism. The carrier-mediated absorption of

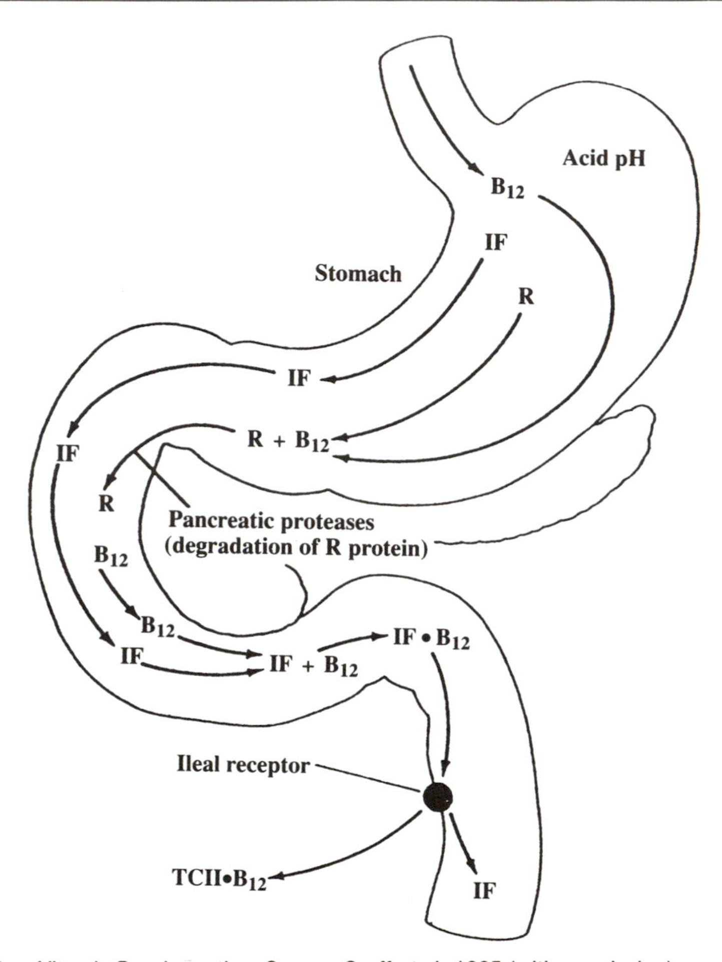

Fig. 9.2. Vitamin B_{12} absorption. Source: Groff *et al.*, 1995 (with permission).

vitamin B_{12} is highly efficient and is important at physiological dose levels (1–3 μg) of the vitamin. Absorption by diffusion, on the other hand, occurs with low efficiency throughout the small intestine and becomes significant only at high dose levels (> 500 μg day^{-1}) of vitamin B_{12}.

In the circulation, vitamin B_{12} binds primarily to three different proteins, which are transcobalamins (TC) I, II, and III. Of these, TC-II, a β-globulin of about 50,000 Da, synthesized in the gut wall, is the most abundant and important for transport of vitamin B_{12} to the circulation (Seetharam and Alpers, 1982). Of the absorbed vitamin B_{12}, 95% is transported in the portal veins combined with this protein; the other 5% is carried by the lymphatics. TC-II appears to be the first protein to which vitamin B_{12} is bound, and within hours of absorption, binding commences with TC-I, an α-globulin. After 24 hours very little of the binding remains with TC-II. The half-lives of TC-II and TC-I molecules combined with vitamin B_{12} are 1.5 hours and 9–10 days, respectively. TC-III may also carry some vitamin B_{12} to the liver. TC-I and TC-III are heterogeneous glycoproteins of about 60,000 Da produced by granulocytic white blood cells; these two proteins are also known as 'R-binders' (Carmel, 1985).

Vitamin B_{12} in the body occurs mostly as the coenzymes, methylcobalamin (predominantly in plasma) and adenosylcobalamin (predominantly in tissues). The total body content in adults is 2–5 mg, most of which is stored in the liver. Vitamin B_{12} is lost from the body in an exponential fashion, the lower the intake and stores the less the excretion (Hall, 1964). Hence the stores are depleted very slowly; in the liver the biological half-life for vitamin B_{12} is estimated to be over 400 days. The long biological half-life of the vitamin provides protection against deprivation for more than a year.

Vitamin B_{12} is excreted in bile, from which 65–75% is absorbed again in the distal ileum (enterohepatic circulation), providing efficient conversion of the vitamin (Herbert, 1987); the remainder, as well as vitamin B_{12} newly synthesized by the intestinal flora, pass out in the faeces. There is very little urinary excretion of vitamin B_{12} at physiological dose level; however, the excretion is markedly increased when the vitamin is taken in pharmacological doses. In humans, the total daily loss of vitamin B_{12} via renal and biliary routes is normally about to 0.1–0.2% of total body reserves (equivalent to 2–5 μg). This amount constitutes the daily dietary requirements for vitamin B_{12}.

Metabolic Role

Vitamin B_{12} participates in a number of enzymatic reactions in microorganisms, but there are only three vitamin B_{12}-dependent enzyme systems of metabolic significance that exist in mammals. These are 5-methyltetrahydrofolate-homocysteine methyl transferase, methylmalonyl CoA mutase, and leucine mutase. The first enzyme system is important in the generation not only of methionine but also of FH_4, a reaction where methylcobalamin functions as a coenzyme (Bannerjee and Mathews, 1990). In the second reaction, 5′-adenosylcobalamin is required for the action of methylmalonyl CoA mutase, which converts methylmalonyl CoA to succinyl CoA in the degradation of propionate. The adenosylcobalamine-dependent leucine mutase is involved in the conversion of L-α-leucine to 3-aminoisocapronate as the first step in the synthesis or degradation of the amino acid.

Methylcobalamin

The majority of naturally occurring folates, when ingested, is converted to 5-CH_3-FH_4, which is stored in the liver and other tissues (see Chapter 8). The CH_3-FH_4 is first converted to FH_4 before entering the cellular pool of active folates. In this conversion, the CH_3 group of CH_3-FH_4 is transferred to homocysteine, generating methionine and FH_4 (Eq. 9.1). This is a reaction catalysed by vitamin B_{12}-dependent 5-CH_3-FH_4 homocysteine methyltransferase (Shane and Stokstad, 1985).

$$CH_3\text{-}FH_4 + \text{Homocysteine} \longrightarrow FH_4 + \text{Methionine} \quad (9.1)$$

Mammals are unable to synthesize homocysteine *de novo*. They use 5-CH_3-FH_4-homocysteine methyltransferase (methionine synthase) to regenerate methionine from homocysteine. This methylation reaction is dependent upon cobalamin, which acts as a methyl carrier from 5-CH_3-FH_4-homocysteine. The 5-CH_3-FH_4-homocysteine methyltransferase is the major, and probably the only, metabolic reaction that can use 5-CH_3-FH_4. If this reaction is blocked due to lack of vitamin B_{12}, 5-CH_3-FH_4 accumulates and the FH_4 needed for other folate requiring reactions is not formed (Fig. 9.3). Consequently, most of the folate in the body is trapped as 5-CH_3-FH_4, becoming metabolically unavailable to the body (Scott and Weir, 1981).

The concept of metabolic availability of folate requiring vitamin B_{12} may provide an explanation of why the haematological damage of vitamin B_{12} deficiency is not distinguishable from that of folate deficiency. In both instances, the haematological damage results from lack of adequate N^5N^{10}-methylene FH_4

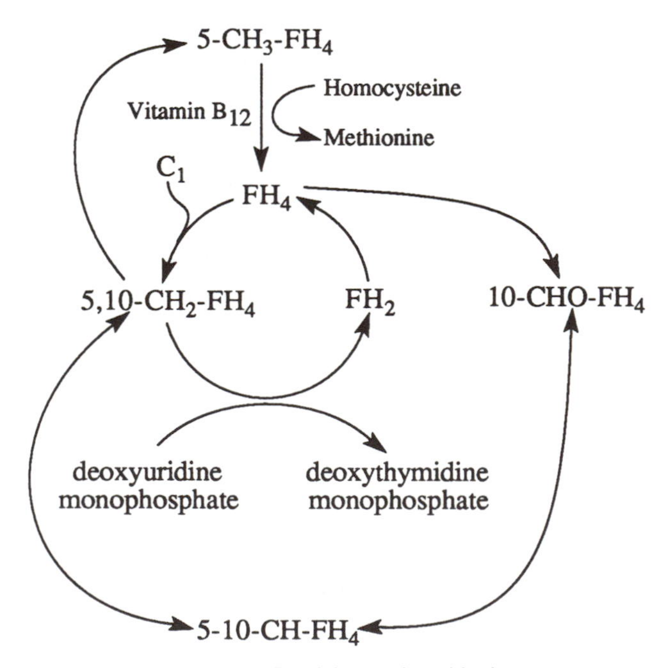

Fig. 9.3. Interrelationship of vitamin B_{12}, folate and methionine.

which delivers its methyl group to deoxyuridylate to synthesize thymidylate and thus to synthesize DNA (Fig. 9.3). In either folate or vitamin B_{12} deficiency, the lack of adequate DNA synthesis causes many haemopoietic cells to die in bone marrow leading to megaloblastosis, characterized by large, oval red and white blood cells (giant germ cells).

Another possible interrelationship of folate and vitamin B_{12} appears to be vitamin B_{12} aiding in the retention of folate by cells (Shane and Stokstad, 1985). Folate is stored in tissues primarily in its polyglutamate form (folyglutamates). The decreased synthesis of folypolyglutamates has been reported to be associated with vitamin B_{12} deficiency. It is possible that the CH_3-FH_4 that accumulates in the presence of inadequate vitamin B_{12} supply is inhibitory to the folyglutamate synthetase.

5′-Adenosylcobalamin

Succinyl CoA is a breakdown product of valine, isoleucine, threonine and methionine. Propionyl CoA is an intermediate of this metabolic pathway; it undergoes a biotin-dependent CO_2 fixation reaction (see Chapter 6) to form methylmalonyl CoA. Subsequently the malonyl CoA is isomerized to form succinyl CoA. This reaction is catalysed by methylmalonyl mutase, for which 5′-adenosylcobalamin acts as a coenzyme. The subsequent oxidation of succinyl CoA in the Krebs' cycle permits complete oxidation of its precursors to CO_2, thereby involving vitamin B_{12} in intermediary metabolism.

The conversion of methylmalonyl CoA to succinyl CoA is interrupted in vitamin B_{12} deficiency. As a consequence, there is an accumulation of methylmalonic acid in the body accompanied by its increased urinary excretory levels. Urinary excretory level of methylmalonic acid (methylmalonic aciduria) is one of the biochemical indices used in diagnosing vitamin B_{12} deficiency and in monitoring therapy.

Signs of Deficiency

Tissues with high turnover rates, such as the red blood cells and the epithelial cells of the gastrointestinal tract, are particularly affected by vitamin B_{12} (or folate) deficiency. The ultimate cause is a decreased metabolic availability of folate coenzyme, 5, 10-methylene FH_4, which is involved in the *de novo* synthesis of DNA (Fig. 9.3).

The clinical signs of vitamin B_{12} deficiency can be related to general symptoms of anaemia, which include pallor of the skin and mucous membrane, accompanied by tiredness, breathlessness, palpitation, angina and anorexia. Disturbances of the alimentary tract leading to diarrhoea and malabsorption are also seen. Other signs include a burning tongue which is often red and smooth, and may show ulcers. The more specific sign of a deficiency of vitamin B_{12} in man is the development of a macrocytic anaemia either alone or in combination with characteristic lesions of the nervous system.

The haematological changes due to vitamin B_{12} deficiency are identical with those described in folate deficiency (see Chapter 8), but usually take much longer to appear. Generally, the abnormal morphological changes in the bone marrow and peripheral blood are manifested at the final stages of chronic vitamin B_{12} deficiency. Unlike folate deficiency, the lack of vitamin B_{12} may cause neurological disorders, characterized by peripheral neuropathy, subacute combined degeneration of the spinal cord, and mental changes (Lindenbaum *et al.*, 1988). Some of the clinical manifestations include numbness and tingling in the hands and feet, diminution of vibration sense, unsteadiness, poor muscular coordination with ataxia, and poor memory.

The underlying cause for the effect of vitamin B_{12} deficiency on the nervous system is not yet clearly understood. Inadequate myelin synthesis is seen in vitamin B_{12} deficiency (Herbert, 1984). The disturbance in methylmalonyl CoA metabolism, a specific abnormality to vitamin B_{12} deficiency, has been suggested to play some role. Using the peripheral nerve tissue taken from vitamin B_{12} deficient animals, abnormal synthesis of total fatty acids and odd chain fatty acids (C-15, C-17) have been demonstrated *in vitro*; this may be related to excessive accumulations of methylmalonic acid (and propionic acid) and to shortage of succinyl CoA. The methylmalonyl CoA that accumulates may inhibit myelin sheath formation by competitive inhibition of malonyl CoA in fatty acid synthesis or by substitution of branched-chain fatty acids for malonyl CoA in the myelin sheath. This concept, however, has been disputed on the basis of the fact that children with an inborn error, where the conversion of methylmalonic acid to succinyl CoA is metabolically disturbed, have not been found to be associated with any neuropathy (Mahoney and Rosenberg, 1980).

In recent years, homocysteine has been attributed to be one of the factors causing neurological disorder in vitamin B_{12} deficiency. As discussed earlier, CH_3-FH_4 transfers a methyl group to vitamin B_{12}, which transfers it to homocysteine, and thereby converts homocysteine to methionine (Eq. 9.1). In vitamin B_{12} deficiency, therefore, homocysteine concentration builds up in the body. Since homocysteine is known to be a neurotoxin (Herbert, 1992), this may account for the neuropathy in vitamin B_{12} deficiency. It is worth noting here that although the reaction catalysed by vitamin B_{12}-dependent methionine synthetase (Eq. 9.1) is an important route for utilizing methyl FH_4, most tissues can form methionine from homocysteine by an alternative vitamin B_{12}-dependent reaction using beteine as $-CH_3$ donor. Since the central nervous system (CNS) has little or no beteine, this may also explain the CNS damage in vitamin B_{12} deficiency. This concept has been further supported by the fact that subacute combined degeneration in monkeys, induced by the anaesthetic gas nitrous oxide, could be prevented by dietary supplementation with methionine (Scott *et al.*, 1981). These findings suggest that neuropathy induced by vitamin B_{12}

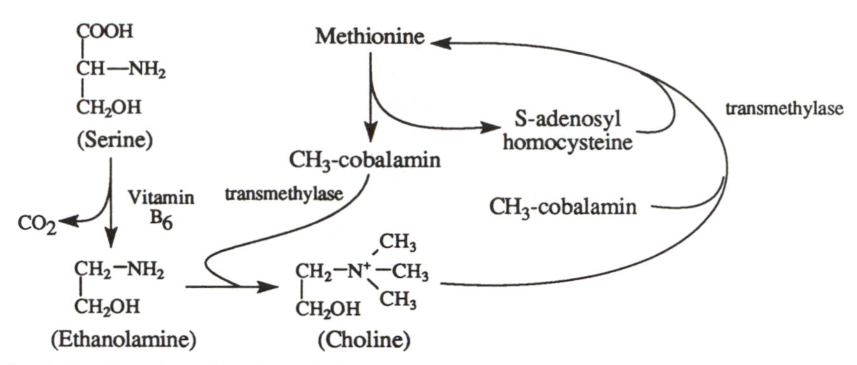

Fig. 9.4. Possible role of vitamin B_{12} in choline and methionine synthesis.

deficiency may be due to inability to synthesize methionine and *S*-adenosylmethionine, resulting in an accumulation of excess homocysteine.

Another possible underlying mechanism by which vitamin B_{12} deficiency could be responsible for neuropathy is through affecting the synthesis of acetylcholine. This neurotransmitter is known as a chemical mediator of parasympathetic activity as well as certain other types in the nervous system. It is produced from choline and acetyl CoA, catalysed by the enzyme choline acetylase (Eq. 9.2).

$$\text{Choline} + \text{Acetyl CoA} \xrightarrow{\text{Choline acetylase}} \text{Acetylcholine} + \text{CoA} \qquad (9.2)$$

An inadequate amount of methionine caused by vitamin B_{12} deficiency may affect choline biosynthesis (Fig. 9.4). The *de novo* synthesis of choline begins with the amino acid serine which is decarboxylated in a vitamin B_6-dependent reaction to ethanolamine. This latter compound is then progressively methylated to choline. This methylation process, in which methionine acts as a methyl donor, therefore, indirectly depends upon vitamin B_{12} and folate for synthesis of methionine and homocysteine, controlling the amount of choline available to the body.

It is noteworthy that macrocytic anaemia may be caused by the deficiency of either folate, vitamin B_{12}, or both. Supplementation with folate in high doses may alleviate the clinical signs of this condition, but it has no effect in preventing the irreversible neurological damage resulting from a continued vitamin B_{12} deficiency. For this reason, supplemental folates should not be used until the underlying cause of a macrocytic anaemia has been established.

Causes of Deficiency

As discussed above, megaloblastic anaemia is the result of a deficiency of vitamin B_{12} or folic acid. Folate deficiency, however, is a more common cause of the anaemia than deficiency of vitamin B_{12}. This is because body stores of folate are more easily depleted than those of vitamin B_{12}. Furthermore, much of the naturally occurring folates are labile and easily destroyed by cooking. Vitamin B_{12}, on the other hand, is relatively more stable. Its excretion is generally proportional to the total vitamin B_{12} content of the body. When there is a low intake, its excretion is reduced through being re-utilized by enterohepatic circulation. Normal body stores of vitamin B_{12} can therefore last for many years even when its intake is completely discontinued. Although rare, there are many situations where vitamin B_{12} deficiency may, however, occur.

Strict vegetarians (vegans)

Any food of plant origin is devoid of vitamin B_{12} and the usual dietary sources are meat and its products and, to a lesser extent, milk. Because of its predominantly animal source, strict vegetarians (eating no food of animal origin including egg and dairy products) may be potentially vulnerable to developing vitamin B_{12} deficiency over a period of many years. Occasional cases of vitamin B_{12} deficiency have been observed in breast-fed infants, whose mothers have been strict vegetarians for several years prior to and during pregnancy. Most vegetarians (or vegans), however, show no evidence of vitamin B_{12} deficiency. This is thought to be because: (i) a small amount of the vitamin is available through contamination of plant food with soil containing vitamin B_{12}; (ii) the body reabsorbs with high efficiency the vitamin excreted daily into the bile from body stores; and (iii) the small intestine of vegetarians harbours microflora that may synthesize significant amounts of absorbable vitamin B_{12} (Albert *et al.*, 1980).

Gastric factor

The most common cause of deficiency of vitamin B_{12} is failure of its absorption due to atrophy of the gastric mucosa. It is thought that an autoimmune process leading to the disappearance of gastric glands and cessation of gastric secretion is contributory, if not fundamental, to this condition (Grasbeck and Salonen, 1976). There appears to be substantial evidence indicating that this is a familial condition. Some patients with atrophic gastritis have antibodies to the parietal cells of the gastric mucosa without any evidence of IF deficiency, and some have antibodies that react with IF, resulting in IF deficiency. Prevalence of the former type is known to be greater among the genetically determined vitamin B_{12}-deficient subjects than the latter type (Table 9.2). Anaemia caused by vitamin B_{12} malabsorption, due to inadequate or absent secretion of the gastric IF, is called pernicious anaemia. This condition was first described by Thomas Addison of Guy's Hospital, England, in 1855.

Table 9.2. Percentage distributions of different types of antibodies among patients with pernicious anaemia.

	IF antibody	Parietal cell antibody
Normal (adult)	0%	11%
Pernicious anaemia	57%	86%

Source: Doniach *et al.* (1963).

In addition to the genetically determined autoimmune pernicious anaemia, there are situations such as chronic gastritis, gastrectomy, iron deficiency, and thyroid dysfunction, where secretion of IF may be affected, leading to vitamin B_{12} deficiency and therefore pernicious anaemia. The term 'addisonian pernicious anaemia' is often used to distinguish true pernicious anaemia (i.e. decreased IF production due to genetic factors) from 'non-addisonian pernicious anaemia', where IF production is affected due to non-genetic factors. Patients with addisonian pernicious anaemia usually have one or two types of serum antibody which react with IF. One type is called type I (or blocking antibody) where the antibody blocks the binding of vitamin B_{12} with IF. The other type, type II (or binding antibody), is where the antibody binds the IF–vitamin B_{12} complex. Pernicious anaemia is frequently associated with gradual progressive gastric atrophy which is a genetically determined and age-dependent phenomenon. Hence, the frequency of the condition increases with increasing age.

Another variety of pernicious anaemia with vitamin B_{12} deficiency occurs very early in the postnatal period, usually before the age of two. This is a rare autosomal recessive disorder, caused by the congenital absence of IF. This condition is not accompanied by absence of gastric acid secretion or by abnormalities in the histological structure of the gastric mucosa including the gastric parietal cells. This syndrome is called 'congenital pernicious anaemia', which does not appear to be related to adult or juvenile pernicious anaemia, and is characterized by a failure to secrete IF and an associated achlorhydria, atrophic gastritis and antibodies to IF or to parietal cells.

Intestinal factor

Despite an adequate secretion of IF and its ability to complex with vitamin B_{12}, there may be a failure of intestinal absorption of IF–vitamin B_{12} complex. Intestinal diverticuli, blind loops caused by pathologic or surgical entero-enteric anastomosis, and impaired mobility tend to permit bacterial growth, and these bacteria not only take up vitamin B_{12} but IF–B_{12} complex as well. Another organism is *Diphyllobothrium latum*, a fish tapeworm which has affinity for vitamin B_{12}. This parasite infects man as a result of eating uncooked infested fish. It occurs commonly in Finland and Japan in particular, but nevertheless only a small number of infested patients develop vitamin B_{12} deficiency.

Small intestinal disorders, such as Crohn's disease, ileitis, tropical sprue, and non-tropical (gluten-sensitive) sprue, appear to affect the ileal mucosa, which alters the brush border structure containing the receptor for IF, and consequently the vitamin B_{12} absorption is markedly depressed. Deficiency of vitamin B_{12} is, therefore, frequently seen especially among the chronic patients with these disorders. Impaired vitamin B_{12} absorption is also seen among those who undergo bypass surgery for massive obesity. This is particularly true if the surgery removes the distal 24 cm of ileum, the main site of vitamin B_{12} absorption. Chronic pancreatitis is another cause of impaired vitamin B_{12} absorption. The pancreatic insufficiency results in loss of proteolytic activity causing failure to digest intestinal R-proteins. As a consequence, vitamin B_{12} is not released for binding by IF. The condition can be corrected by oral administration of pancreas powder or pancreatic proteases.

A variety of drugs, including p-aminosalicylic acid (PAS), colchicin, antacids, and alcohol, when taken in excess, are known to affect vitamin B_{12} absorption. The underlying mechanism for their interference with vitamin B_{12} absorption is not known. They may affect the intracellular transport of the vitamin, inhibit the IF secretion, or interfere with the formation of IF–vitamin B_{12} complex (for details, see Chapter 18).

Abnormality of vitamin B_{12} transport

Transcobalamin (TC) II is a carrier protein for vitamin B_{12} in plasma, and it facilitates its uptake into the cells by attaching to a specific receptor on the cell membrane. Rare cases with absence of this protein accompanied by a severe megaloblastic anaemia have been reported. These cases appear to be responsive to massive doses (1000 μg, 2–3 times per week) of vitamin B_{12}. This may be because of the fact that in the absence of TC II, vitamin B_{12} may enter the cell by passive diffusion, and this can be achieved only if the vitamin is taken in large doses.

Derangements of metabolism

Methylmalonic acid (MMA) is barely detectable in the urine of healthy individuals, but it is excreted in large amounts in the presence of vitamin B_{12} deficiency and by individuals with inherited methylmalonic aciduria. The abnormality in MMA metabolism may be ascribed to defective synthesis of adenosyl cobalamin, which is an essential cofactor with the enzyme methylmalonyl mutase in catalysing conversion of methylmalonyl CoA to succinyl CoA (Fig. 9.5).

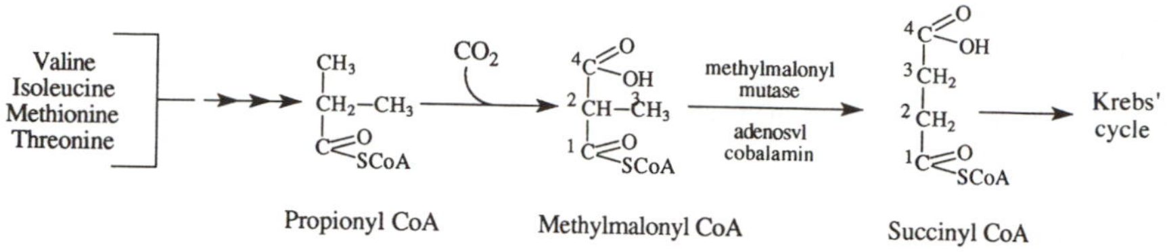

Fig. 9.5. Vitamin B_{12}-dependent isomerization of methylmalonyl CoA to succinyl CoA.

Requirements

An average size of the saturated body pool of vitamin B_{12} in adults is estimated to be 4 mg, the majority being in the liver. Loss of this pool occurs via the faeces, urine and skin; it is generally lost in an exponential manner at a rate of only 0.05–0.2% of the pool per day, irrespective of the pool size (Herbert, 1987). The biological half-life of vitamin B_{12} is estimated to be from 480 to 1284 days. Nutritional equilibrium for vitamin B_{12} can be maintained on a wide range of intakes. An average diet in most industrialized nations supplies between 4–7 μg day^{-1}, with individual diets ranging from 1 to 100 μg day^{-1} (Chung *et al.*, 1961). The intake in many developing countries and by vegetarians often falls below 0.5 μg day^{-1} (Armstrong *et al.*, 1974; FAO/WHO, 1988) with no clinical manifestations of vitamin B_{12} deficiency.

On the basis of body pool size of vitamin B_{12}, its turnover rate, and amounts in the diets of healthy subjects, the daily recommended intake of the vitamin in the United States and in Canada is set at 2 μg day^{-1} for adults (Food and Nutrition Board, 1989; Health and Welfare Canada, 1990). This allowance is increased to 3 μg for the pregnant woman and 2.5 μg for the nursing mother.

When there is vitamin B_{12} deficiency, there are two objectives of its treatment. These are: (i) to alleviate all clinical and biochemical signs; and (ii) to replenish body stores of vitamin B_{12}. Pernicious anaemia is usually treated with vitamin B_{12} at a dose level of 1000 μg given subcutaneously, followed by 500 μg every other day for 10 days and then either 200 μg monthly or 1 μg daily. Patients with sprue usually respond to 25–50 μg of vitamin B_{12} given daily through the parenteral route. Sometimes the best results can be obtained if vitamin B_{12} is given along with folic acid.

Assessment of Vitamin B_{12} Status

Serum concentrations

Vitamin B_{12} deficiency is often accompanied by folate deficiency with clinical manifestations of macrocytic anaemia, but without a sign of neurological disorder. It is important that patients with the anaemia are tested for deficiencies

Table 9.3. Relationships between circulatory levels of vitamin B_{12} and folate with their deficiency states.

Vitamin status	Serum vitamin B_{12}	Serum folate	Red cell folate
Normal	Normal	Normal	Normal
B_{12} deficiency	Normal	Normal/High	Low
Folate deficiency	Normal	Low	Low
Deficiency of both	Low	Low	Low

Source: Herbert (1980).

Table 9.4. Guidelines for the interpretation of serum vitamin B_{12} and urinary methylmalonic acid concentrations.

	Serum B_{12} (pg ml^{-1})	Methylmalonic acid (mg per 24 h urine)
Normal	200–900	1.5–2.0
B_{12} deficiency	< 100	> 300

of these two vitamins. Serum vitamin B_{12}, and serum and red cell folate are three laboratory assays that could be used effectively for the differential diagnosis (Table 9.3). When there is a deficiency of vitamin B_{12}, its serum concentration is low along with low RBC folate but normal or high serum folate. The high serum folate and the low RBC folate are generally the reflections of failure in proper utilization of the circulating 5-CH_3-FH_4 and reduced uptake of folate by erythrocytes, respectively.

Methylmalonic acid excretion

Another biochemical test that can be used to identify vitamin B_{12} deficiency is the determination of methylmalonic acid (MMA) levels in the urine. This substance is present in urine of normal individuals in very small amounts, but there is a profound increase in its excretion level in the presence of a deficiency of vitamin B_{12} (Table 9.4). The increase can be even further accentuated by administering a loading dose (5–10 g) of valine (valine loading test). The methylmalonicaciduria seen in a vitamin B_{12} deficiency is suggestive of a lack of 5′-adenosylcobalamine, which is required for MMA metabolism as indicated in Fig. 9.5. MMA in urine can be accurately measured by gas–liquid chromatography (Cox and White, 1962), but the method is technically difficult and time consuming. Hence, the measurement of MMA has not yet been routinely used in assessing vitamin B_{12} status.

Schilling test

In order to confirm the diagnosis of pernicious anaemia and to reveal its underlying pathogenesis, it is necessary that the vitamin B_{12} absorption test is carried out. The most popular method for testing this function is the 'Schilling test', which was first described by Schilling in 1953. This test involves three steps, as required. In step I, vitamin B_{12} labelled with ^{57}Co (0.5–1.0 μCi) is given orally to a fasting patient. At the same time 1000 μg of non-radioactive vitamin B_{12} is given intramuscularly in order to saturate the protein binding sites in the blood for the vitamin, thus ensuring that any radioactivity absorbed is readily excreted in the urine and not taken up by the tissues. Step II of the Schilling test can be performed 72 h later and consists of repeating step I, except that the labelled vitamin B_{12} is administered orally with a commercial preparation of IF capsule. Step III of the test is rarely used; patients are first treated with a gastrointestinally active antibiotic (such as tetracycline) for 5 days before repeating step II of the test in an attempt to eliminate intestinal vitamin B_{12} utilizing microorganisms. Subsequent to each part of the test, the urine is collected for 24 h, and the radioactivity is measured using a gamma counter.

In step I of the Schilling test, less than 5% of the administered labelled vitamin B_{12} is excreted in the urine of patients with pernicious anaemia compared with greater than 15% in normal subjects. Such interpretations are valid provided that there is no vomiting involved during the test, urine collection is complete, and renal glomerular function is normal. If the reduced vitamin B_{12} absorption is caused by IF deficiency, urinary excretion should be restored to near normal values in step II of the test. If, however, the basis for the malabsorption lies in bacterial or parasitic consumption of the vitamin B_{12}, the test results will show normal values only in step III of the Schilling test. Interpretive guidelines of the Schilling test are summarized in Table 9.5.

The majority of patients with adult-onset pernicious anaemia have circulating antibody to IF. Hence, the presence of serum antibody to IF is thought to be diagnostic for actual or latent pernicious anaemia (Fairbanks *et al.*, 1983). Indeed, the detection of the presence of IF antibody in serum along with the measurement of serum vitamin B_{12}, are more routinely being carried out than the Schilling test in many laboratories. A positive test for the antibody in a

Table 9.5. Summary of the Schilling test.

^{57}Co-B_{12} absorption	Megaloblastic anaemia	Pernicious anaemia	Intestinal malabsorption due to microorganisms
^{57}Co-B_{12} alone	Normal	Subnormal	Subnormal
^{57}Co-B_{12} + IF	Normal	Improvement	No improvement
^{57}Co-B_{12} after antibiotics	Normal	–	Improvement

patient with low serum vitamin B_{12} concentration is virtually diagnostic of pernicious anaemia.

Although the Schilling test is generally considered to be the most detailed diagnostic test for vitamin B_{12} deficiency, its normal values do not always rule out deficiency of the vitamin. This is particularly the case with elderly people who may be deficient in vitamin B_{12}, not because of the lack of gastric IF but of gastric acid and enzymes, making them unable to split vitamin B_{12} from its peptide linkages in foods. Their Schilling tests may appear normal (because the test is performed with crystalline vitamin B_{12}) along with a negative test for IF antibody, and yet they may have low serum vitamin B_{12} (Doscherholmen *et al.*, 1977).

Deoxyuridine (dU) suppression test

This is a sensitive *in vitro* test which is used to assess folate and/or vitamin B_{12} status. It can be a reliable tool to distinguish the deficiency of these two vitamins. This test can be performed in bone marrow cells, lymphocytes or whole blood (Das *et al.*, 1980).

Folate is required for the methylation of deoxyuridine (dU) to produce thymidine (Eq. 9.3).

$$\text{dU} \xrightarrow{5,10\text{-}CH_3\text{-}FH_4} \text{Thymidylate} \longrightarrow \text{DNA} \qquad (9.3)$$

In folate deficiency, this methylation step is expected to be impaired. This principle is used in the dU suppression test, where cells are incubated with and without non-labelled dU prior to the addition of $[^3H]$ thymidine. The uptake of the latter for each culture is then counted and the results expressed as a percentage of the uptake without dU. This method can be adapted to differentiate folate and vitamin B_{12} deficiencies, by adding cobalamin or CH_3-FH_4 to the cell cultures. The uptake of $[^3H]$thymidine radioactivity is then compared in the absence or presence of the *in vitro* addition of cobalamin or CH_3-FH_4. The dU depression values for both folate and vitamin B_{12} deficient bone marrow cultures are known to be $> 20\%$. Cobalamin-deficient marrow is not corrected by CH_3-FH_4, and folate-deficient marrow is not corrected by cobalamin.

References

Albert, M.J., Mathan, V.I. and Kaber, S.J. (1980) Vitamin B_{12} synthesis by human small intestinal bacteria. *Nature* 283, 781–782.

Armstrong, B.K., Davis, R.E., Nicol, D.J., Van Merwyk, A.J. and Larwood, C.J. (1974) Hematological vitamin B_{12} and folate studies on Seventh Day Adventist vegetarians. *American Journal of Clinical Nutrition* 27, 712–718.

Bannerjee, R.V. and Mathews, R.G. (1990) Cobalamin-dependent methionine synthase. *FASEB Journal* 4, 1450–1459.

Carmel, R. (1985) The distribution of endogenous cobalamin among cobalamin-binding proteins in the blood in normal and abnormal states. *American Journal of Clinical Nutrition* 41, 713–719.

Chung, A.S.M., Pearson, W.N., Darby, W.J., Miller, O.N. and Goldsmith, G.A. (1961) Folic acid, vitamin B_6, pantothenic acid and vitamin B_{12} in human dietaries. *American Journal of Clinical Nutrition* 9, 573–582.

Cox, E.V. and White, A.M. (1962) Methylmalonic acid excretion: an index of vitamin B_{12} deficiency. *Lancet* 2, 853–856.

Dagnelie, P.C., Van Staveren, W.A. and Van den Berg, H. (1991) Vitamin B_{12} from algae appears not to be bioavailable. *American Journal of Clinical Nutrition* 53, 695–697.

Das, K.C., Manusselis, C. and Herbert, V. (1980) Simplifying lymphocyte culture and the deoxyuridine suppression test by using whole blood (0.1 ml) instead of separated lymphocytes. *Clinical Chemistry* 26, 72–77.

Doniach, D., Roitt, I.M. and Taylor, K.B. (1963) Autoimmune phenomena in pernicious anaemia. *British Medical Journal* 1, 1374–1379.

Doscherholmen, A., Ripley, D., Chang, S., Ripley, P., Chang, S. and Silvis, S.E. (1977) Influence of age and stomach functions on serum vitamin B-12 concentration. *Scandavanian Journal of Gastroenterology* 12, 313–319.

Fairbanks, V.F., Lennon, V.A., Kokmen, E. and Howard, F.M. (1983) Tests for pernicious anaemia: Serum intrinsic factor blocking antibody. *Mayo Clinic Proceedings* 58, 203–204.

FAO/WHO (1988) Report of a joint FAO/WHO expert consultation. Requirements of vitamin A, iron, folate and vitamin B12. *FAO Food and Nutrition Series* No. 23.

Food and Nutrition Board (1989) *Recommended Dietary Allowances*, 10th edn. National Academy Press, National Research Council, Washington, D.C.

Grasbeck, R. (1984) Biochemistry and clinical chemistry of vitamin B_{12} transport and the related diseases. *Clinical Biology* 17, 99–107.

Grasbeck, R. and Salonen, E.M. (1976) Vitamin B_{12}. *Progress Food Nutrition Science* 2, 193–231.

Groff, J.L., Groffen, S.S. and Hunt, S.M. (1995) Vitamin B_{12} (cobalamins). In: *Advances in Nutrition and Human Metabolism*. West Publishing, St. Paul, pp. 270–276,

Hall, C.A. (1964) Long-term excretion of Co^{57}-vitamin B_{12} and turnover within the plasma. *American Journal of Clinical Nutrition* 14, 156–162.

Health and Welfare Canada (1990) *Nutrition Recommendation: The Report of the Scientific Review Committee.* Canadian Government Publishing Centre, Ottawa.

Herbert, V. (1980) The nutritional anaemias. *Hospital Practice* 15, 65–89.

Herbert, V. (1984) Vitamin B_{12}. In: *Present Knowledge in Nutrition*. Nutrition Foundation, Washington, D.C., pp. 347–364.

Herbert, V. (1987) Recommended dietary intakes (RDI) of vitamin B_{12} in humans. *American Journal of Clinical Nutrition* 45, 671–678.

Herbert, V. (1992) Folate and neural tube defects. *Nutrition Today* 27, 30–33.

Lindenbaum, J., Healton, E.B. and Savage, D.G. (1988) Neuropsychiatric disorders caused by cobalamin deficiency in the absence of anaemia or macrocytosis. *New England Journal of Medicine* 318, 1720–1728.

Mahoney, M.J. and Rosenberg, L.E. (1980) Inherited defects of B_{12} metabolism. *American Journal of Medicine* 48, 584–593.

Rickes, E.L., Brink, N.G., Koniuszy, F.R., Wood, T.R. and Folkers, K. (1948) Crystalline vitamin B_{12}. *Science* 107, 396–397.

Rose, R.C., Hoyumpa, A.M. Jr., Allen, R.H., Middleton, H.M. III, Henderson, L.M. and Rosenberg, I.H. (1984) Transport and metabolism of water soluble vitamins in intestine and kidney. *Federation Proceedings* 43, 2423–2429.

Rothenberg, S.P. and Cotter, R. (1978) Nutrient deficiencies in man: Vitamin B_{12}. In: Rechcigl, M.D. Jr., (ed.) *CRC Handbook Series in Nutrition and Food*, CRC Press, Florida, pp. 69–88.

Schilling, R.F. (1953) Intrinsic factor studies ll. The effect of gastric juice on the urinary excretion of radioactivity after the oral administration of radioactive vitamin B_{12}. *Journal of Laboratory Clinical Medicine* 42, 860–866.

Scott, J.M. and Weir, D.G. (1981) The methyl-folate trap. *Lancet* 2, 337–340.

Scott, J.M., Dinn, J.J., Wilson, P. and Weir, D.G. (1981) Pathogenesis of subacute combined degeneration: A result of methyl group deficiency. *Lancet* 2, 334–337.

Seetharam, B. and Alpers, D.H. (1982) Absorption and transport of cobalamin (vitamin B_{12}). *Annual Review of Nutrition* 2, 343–369.

Seetharam, B. and Alpers, D.H. (1985) Cellular uptake of cobalamin. *Nutrition Reviews* 43, 97–102.

Shane, B. and Stokstad, E.L.B. (1985) Vitamin B_{12}-folate interrelationships. *Annual Review of Nutrition* 5, 115–141.

Stadtman, T.C. (1971) Vitamin B_{12}. *Science* 171, 859–867.

Vitamin C (Ascorbic Acid) 10

Vitamin C comprises essentially two compounds, L-ascorbic acid (mol. wt 176), a strong reducing agent, and its oxidized derivative L-dehydroascorbic acid. Although most vitamin C in body fluids and tissues is in its reduced form, both ascorbic acid and dehydroascorbic acid have biological activity, and are interconvertible by an oxidation and reduction reaction (Fig. 10.1). Some of the enzymes responsible for these interconversions include glutathione dehydrogenase and ascorbate oxidase (Basu and Schorah, 1982).

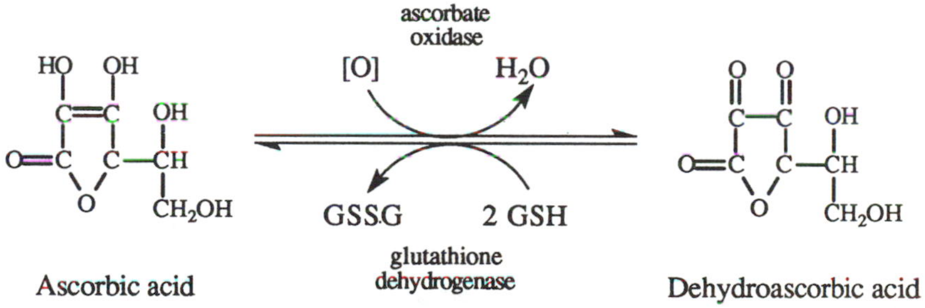

ascorbate oxidase

[O] H_2O

GSSG 2 GSH

glutathione dehydrogenase

Ascorbic acid

Dehydroascorbic acid

Fig. 10.1. Interconvertibility of ascorbic acid by oxidation and reduction.

Biosynthesis

Most plants and animals have the ability to synthesize vitamin C from D-glucose or D-galactose via the glucuronic acid pathway (Fig. 10.2). In the first phase of its synthesis, glucose is converted through several stages to D-glucuronic acid, which is then reduced to L-gulonate. Subsequently, L-gulonate lactonizes to form L-gulono-γ-lactone, which is oxidized to 2-keto-L-gulonolactone. The rate limiting enzyme for this oxidation step is L-gulono-γ-lactone oxidase. The oxidized product in its subsequent step is spontaneously isomerized to form L-ascorbic acid.

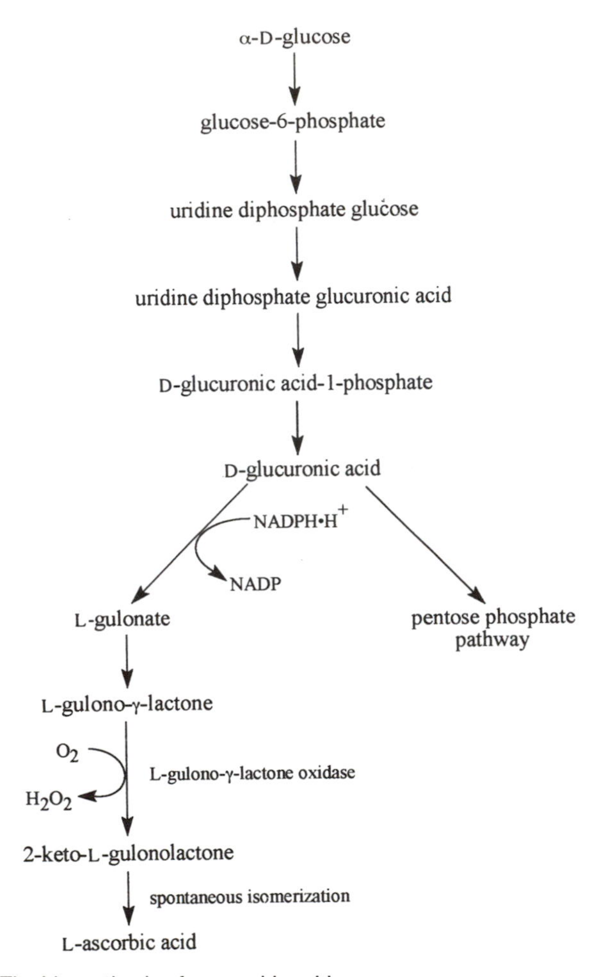

Fig. 10.2. The biosynthesis of L-ascorbic acid.

The only species unable to synthesize vitamin C are primates, including humans, guinea pigs, the red-vented bulbul, the fruit-eating bat (*Pteropus medius*), the rainbow trout, and the coho salmon. They lack the enzyme L-gulono-γ-lactone oxidase, necessary for the conversion of 2-keto-L-gulonolactone to L-ascorbate. Hence, these species become dependent upon exogenous sources for the vitamin (Chatterjee, 1978).

Occurrence in Foods

Vitamin C is found almost exclusively in foods of plant origin. Aside from kidney, no other animal food is considered a significant source. In the United States,

Canada, and most European countries, ascorbic acid is used as a dietary supplement and chemical preservative. It is added in soft drinks as an antioxidant for flavouring, in meat and its products for curing, and in flour to improve baking quality. Hence, food such as cereals, cakes, confectionery, fish and meat products, and soft drinks become an important dietary source of vitamin C. Generally speaking cow's milk is a poor source of vitamin C for the infant. Although human milk is also not a significant source of the vitamin, it contains 3–4 times as much vitamin C as cow's milk.

Particularly rich sources of vitamin C are the West Indian cherry (*Acerola*) and the rose hip, while moderately rich sources of the vitamin are blackcurrants, oranges, lemons, strawberries, kiwi fruit, most green leafy vegetables, and potatoes, particularly new ones, are important on account of the large amount generally eaten. The amount of vitamin C naturally present in plant foods is determined by various factors, such as part and type of the plant, and stage of maturity. Thus, the head of broccoli contains more vitamin C (158 mg/100 g) than its stem (110 mg/100 g). However, stems appear to retain more than 80% of their vitamin C content during a 10 min cooking period whereas heads tend to retain less than 60%. In fruits, vitamin C accumulates during the period up to the point when they are ripe. Therefore, the longer the fruit remains on the vine or tree, the more vitamin C it will contain. In contrast, however, immature seeds such as peas and beans contain more vitamin C than when they are fully mature.

Vitamin Losses

Vitamin C is readily lost in cooking because of its water solubility. Whenever plant foods are eaten raw, the availability of this vitamin is generally higher. Thus, fresh fruit and salad are the most reliable sources for vitamin C in our diet. However, storage of fresh food for considerable periods will deplete the level of vitamin C significantly. For instance, during the summer months when potatoes are freshly picked, their vitamin C content is higher than 30 mg/100 g, while the vitamin content of the same potatoes, stored until spring, is reduced to as low as 7–8 mg/100 g.

Vitamin C is susceptible to oxidation to dehydroascorbate, which is irreversibly degraded further by hydrolytic opening of the lactone ring (Fig. 10.3). These reactions occur in the presence of oxygen. The rate of oxidation is enhanced in the presence of factors such as alkali, heat, light, and the mineral, copper. Thus, addition of baking soda to vegetables, excessive cooking, storage at room temperature or contamination with copper (from cooking utensils) will promote the oxidation of vitamin C. A shorter cooking time and limited exposure to air during preparation help reduce the vitamin loss. Any method that reduces the surface area exposed to air minimizes the loss. Thus, cabbage loses vitamin C faster when shredded. Many plants contain a copper-containing oxidase

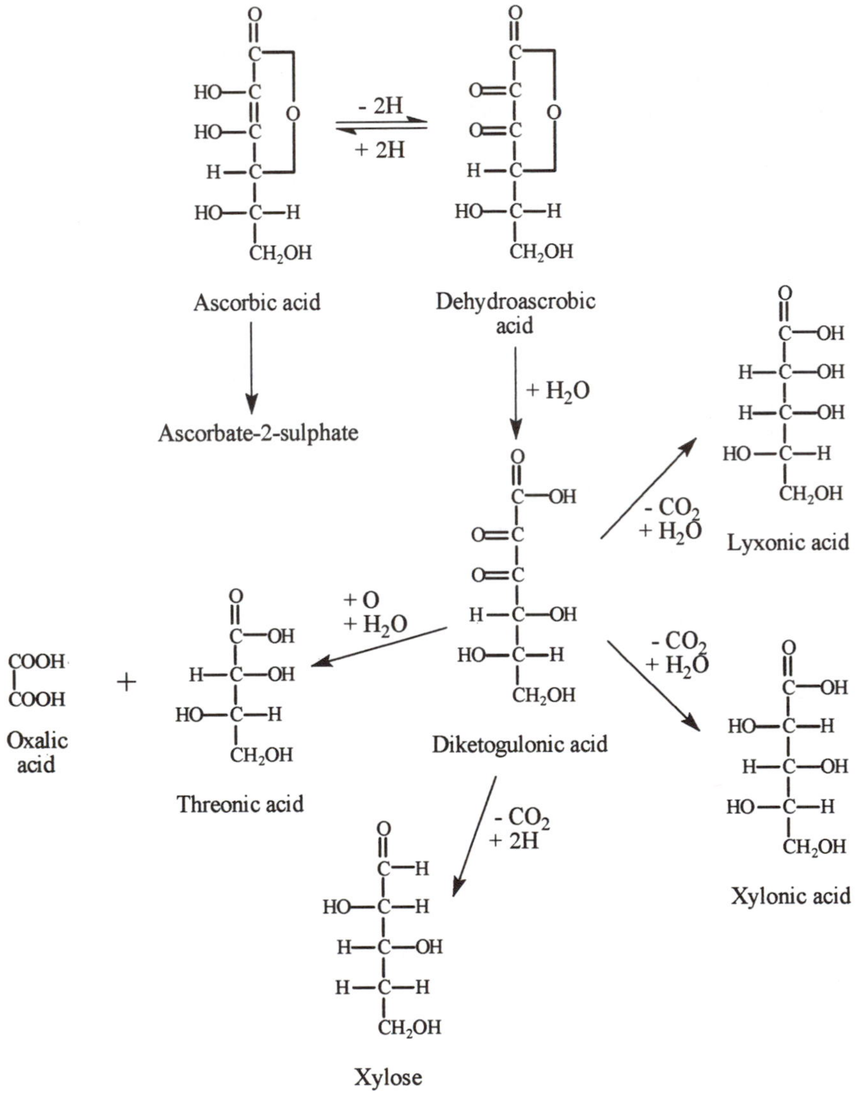

Ascorbic acid

$- 2H$ / $+ 2H$

Dehydroascrobic acid

Ascorbate-2-sulphate

$+ H_2O$

Diketogulonic acid

$- CO_2$ $+ H_2O$

Lyxonic acid

$- CO_2$ $+ H_2O$

Xylonic acid

$+ O$ $+ H_2O$

Oxalic acid + Threonic acid

$- CO_2$ $+ 2H$

Xylose

Fig. 10.3. Degradation of ascorbic acid. Source: Basu and Schorah, 1982.

enzyme that facilitates the oxidation of the vitamin in the presence of air. This enzyme appears to have no direct contact with vitamin C in the intact plant. It is, however, released from plant cells when leaves or fruits are damaged by drying, bruising or cutting with a blunt knife. This is of particular concern in those fruits, such as apples, stone-fruits and pineapples, which have high levels of activity of the enzyme.

The stability of vitamin C in the fruit juice is also determined by its nature (Henshall, 1981). Citrus juices are more susceptible to losses of the vitamin by trace elements than blackcurrant juice. The underlying basis for this is thought to be the type of flavonoids that are present in the fruits. Flavonoids act as inhibitors of oxidation of vitamin C through complexing with metals. The flavonoids of citrus fruits do not possess the 3-hydroxy-4-carboxyl group in the pyrone ring or the 3′,4′-dihydroxy group in the B-ring which are necessary for these compounds to complex with metal ions.

Absorption

The absorption of vitamin C in humans occurs in the buccal mucosa, stomach and small intestine. Buccal absorption is believed to be mediated by passive diffusion through the membrane of the buccal mucosal cells. The rate and extent of diffusion are determined by the initial concentration of vitamin C in the buccal cells and by its rate of passage from the cells into the blood in the mucosal capillaries.

Gastrointestinal absorption of vitamin C is rapid and efficient, and an active carrier-mediated transport system has been suggested, especially at low concentrations. This active absorption mechanism becomes saturated when the mucosal concentration of the vitamin is greater than 6 mmol l^{-1}. This may account for the fact that the proportion of dietary vitamin C absorbed decreases with increasing intake of the vitamin (Hornig *et al.*, 1980).

Distribution in the Body

After absorption, vitamin C rapidly equilibrates in intra- and extracellular compartments. Although no particular organ acts as a storage reservoir for the vitamin, tissues such as the pituitary and adrenal glands, eye lens and leucocytes are concentrators of vitamin C (Table 10.1).

Vitamin C exists in blood and tissues mainly in the reduced form; its oxidized form is generally less than 10%. The average half-life of the vitamin in an adult human is about 20 days, with a turnover of 1 mg $kg^{-1}day^{-1}$ and a total body pool size of 1500 mg (Basu and Schorah, 1982). The daily 'utilization breakdown' of vitamin C is believed to be constant: 0.2 mg kg^{-1} fat-free weight. Thus a man with a lean body mass of 70 kg requires 14 mg vitamin C daily to maintain the body pool. If as much as 70% of dietary vitamin C is absorbed, the daily requirement would then be 20 mg. The requirement of the vitamin in women is likely to be lower than in men of the same body weight since the average body fat content is higher in females.

There appears to be a direct relationship between clinical signs of vitamin C deficiency and its pool size. Frank signs of deficiency are generally observed when the body pool of the vitamin is depleted to a level of 300 mg or less, and

Table 10.1. Plasma and tissue distribution of vitamin C in an adult human.

Tissue	Vitamin C (mg/100 g wet tissue)
Pituitary gland	40–50
Adrenal glands	30–40
Eye lens	25–31
Brain	13–15
Liver	10–16
Spleen	10–15
Kidneys	5–15
Heart muscle	5–15
Lungs	7
Skeletal muscle	3
Testes	3
Thyroid	2
Leucocytes	35
Plasma	0.4–1.0

Source: Adapted from Basu and Schorah (1982).

the clinical signs tend to disappear when repletion takes the vitamin C pool above a level of 300 mg.

Metabolism and Excretion

Catabolism of vitamin C in humans occurs by the irreversible hydrolysis of dehydroascorbic acid to diketogulonic acid, followed by oxidation to oxalic and threonic acids (Fig. 10.3). These metabolites, together with some ascorbate-2-sulphate, are excreted in the urine along with unmetabolized ascorbic acid. Excretion of vitamin C in man can also occur in the forms of lyxonic acid, xylonic acid, and xylose, which are the breakdown products of diketogulonic acid following decarboxylation. In addition, a small portion of vitamin C (usually less than 2%) is metabolized to CO_2, and is exhaled during respiration. In guinea pigs, on the other hand, 60–70% of the vitamin is excreted in the breath as CO_2. It appears, therefore, that there is a fundamental difference in the elimination of vitamin C in humans from that of guinea pigs.

Although the respiratory route in humans is not a major pathway for vitamin C elimination, a number of factors are known to accelerate the rate of oxidation of vitamin C to CO_2. Thus the level of exhalation of CO_2 increases as the intake level of the vitamin increases (Kallner *et al.*, 1985). A presystemic effect as a result of microbiological degradation of unabsorbed vitamin C in the gut rather than its hepatic degradation has been suggested as the underlying

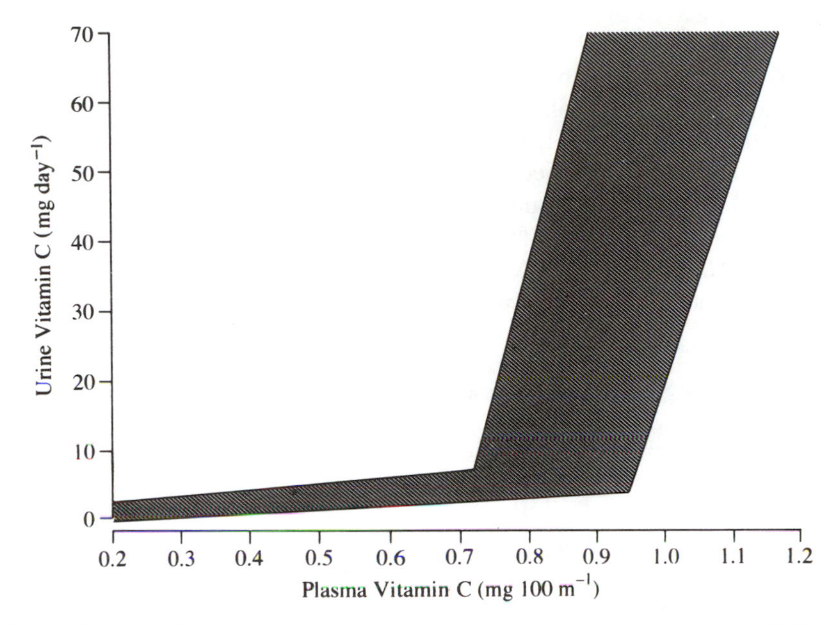

Fig. 10.4. A relationship between plasma and urinary concentrations of vitamin C. The shaded area indicates the approximate 95% range. Source: Basu and Schorah, 1982.

basis for the dose-related effects. Iron toxicity is another factor that is thought to accelerate oxidation of vitamin C to CO_2. A rapid rate of oxidative catabolism of this vitamin through the decarboxylation process has been noted among patients with haemosiderosis. The mechanism by which iron accumulation induces alteration in the metabolism of vitamin C awaits clarification.

The overall metabolism of vitamin C is affected by the level of its intake. At a physiological level (30 mg), less than 10% is excreted in the urine as ascorbic acid and more than 90% as metabolites, whereas a reverse is seen when a large dose level of vitamin C (1–2 g) is ingested. The capacity of kidney tubules for reabsorption saturates at plasma concentrations of vitamin C below 0.8 mg dl^{-1} (Fig. 10.4), and most is lost in the urine within 24 hours.

Biochemical Role

Vitamin C is a strong reducing agent, and hence has a general importance as an antioxidant, affecting the body's 'redox potential'. It has many diverse biochemical functions that are a consequence of its ability to donate one or two electrons (Levine, 1986).

Free-radical scavenger

Ascorbic acid is oxidized to dehydroascorbate through a short-lived intermediate, ascorbate free-radical, which is also called monodehydroascorbic acid. This intermediate is generally regarded as innocuous. In fact, it forms a part of the body's antioxidant defences against reactive oxygen species and free radicals, and thereby prevents tissue damage (Stadtman, 1991). Thus, ascorbic acid (AH^-) reacts with a variety of free radicals and oxygen species as illustrated in Eqns 10.1–10.4, in which $A^{\cdot-}$ = monodehydroascorbic acid; A = dehydroascorbic acid; and $R^{\cdot}$ = alkyl radical.

$$AH^- + {}^{\cdot}OH \longrightarrow H_2O + A^{\cdot-} \quad (10.1)$$

$$AH^- + O_2^- + H^+ \longrightarrow H_2O_2 + A^{\cdot-} \quad (10.2)$$

$$AH^- + R^{\cdot} \longrightarrow RH + A^{\cdot-} \quad (10.3)$$

$$AH^- + H_2O_2 + H^+ \longrightarrow 2\ H_2O + A \quad (10.4)$$

The antioxidant action of ascorbic acid is potentiated by the presence of other reducing agents, such as glutathione (GSH) and NADH, which assist with the regeneration of ascorbic acid (AH^-) from its oxidation products ($A^{\cdot-}$) as shown in Eqns 10.5 and 10.6.

$$2A^{\cdot-} + 2\ GHS \longrightarrow GSSG + 2AH^- \quad (10.5)$$

$$2A^{\cdot-} + NADH + H^+ \longrightarrow NAD^+ + 2AH^- \quad (10.6)$$

Role as a cofactor

Ascorbic acid plays an important role in many reactions involving oxygenases (Englard and Seifter, 1986). These reactions also require molecular oxygen and Fe^{2+} or Cu^{2+} as a cofactor. Essentially, ascorbic acid plays either of two roles: (i) as a direct source of electrons for reduction of oxygen; or (ii) as a protective agent for maintaining Fe and Cu in their reducing states. The oxygenases are classified according to the type of reaction they catalyse. There is a mono-oxygenase where one atom of O_2 is incorporated into the substrate (Sub), the other being reduced to water with the help of a hydrogen donor which can be a reducing agent such as ascorbate (Asc) (Eq. 10.7).

$$\text{Sub} + \text{Asc} + O_2 \xrightarrow{Cu^{2+}} \text{OH-Sub} + \text{Dehydroasc} + H_2O \quad (10.7)$$

The other type of oxygenase is the dioxygenase where both molecules of oxygen are incorporated into two separate substrates, one being α-ketoglutarate (αkg).

$$\alpha kg + Sub_2 + O_2 \xrightarrow[\text{Ascorbate}]{Fe^{2+}} \text{Succinate} + CO_2 + OH\text{-}Sub_2 \tag{10.8}$$

Mono-oxygenase

A high concentration of vitamin C is found in the adrenal medulla of mammals. It is a cofactor for dopamine β-mono-oxygenase (dopamine β-hydroxylase), and peptidylglycine α-amidating mono-oxygenase systems which catalyse the synthesis of noradrenaline (Fig. 10.5) and a variety of α-amidated peptides (Eq. 10.9), respectively. In these reactions, ascorbic acid (Asc) appears to be a direct hydrogen donor to the substrate, 3,4-dihydroxyphenylethylamine (dopamine), or peptide with C-terminal glycine. The hydrogen from ascorbic acid is used to form water.

$$\text{Peptidyl-}\alpha\text{-OH-glycine} \xrightarrow[\text{mono-oxygenase}]{\text{Peptidylglycine } \alpha\text{-amidating}} \text{Amidated peptide} + \text{Glyoxalate} \tag{10.9}$$

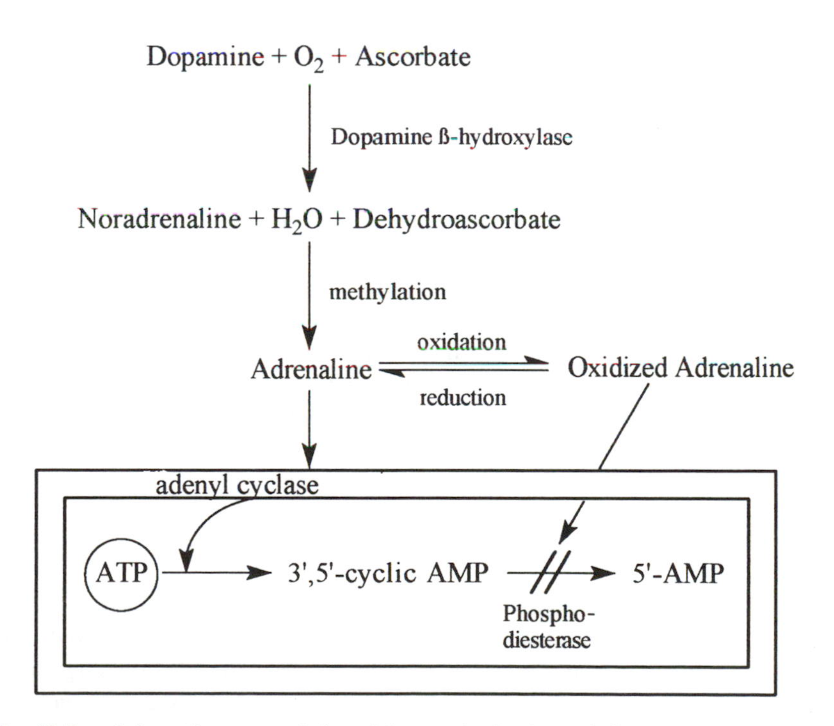

Fig. 10.5. Schematic representation of the synthesis of catecholamines and the formation of cyclic AMP.

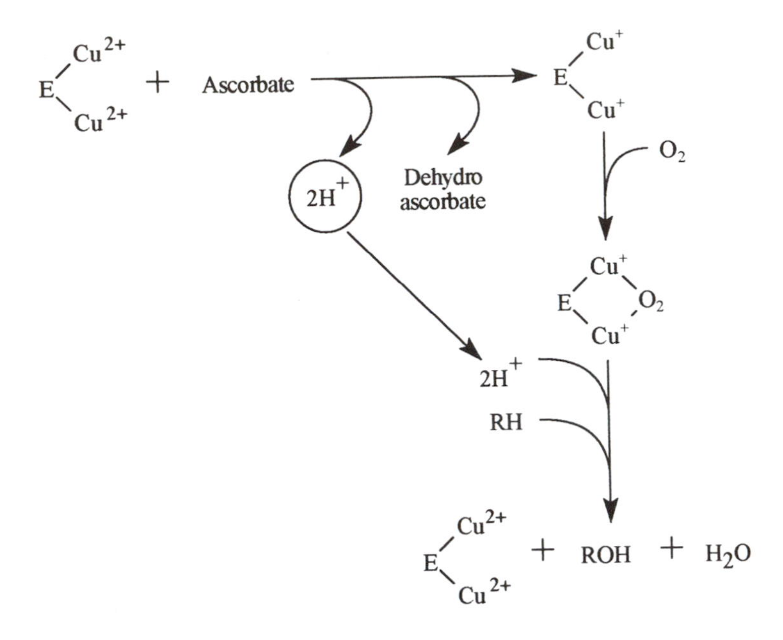

Fig. 10.6. Copper as an integral part of the function of vitamin C (E: enzyme; RH: substrate; ROH: hydroxylated substrate).

Both mono-oxygenase systems contain copper, which accepts electrons from ascorbate as it is reduced to the cuprous ion, Cu^+ (Fig. 10.6). These electrons are subsequently transferred to oxygen and then to the substrate, dopamine or peptidyl-α-OH-glycine (RH) to give the hydroxylated product, noradrenaline or peptide-NH_2 (ROH).

The protein-bound copper is also reoxidized to the cupric (Cu^{2+}) state during this reaction. The role of ascorbate in catecholamine and amidated peptide synthesis is to reduce the essential metal on the enzyme so that the reduced enzyme may mediate in the hydroxylation reaction.

Copper appears to be an integral part of the functioning of vitamin C. Hence, the clinical or biochemical deficiency of this vitamin may not be associated with only vitamin C intake *per se* but also copper deficiency.

Dioxygenases

A variety of reactions involving ascorbate-dependent dioxygenases are listed in Table 10.2. In reaction 1 involving *p*-hydroxyphenylpyruvate hydroxylase, both atoms of a dioxygen molecule are incorporated into a single product, homogentisate. Unlike this reaction, others (reactions 2–6) involving dioxygenases do, however, require α-ketoglutarate as a cosubstrate. In these reactions, one atom of oxygen is incorporated into succinate and one into the

Table 10.2. Enzymatic reactions involving ascorbic acid - dependent dioxygenases.

Reaction	Substrate	Enzyme	Product
1	*p*-OH-phenylpyruvate hydroxylase[1]	*p*-OH-phenylpyruvate	Homogentisate
2	Peptidyl L-proline	Prolyl 4-hydroxylase[2]	Peptidyl 4-transhydroxyl-L-proline
3	Peptidyl L-proline	Prolyl 3-hydroxylase[2]	Peptidyl 3-transhydroxyl-L-proline
4	Peptidyl L-lysine	Lysyl hydroxylase[2]	Peptidyl 5-erythrohydroxy-L-lysine
5	6-*N*-Trimethyl L-lysine	6-*N*-trimethyl L-lysine hydroxylase[2]	Erythro-3-hydroxy-6-*N*-trimethyl-L-lysine
6	4-*N*-Trimethyl aminobutyrate	4-*N*-Trimethyl amino-butyrate hydroxylase[2]	3-Hydroxy-4-*N*-trimethyl amino-butyrate

[1]α-Ketoglutarate is not required as a cosubstrate.
[2]α-Ketoglutarate is required as a cosubstrate.
Source: modified from Englard and Seifter (1986).

product of oxidation of the specific substrate (Eq. 10.8). There is a similarity in all of the dioxygenases (reactions 1–6) in that they all contain iron, and the iron must be in the ferrous state for activity. It is possible that ascorbate is an absolute requirement for (i) maintaining iron in its reduced state, and (ii) prevention of oxidation of sulphydryl groups in the enzymes.

Prolyl and lysyl hydroxylases (Table 10.2, reactions 2–4) are responsible for the hydroxylation of prolyl and lysyl residues in procollagen to hydroxyprolyl and hydroxylysyl residues. For the collagen molecule to aggregate into its triple-helix configuration, the proline and lysine residues on newly synthesized collagen must be hydroxylated. Formation of the triple-helix is important because it is in this configuration that the procollagen is secreted from the fibroblast. Defective hydroxylation within the synthesizing fibroblast gives rise to the formation of an abnormal collagen precursor, called protocollagen. This substance may not be extruded from the cell as is the normal precursor, tropocollagen. If it is extruded then it polymerizes into an abnormal collagen which reduces the tensile strength of tissues. In the absence of hydroxylations of prolyl and lysyl residues the normal maturation of collagen is interrupted. Hence as a consequence of ascorbic acid deficiency, a markedly abnormal collagen will be synthesized, with a shorter half-life than normal. Basement membranes in tissues such as skin, connective tissues, blood capillaries, and bone matrix which are rich in collagen will be markedly affected in ascorbate deficiency (Prockop *et al.*, 1979).

Reduced carnitine levels are also the sequel to ascorbic acid deprivation (Rebouche, 1991). Carnitine plays a role in transporting long-chain fatty acyl groups into the inner mitochondrial membrane where oxidation, yielding energy, takes place. Trimethyl L-lysine and trimethyl aminobutyrate hydroxylases (Table 10.2, reactions 5–6), responsible for the synthesis of carnitine from trimethyl lysine are, dioxygenases that require not only ascorbate but also α-ketoglutarate. These enzymes are identical to prolyl and lysyl hydroxylases. Carnitine concentrations in various tissues and its synthesizing enzyme activities in the liver or kidney appear to be markedly decreased in scorbutic guinea pigs (Rebouche, 1991).

Mixed-function oxygenase

Most foreign compounds including therapeutic agents are lipid soluble, and hence modified prior to excretion. This modification is usually carried out by an enzyme system, mixed-function oxygenase (MFO), which is found predominantly in hepatic microsomes but also in the extrahepatic tissues (Basu and Schorah, 1982). This enzyme system requires a number of components. These include enzymes for hydroxylating substrates, flavoproteins, cytochrome P450 (a haemoprotein), molecular oxygen and a reducing agent. There is considerable evidence that ascorbic acid is one of the factors capable of modifying this MFO system with consequent effects upon the body's response to many environmental chemicals.

Phenylalanine hydroxylase, a mixed-function oxygenase involved in the conversion of phenylalanine to tyrosine, and cholesterol-7α-hydroxylase, the rate-limiting enzyme for cholesterol degradation, are also thought to belong to the MFO system. These hydroxylation reactions as well as the conversion of tryptophan to 5-hydroxytryptamine (serotonin) appear to require ascorbic acid. How the vitamin participates in these reactions remains to be determined. The relationship between ascorbic acid and the MFO system is dealt with further in Chapter 18.

Other reducing roles

Cyclic nucleotides

Ascorbic acid is required for the synthesis of noradrenaline, which is converted to adrenaline following methylation (Fig. 10.5). Under physiological conditions adrenaline tends to be oxidized, but ascorbic acid prevents it. Adrenaline potentiates the adenyl cyclase which in turn activates the formation of cAMP from ATP. Ascorbic acid also reduces the breakdown of cAMP to 5′-AMP by inhibiting the enzyme phosphodiesterase. Thus, ascorbic acid may increase the tissue level of cAMP by stimulating its synthesis and by decreasing its degradation.

There is evidence that vitamin C intake can relieve both experimental and clinical asthma. It is thought that this effect is mediated through increasing cAMP levels and hence leading to a decreased histamine release (Lewin, 1976).

Iron absorption and metabolism

Non-haem iron (from plant food) usually constitutes more than 90% of the dietary iron; its absorbability is considerably less than that of haem iron (from animal food). Ascorbic acid is a potent enhancer of non-haem iron absorption from food (natural iron) and also of iron added in food. The enhancing effect is strongly dose related, and is different for different meals probably due to varying content of inhibitors in the meals (Fig. 10.7). Non-haem iron present in food is usually in its ferric (Fe^{3+}) state. Inorganic iron is, however, absorbed from the intestinal lumen as ferrous iron (Fe^{2+}), with very little uptake in its ferric state. Ascorbic acid in the gut is thought to keep iron in its reduced form preventing the formation of insoluble ferric hydroxide, and hence aids absorption. It is not only the reducing but also the chelating property of this vitamin with metal ions that may account for its enhancing effect on iron absorption.

Ascorbic acid may also be involved in the transfer of iron into the blood as well as in the mobilization from its stores. In the circulation, iron is generally in its oxidized form bound to transferrin, whereas the reduced form of iron is bound to ferritin in the liver. In ascorbate deficiency, increasing quantities of iron are deposited in the tissues, and circulatory levels of iron and ferritin bear little relationship to tissue iron reserves. The correlation can be restored by repletion of tissue levels of ascorbic acid (Roesser *et al.*, 1980).

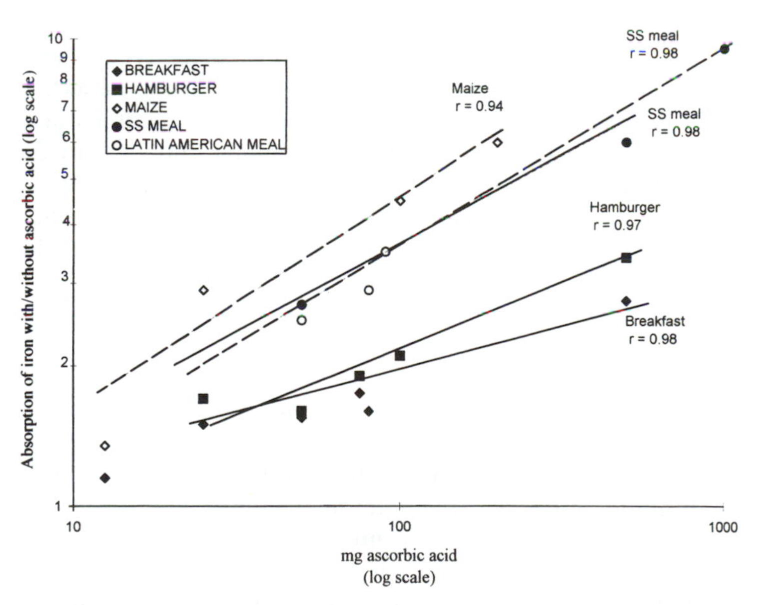

Fig. 10.7. Effect of different meals with or without ascorbic acid on the absorption of iron. Source: Hallberg and Rossander (1982).

Activation of folic acid

The coenzymatically active forms of folic acid are the tetrahydro derivatives, which are produced through its reduction (Eq. 10.10). This reduction process is catalysed by NADPH-dependent reductase enzymes (E_1 and E_2). In this reaction, folic acid (F) is first reduced to dihydrofolate (FH_2) and then to its active form, tetrahydrofolate (FH_4).

$$F \underset{\text{NADPH.H}^+}{\overset{\text{Asc, E}_1}{\rightleftharpoons}} FH_2 \underset{\text{NADPH.H}^+}{\overset{\text{Asc, E}_2}{\rightleftharpoons}} FH_4 \quad (10.10)$$

This two-step reaction is, however, reversible, and hence subject to oxidative destruction. Ascorbic acid (Asc), owing to its powerful reducing capacity, protects folate reductases, and therefore the active form of folic acid.

Nitrite scavenger

Ascorbic acid, owing to its reducing property, is known to function in other cofunctional capacities. It thus appears to have potential importance as a nitrite scavenger.

The principal source of nitrite is nitrate from food. Following absorption, part of the nitrate reaches the buccal cavity where microorganisms convert it to nitrite. Consequently, nitrite reaches to the stomach where it can react with amines or amides (Tannenbaum *et al.*, 1978) forming nitroso compounds (Eq. 10.11).

$$(R_1,R_2)\text{-N-H} + HNO_2 \longrightarrow (R_1,R_2)\text{-N-N=O} + H_2O \quad (10.11)$$

Nitrostable amines or amides are generally found in many types of foods; they can also be synthesized in the body (Zeisel *et al.*, 1985). Nitroso compounds have been found to be both hepatotoxic and carcinogenic in experimental animals. There appears to be substantial evidence suggesting that vitamin C has the potential to block the formation of nitrosamines, because the vitamin C–nitrite reaction is more rapid than the amine–nitrite reaction (Tannenbaum *et al.*, 1991). The vitamin reduces nitrite, being oxidized to dehydroascorbic acid in the process, thereby inhibiting the reaction of nitrite with amines (Eq. 10.12).

$$2\,HNO_2 + \text{Asc} \longrightarrow 2NO + \text{Dehydroasc} + 2\,H_2O \quad (10.12)$$

Immune function

There are two types of mechanisms by which the immune response is generally initiated: humoral antibodies (the immunoglobulins), and the cell mediated response. The production of humoral antibodies is essentially a function of the B lymphocytes, which are derived from the bone marrow. Cell-mediated immune responses are mediated by the T lymphocytes which come from the thymus.

The most obvious link between immunity and vitamin C is that the cells involved in the immune response contain very high concentrations of vitamin C of the order of 40–60 times the concentration found in the plasma (Table 10.1). These high concentrations within the leucocytes are rapidly depleted by acute disease, infection and trauma. Furthermore, pregnancy, ageing, and corticosteroid therapy, together with chronic disorders such as cancer and diabetes mellitus, are associated with depression of immunity; these states, too, are accompanied by low levels of both plasma and leucocyte vitamin C.

The implication from these observations is that vitamin C might be involved in some way in immunological processes. There is little evidence, however, supporting its role in the humoral side of the immune response, but studies aimed at investigating lymphocyte function and phagocytic activity of the neutrophil have proved more positive (Anderson, 1981).

Overall there appears to be appreciable evidence that vitamin C is involved in immune response through participating in the maintenance of phagocytosis. It promotes chemotaxis making bactericidal polymorphonuclear leucocytes move against microbes. It also reduces allergic reactions and raises interferon production, which is essential to combat viral infections. Most recent interest has, however, concentrated on the effectiveness of vitamin C in combating respiratory and other viral infections. This area will be considered further in Chapter 17.

Deficiency

Vitamin C appears to play an important role in numerous biological systems. Its functions include synthesis of collagen, neurotransmitters, and carnitine. It plays a major role as an antioxidant and free radical scavenger and it is involved in the detoxification of many foreign compounds. It is also an important factor for the utilization of other nutrients, such as iron and folacin, and for the function of the immune system. Because of the involvement of vitamin C in multiple functions, the early signs of deficiency are relatively non-specific. These signs often include fatigue, weakness, shortness of breath, aching bones, joints and muscles, and loss of appetite. These are, however, followed by more specific symptoms, such as swollen, bleeding and sensitive gums, hardening and roughness around hair follicles (hyperkeratosis), petechial haemorrhages under the skin and delayed wound healing. These clinical signs of scurvy are due to inhibition of collagen synthesis leading to failure to maintain the cellular structure of the supporting tissues of mesenchymal origin, such as bone, dentine, cartilage, and connective tissues. The dentine becomes porous, alveolar bone becomes osteoporotic, and teeth loosen and fall out. These conditions are associated with severe pain and immobility. This may be the reason why infants with scurvy adopt a typical 'frog-leg' position, this being the least painful for them.

If vitamin C is withdrawn from the diet, it takes 100–160 days for the clinical manifestations of scurvy to develop (Table 10.3). During deficiency, the vitamin pool is depleted at a daily rate of about 2.6% of the existing pool; symptoms of mild scurvy are evident when the pool is less than 300 mg.

In vitamin C deficiency the fibroblasts proliferate but they remain immature and fail to synthesize collagen molecules, the 'building blocks' of tissue repair. Because of the impaired fibroblastic activity, wound healing is poor. In vitamin C deficiency the cartilage cells of the epiphyseal plate at the diaphyseal end of the long bones continue to proliferate and line up in rows. The cartilage between the rows is calcified, where osteoblasts do not migrate, resulting in compressed and brittle bone.

Additional clinical manifestations observed in vitamin C deficiency include behavioural changes, such as apathy, depression and emotional disturbances. These signs may be the consequences of decreased synthesis of catecholamines, serotonin, and α-amidated peptide, a neurotransmitter hormone. In vitamin C deficiency, there is a loss of blood associated with petechiae, perifollicular haemorrhages, and bleeding gums. Vitamin C is an important factor for iron absorption and utilization; it also plays an important role in stabilizing the folate reductase reaction. In patients with vitamin C deficiency, marginal folate

Table 10.3. The consequences of withdrawing vitamin C from the diet in adults.

	Plasma ascorbic acid		Buffy coat ascorbic acid in leucocytes (WC)		Body pool of ascorbic acid		
Day	(mg l^{-1})	(μmol l^{-1})	(μg 10^{-8})	(nmol 10^{-8})	(g)	(nmol)	Clinical state
0	8–15	45–85	21–57	119–323	0.6–1.5	3.4–8.5	
20	3	17	10–38	57–216			
40	1–3	6–17	2–10	11–57	0.3–0.6	1.7–3.4	Subclinical deficiency
60	< 1	< 6	< 5	< 28	0.3–0.6	1.7–3.4	
80	< 1	< 6	< 5	< 28	0.3–0.6	1.7–3.4	
100	< 1	< 6	< 5	< 28	0.3–0.6	1.7–3.4	
120	< 1	< 6	< 2	< 11	< 0.3	< 1.7	Peri-follicular hyperkeratosis
140	< 1	< 6	< 2	< 11	< 0.3	< 1.7	
160	< 1	< 6	< 2	< 11	< 0.3	< 1.7	Petechiae and ecchymosses of the skin; failure of wounds to heal
180	< 1	< 6	< 2	< 11	< 0.1–0.3	0.6–1.7	Gingival changes
200	< 1	< 6	< 2	< 11	< 0.1	< 0.6	Dyspnoea, oedema and very rapid progression

and iron intakes may precipitate megaloblastic and hypochromic anaemia, respectively.

Requirements

A daily intake of 40–60 mg of vitamin C is believed to maintain the saturated state with a 1500 mg or greater body pool size. A minimum of 10 mg day^{-1} will prevent scurvy. The recommended intake of vitamin C, however, varies in different parts of the world (Table 10.4). Its recommendation has been subject to frequent changes even within a country. Thus in the United States the adults' recommended dietary allowance (RDA) for vitamin C has changed from 75 mg day^{-1} in 1943 to 45 mg day^{-1} in 1974, and currently it is 60 mg (National Research Council, 1989). Our inadequate knowledge of what tissue concentration is required for vitamin C to be maximally effective in carrying out its basic functions, its instability, bioavailability and metabolic turnover, are primarily responsible for the differences that exist in its recommended intakes.

Vitamin C status in man depends on dietary intake, metabolic demands, and renal clearance. Dietary intake can be influenced by social and economic conditions, and by the age of the individual. Metabolic changes affect the vitamin C demand of the tissues as in pregnancy, infections, alcoholism, smoking, and other stress conditions.

Scurvy is now a rare condition. Outbreaks occur in poor nomadic populations in arid or semi-desert districts when there is a threat of famine or a long standing drought. A marginal state of vitamin C deficiency, however, can be seen more frequently in certain population groups. The 'at risk' groups are often elderly people, food faddists, alcoholics, smokers, and persons living in institutions and patients with psychiatric disorders.

Table 10.4. Recommended intakes of vitamin C for an adult (male) in different parts of the world.

Countries	Recommended intake (mg day^{-1})
United Kingdom	30
Canada	40
Denmark	45
Netherlands	50
Sweden	60
United States	60
West Germany	75
Republic of Russia	75

Elderly people

Both plasma and leucocyte concentrations of vitamin C have long been known to fall with increasing age in both males and females. Studies based on urinary excretion of the vitamin following oral loads, have suggested that older people may have a higher requirement for vitamin C, due to impaired absorption or increased utilization (Irwin and Hutchins, 1976). However, there is substantial evidence to suggest that inadequate dietary intake may be the principal cause of the low vitamin C status in the elderly (Basu and Schorah, 1982). It may be that as we age we eat less and prefer foods that contain little vitamin C. Alternatively, it may be that today's elderly population acquired their habits at a time when vitamin C-containing foods were less popular than they have been in recent years.

Vitamin C intake is much of a concern especially in institutions where there is an additional loss of the vitamin in large-scale institutional cooking. The delay in delivery of the meal to the recipient and an inadequate supply of fruit and fruit juices are possible contributing factors. Similar factors may be responsible for the poor vitamin C status of the elderly receiving meals-on-wheels. Some studies have indicated that vitamin C status of about 5% of the elderly population is dangerously low, and that of a much higher proportion is marginal.

The high incidence of inadequate vitamin C status among the elderly has been reported from studies carried out in countries outside North America. The intake of the vitamin is generally sufficient for all age groups among Americans and Canadians. However, in spite of this adequate intake of vitamin C, elderly subjects, especially males, appear to have low serum levels of the vitamin. According to Garry and his associates (1987) the elderly should receive vitamin C to the amounts that would allow serum concentrations to be maintained close to 1.0 mg dl^{-1}. The daily intakes required to maintain this serum level would be approximately 125 mg for healthy males and 75 mg for females. This prudent approach would assume adequate body reserves of vitamin C in the elderly.

Alcoholics

Decreased intakes of vitamin C accompanied by low plasma and leucocyte levels have been reported among alcoholics with or without any obvious liver disease (Bonjour, 1979). It is not, however, known if these effects are due to a direct effect of alcohol consumption. Both animal and human studies have indicated that vitamin C may be useful in alcohol detoxification and hence in protecting against liver disease.

Smokers

Decreased plasma and leucocyte levels of vitamin C are found among smokers. Using controlled doses of labelled vitamin C in a pharmacokinetic approach, smoking has also been reported to be associated with a reduced absorption rate and biological half-life of the vitamin (Kallner *et al.*, 1981). In addition, this group of the population has a higher turnover of vitamin C and therefore an increased requirement compared with non-smokers. The daily metabolic turnover appears to approach saturation at 70–90 mg of vitamin C in smokers as opposed to only 40–50 mg in non-smokers (Fig. 10.8). In order to reach this saturation level a total turnover of 60 mg day^{-1} of vitamin C is required in non-smokers, whereas the smokers require 90 mg day^{-1} to achieve this saturation level. Both in smokers and non-smokers the saturation level of the metabolic turnover corresponds to a plasma steady state concentration of vitamin C of 0.8–0.9 mg dl^{-1}.

According to the recent recommendation in Canada (Health and Welfare Canada, 1990) the intakes of vitamin C for heavy smokers should be increased by as much as 50%. Thus, in smokers the recommended intakes of the vitamin are 60 mg and 45 mg day^{-1} for men and women, respectively (non-smokers: male, 40 mg; female, 30 mg day^{-1}). In the United States the RDA of vitamin C for smokers was recently increased from 60 mg to 100 mg day^{-1} (National Research Council, 1989).

The requirement of vitamin C for smokers is difficult to determine. It is difficult to measure the effect of 'stress' or perturbation of homeostasis. Furthermore, we do not know what clinical or biochemical measure best reflects

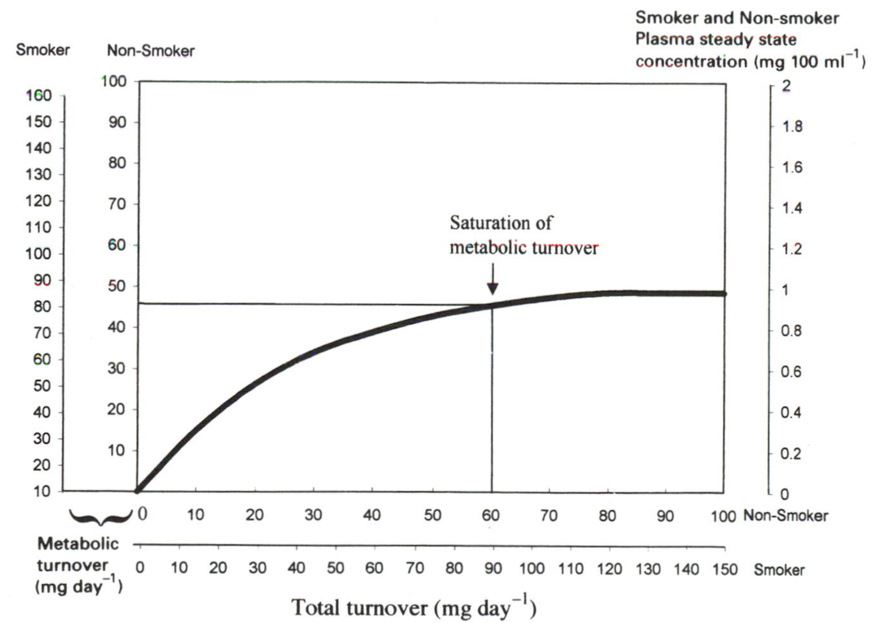

Fig. 10.8. The saturation levels of the metabolic turnover in smokers and in non-smokers.

optimal vitamin C intake. The assays have been used empirically, and hence many of the data conflict. The requirement of vitamin C for smokers is still under re-evaluation. Schectman *et al.* (1991) estimated additional vitamin C needs for smokers from comparisons of dietary intake and serum levels of vitamin C. Data were analysed from the Second National Health and Nutrition Examination Survey (NHANES II), which surveyed nutritional status, including vitamin C, of a nationwide population sample of over 11,000 adult individuals. They found that smokers had to consume more than 200 mg of vitamin C per day to have serum vitamin C concentrations similar to those of non-smokers meeting the RDA (60 mg day^{-1}).

Stress

The adrenal cortex is rich in vitamin C. Stimulation of the adrenal cortex by adrenocorticotropic hormone (ACTH) or by adrenaline leads to depletion of vitamin C. Depletion of adrenal cortex vitamin C is commonly used as an index of the effect of various stressful stimuli in experimental animals, such as infectious disease, physical trauma, and reaction to shock of various kinds.

In humans, evidence that vitamin C reserves are depleted in acute or chronic disease is overwhelming (Basu and Schorah, 1982). Metabolic changes and dietary restrictions both probably contribute to different extents in different conditions.

Pregnancy is a physiological condition which affects vitamin C reserves. During pregnancy, plasma vitamin C levels fall. Whether this is due to physiological responses to pregnancy or to increased demands of pregnancy is uncertain. Vitamin C normally crosses the placental barrier against a concentration gradient, resulting in fetal levels 50% greater than maternal levels at term. Human milk is reasonably rich in vitamin C (30–55 mg l^{-1}) and so lactation can potentially lead to significant losses of maternal vitamin C, of the order of 32 mg day^{-1}. For pregnant and lactating women, an additional allowance of vitamin C is therefore recommended in order to assure satisfactory intakes of the vitamin for the mother as well as the baby.

Exposure to foreign compounds

There is evidence that a wide variety of drugs may affect the requirement for vitamin C. Many of these drugs reduce absorption, increase excretion, or interfere with the utilization of vitamin C. Drug-induced vitamin C deficiency is dealt with in Chapter 18.

Assessment of Vitamin C Status

Biological tissues and fluids

Plasma concentrations are most commonly measured to assess body reserves of vitamin C. These levels, however, are reflected by the preceding intake or depleted by lower intakes of vitamin C while its tissue reserves are adequate. In guinea pigs, a zero plasma vitamin C level could be associated with a wide range of tissue saturation from 0 to 50%. Thus, plasma values are no more than an indicator of vitamin C status. Many studies have indicated that in human populations the plasma vitamin C can be reduced to less than 0.2 mg dl^{-1} when intake is less than 20 mg day^{-1} (Jacob *et al.*, 1987). On the other hand the ability of the renal tubule to reabsorb vitamin C is decreased when plasma levels rise to 0.75–1.0 mg dl^{-1} (Basu and Schorah, 1982). This explains why plasma vitamin C concentrations rarely exceed 1.4 mg dl^{-1} despite very large intakes of the vitamin.

Urinary vitamin C is frequently used to assess the body saturation index on the basis that less than 15 mg day^{-1} is said to indicate deficiency of this vitamin. There is, however, a risk that the urinary level of vitamin C may be reflected by its preceding intake. The vitamin C saturation test is sometimes used as an alternative as well as an improved index. In this test, vitamin C concentrations are determined in urine samples collected at hourly intervals for 5 hours following administration of a loading dose (0.5–2.0 g) of the vitamin. The rationale of this test is based on the hypothesis that following the administration of vitamin C, the surplus is excreted in the urine only when the tissues have become saturated. If tissues are saturated the urinary level of the vitamin will reach its maximum level by 3–4 hours following a loading dose.

Leucocytes contain high concentrations of vitamin C. The concentrations are higher than those in either plasma, whole blood, or erythrocytes. White blood cell vitamin C content is not affected by the preceding intake, and it reaches its lowest level almost simultaneously with the appearance of clinical signs of scurvy. Thus the leucocyte vitamin C level is thought to be the method of choice for the assessment of tissue stores. A leucocyte vitamin C level of less than 10 μg per 10^8 cells is generally regarded as deficient (Table 10.5).

Although the leucocyte vitamin C is a better index of vitamin C status, it is subject to many disadvantages. Preparation of the leucocytes for vitamin C assay is difficult, and relatively large blood samples (2–5 ml) are required. In addition, leucocyte vitamin C measurements are usually performed on the buffy coat layer from whole blood, which contains granulocytes, mononuclear cells, and platelets. All these fractions contain variable amounts of vitamin C. Since the ratio of these fractions can vary widely among individuals, inconsistencies in leucocyte vitamin C concentrations may be observed when buffy coat samples are used to assess vitamin C status.

Table 10.5. Nutritional status of vitamin C and its concentrations in leucocytes.

Vitamin C Status	Leucocyte vitamin C[1] (μg 10^{-8} cells)
Adequate	> 20
Marginal	10–20
Deficient	< 10

[1]Modified from Sauberlich (1981).

Analytical techniques

A variety of colorimetric techniques are available to assay the vitamin C contents of biological materials. These include the oxidation of ascorbic acid with 2,6-dichlorophenolindophenol or the reduction of dehydroascorbic acid with 2,4-dinitrophenylhydrazine. Another class of colorimetric reactions involves the reduction of ferric to ferrous by ascorbic acid followed by addition of 2,2′-dipyridine, a chelating agent, to form a stable coloured ferrous complex (Omaye *et al.*, 1979).

The colorimetric assays, however, lack sensitivity and specificity for biological samples and may not account for instability of dehydroascorbic acid. In recent years, high-performance liquid chromatographic (HPLC) methods for the detection of dehydroascorbic acid have been developed (Dhariwal *et al.*, 1990). The HPLC methods, especially with electrochemical detection, appear to be highly sensitive and specific for dehydroascorbic acid in biological samples.

References

Anderson R. (1981) Ascorbic acid and immune functions: mechanism of immunostimulation. In: Counsell, J.N. and Hornig, D.H. (eds.) *Vitamin C*, Applied Science Publishers, London, pp. 249–272.

Basu T.K. and Schorah, C.J. (1982) In: *Vitamin C in Health and Disease*. Croom Helm, London.

Bonjour J.P. (1979) Vitamins and alcoholism: I. Ascorbic acid. *International Journal of Vitamin and Nutrition Research* 49, 434–441.

Chatterjee I.B. (1978) Ascorbic acid metabolism. *World Review of Nutrition Diet* 30, 69–87.

Dhariwal, K.R., Washko, P.W. and Levine, M. (1990) Determination of dehydroascorbic acid using high performance liquid chromatography with coulometric electrochemical detection. *Analytical Biochemistry* 189, 18–23.

Englard, S. and Seifter, S. (1986) The biochemical function of ascorbic acid. *Annual Review of Nutrition* 6, 365–406.

Garry, P.J., Vanderjagt, D.J. and Hunt, W.C. (1987) Ascorbic acid intakes and plasma levels in healthy elderly. *Annals of the New York Academy of Sciences* 498, 90–99.

Hallberg, L. and Rossander, L. (1982) Absorption of iron from Western-type of various dinner meals. *American Journal of Clinical Nutrition* 35, 502–509.

Health and Welfare Canada (1990) In: *Nutrition Recommendations*. Canadian Government Publishing Centre, Ottawa.

Henshall, J.D. (1981) Ascorbic acid in fruit juices and beverages. In: Counsell, J.N. and Hornig, D.H. (eds.) *Vitamin C*, Applied Science Publishers, London, pp.123–137.

Hornig, D., Vulleumier, J.P. and Hartman, D. (1980) Absorption of large, single, oral intakes of ascorbic acid. *International Journal of Vitamin Nutrition and Research* 50, 309–314.

Irwin, M.I. and Hutchins, B.K. (1976) A conspectus of research on vitamin C requirements of man. *Journal of Nutrition* 106, 821–879.

Jacob, R.A., Skala, J.H. and Omaye, S.T. (1987) Biochemical indices of human vitamin C status. *American Journal of Clinical Nutrition* 46, 818–826.

Kallner, A., Hartmann, D. and Hornig, D. (1981) On the requirements of ascorbic acid in man: steady-state turnover and body pool in smokers. *American Journal of Clinical Nutrition* 34, 1347–1355.

Kallner, A., Hornig, D. and Pellikka, R. (1985) Formation of carbon dioxide from ascorbic acid in man. *American Journal of Clinical Nutrition* 41, 609–613.

Levine, M. (1986) New concepts in the biology and biochemistry of ascorbic acid. *New England Journal of Medicine* 314, 892–902.

Lewin, S. (1976) In: *Vitamin C: Its molecular biology and medical potential*. Academic Press, New York.

National Research Council (1989) *Recommended Dietary Allowances* 10th edn. National Academy Press, Washington, D.C.

Omaye, S.T., Turnbull, J.D. and Sauberlich, H.E. (1979) Selected methods for the determination of ascorbic acid in animal cells, tissues, and fluids. *Methods in Enzymology* 62, 3–11.

Prockop, D.J., Kivirrikko, K.I., Tuderman, I. and Guzman, N.A. (1979) The biosynthesis of collagen and its disorders. *New England Journal of Medicine* 301, 77–85.

Rebouche, C.J. (1991) Ascorbic acid and carnitine biosynthesis. *American Journal of Clinical Nutrition*, 54, 1147S–1152S.

Roesser, H.P., Holliday, J.W., Sizemore, D.J., Nikles, A. and Willgoss, D. (1980) Serum ferritin in ascorbic acid deficiency. *British Journal of Haematology* 45, 459–466.

Sauberlich, H.E. (1981) Ascorbic acid (vitamin C) In: Labbe, R.F. (ed.) *Symposium on Laboratory Assessment of Nutritional Status* Clinics in Laboratory Medicine 1, 673–684.

Schectman, G., Byrd, J.C. and Hoffmann, R. (1991) Ascorbic acid requirements for smokers: analysis of a population survey. *American Journal of Clinical Nutrition* 53, 1466–1470.

Stadtman, E.R. (1991) Ascorbate and oxidative inactivation of proteins. *American Journal of Clinical Nutrition* 154, 1125S–1128S.

Tannenbaum, S.R, Fett, D., Young, V.R., Land, P.D. and Bruce, W.R. (1978) Nitrite and nitrate are formed by endogenous synthesis in the human intestine. *Science* 200, 1487–1489.

Tannenbaum, S.R., Wishnok, J.S. and Cynthia, D. (1991) Inhibition of nitrosamine formation by ascorbic acid. *American Journal of Clinical Nutrition* 53, 247S–250S.

Zeise, S.H., DaCosta, K.A. and Fox, J.G. (1985) Endogenous formation of dimethylamine. *Biochemical Journal* 232, 403–408.

Vitamin A 11

Vitamin A was the 'first' fat-soluble vitamin to be recognized. The historical development of our knowledge of this vitamin goes back to 1500 BC. Night blindness, a condition associated with an impairment of vision in dim light, is caused by lack of dietary vitamin A. The description of this condition has been found in ancient Chinese, Egyptian, and Greek literature. Since the beginning of this century there have been major advances in our knowledge about vitamin A, especially in the area of its chemistry, sources, metabolism, and biological roles. Despite our knowledge of the vitamin since ancient time, its deficiency remains one of the major public health problems in the world today.

Vitamin A occurs physiologically as the alcohol (retinol), the aldehyde (retinaldehyde), the acid (retinoic acid), and the ester (retinyl ester). The term 'vitamin A' refers to these naturally occurring compounds as well as to carotenoids, the provitamin A (see Chapter 12) while the term 'retinoids' includes both naturally occurring forms of vitamin A and its many synthetic analogues with or without vitamin A activity (IUPAC-IUB, 1982). It appears that the retinoids, in addition to their role in vision, are involved in cell growth, reproduction, the immune system, and the integrity of epithelial cells. Furthermore, these biological potentials have brought them under clinical investigation for clinical application as therapeutic agents in many pathological conditions (see Chapter 19).

Chemistry

Vitamin A is a long-chain primary alcohol and is known to exist in a number of isomeric forms. It exists in the pure state as pale yellow crystals soluble in most organic solvents and in fats. Vitamin A occurs in nature largely as retinol in its all-*trans* form (Fig. 11.1), which has three important structural characteristics. These are: (i) a β-ionone ring, the hydrophobic head; (ii) a conjugated isoprenoid side chain, which is subject to isomerization in the presence of light; and (iii) a polar terminal group, which can be enzymatically or chemically modified to

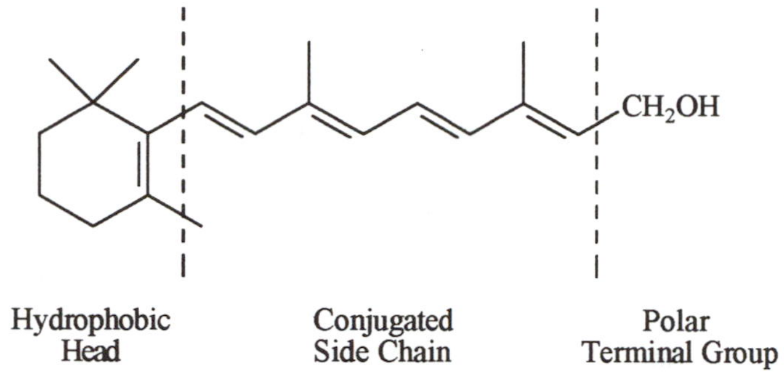

Hydrophobic Head | Conjugated Side Chain | Polar Terminal Group

CH_2OH

Fig. 11.1. Structure of all-*trans* retinol.

become an ester as in retinol palmitate, or an aldehyde as in retinal or be oxidized to a polar metabolite as in retinoic acid.

The all-*trans* retinol molecule (mol. wt 286) has an absorption maximum at 325 nm. Because of its structural configuration, it readily undergoes oxidation in air and isomerizes when exposed to light. Oxidation is the most important cause of its destruction. Vitamin E, due to its antioxidant property, is thought to prevent oxidative destruction of vitamin A. It is also important to know that vitamin A is stored in the animal body primarily in combination with palmitic acid as retinyl palmitate, which is relatively more stable than its other isomeric forms.

Vitamin A is also considered to be sensitive to ionizing radiation. The industrial processing of food, such as pasteurization, sterilization, or dehydration, causes only a very little loss of this vitamin. Generally, vitamin A is stable in heat, acid and alkali; hence the various methods of cooking have very little effect on the loss of this vitamin.

Sources

Dietary compounds that exhibit vitamin A activity are the retinoids which are supplied by preformed vitamin A found exclusively in animal products such as liver, fish oils, dairy products, and eggs. Margarine is supplemented with vitamin A in many countries including the United Kingdom, Canada and the United States; this makes margarine an exceptional food, which although it originates from plant seeds contains preformed vitamin A. However, a variety of carotenoids are widely distributed in the vegetable kingdom, which can be the precursors of vitamin A (see Chapter 12).

The principal sources of vitamin A in the diets of developing countries are generally carotenoids, coming from fruits, green leafy and yellow vegetables, and in some cases red palm oil (see Chapter 12). These plant sources may contribute more than 80% of the total vitamin A intake in their diets. In most of the western world, on the other hand, plant foods contribute less than 20%

to the vitamin A intake; the rest generally comes from animal and fortified food sources.

Vitamin A Activity

Formerly, vitamin A activity in food was expressed in terms of international units (IU). This method of estimation, however, does not take into account the fact that its provitamins are subject to variable activity and absorption. Indeed, foods have to be specified in terms of both preformed and provitamin A. Based on recommendations of the Food and Agriculture Organization (FAO) and the World Health Organization (WHO), the Food and Nutrition Board of the United States (Food and Nutrition Board, 1980) decided that the content of food and the daily requirement should be based on 'retinol equivalent' (RE). The RE is expressed as retinol by its weight and the carotenoids by the weight of retinol to which they would be converted in the body (see also Chapter 12). Thus:

One RE = 1 μg retinol
= 6 μg β-carotene
= 12 μg other carotenoids with provitamin activity
= 3.33 IU vitamin A activity from retinol
= 10 IU vitamin A activity from β-carotene

Metabolism

Absorption

Dietary vitamin A is nearly always present as long-chain fatty acid esters. Its absorption efficiency is normally 80–90%, with only a small reduction in efficiency at high doses. The retinyl ester is hydrolysed in the lumen of the intestine by digestive lipases and esterases in the small intestine (Fig. 11.2). Bile and pancreatic secretions are important for release of the vitamin as retinol. However, recent studies (Rigtrup and Ong, 1992; Rigtrup *et al.*, 1994) have shown that retinyl esters can be hydrolysed in the absence of pancreatic secretions by two enzyme systems located on the brush-border membrane (BBM). These include a pancreatic origin lipase (also called cholesterol ester hydrolase) which is responsible for hydrolysis of short-chain retinyl esters and an enzyme intrinsic to the BBM which hydrolyses long-chain retinyl esters. Dietary retinyl esters are first hydrolysed to retinol in the intestinal lumen before they are absorbed into the intestinal mucosa (enterocytes). Carotenes with vitamin A activity are also absorbed with the lipids of the diet and then converted to retinol primarily in the intestinal mucosa (see Chapter 12). The resulting retinol (newly absorbed + newly synthesized from carotenes) in the mucosal cell is re-esterified mainly with palmitic acid before incorporation into chylomicrons together with triacylglycerols.

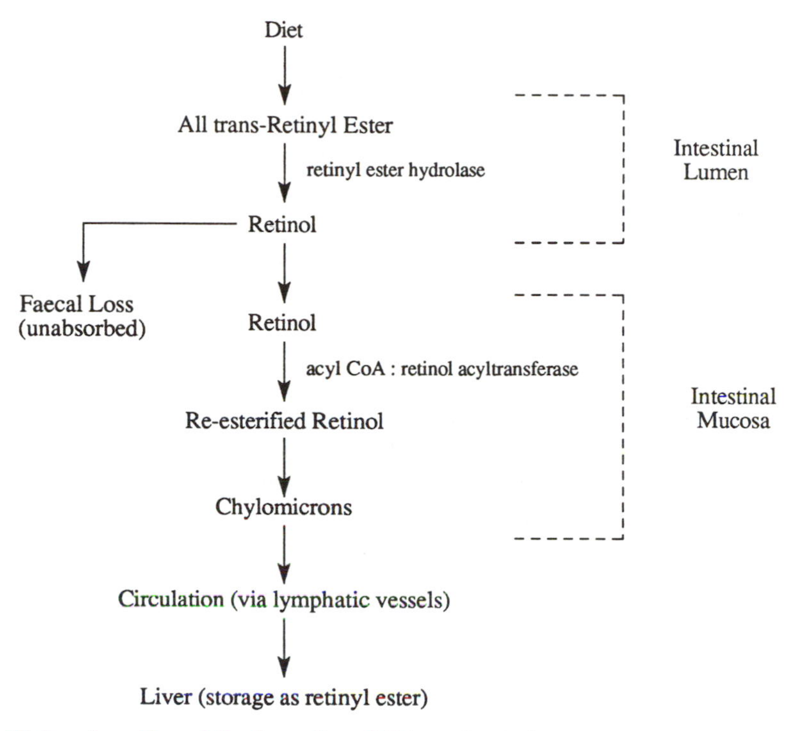

Fig. 11.2. An outline of the absorption of dietary vitamin A.

The mechanism of esterification of retinol appears to involve two enzymes located in absorptive cells of the jejunal mucosa: acyl CoA : retinol acyltransferase (ARAT) and lecithin : retinol acyltransferase (LRAT). It is thought that LRAT is responsible for esterification of retinol bound to cellular retinol binding protein II; this occurs especially during absorption of a normal load of retinol (Norum and Blomhoff, 1992). On the other hand, ARAT esterifies only the free retinol, i.e. when large doses are absorbed and cellular proteins are saturated. The retinyl esters, in association with chylomicrons, are then transported by the lymph into the circulation and then to the liver for metabolism and storage.

Storage and transport

Most of the newly absorbed retinyl esters is taken up by the hepatic parenchymal cells (hepatocytes), the predominant cell type of the liver. Within these cells, chylomicron remnants are degraded by lysosomal enzymes. The retinyl esters are then hydrolysed at the plasma membrane, possibly by retinyl ester hydrolase (Blomhoff, 1994). Subsequently the free retinol is thought to be transferred from the parenchymal cells to stellate cells. Within these cells, retinol is re-esterified by a reaction similar to that of the intestional microsomal ARAT or LRAT. Most of the vitamin A (about 90%) in the human body is thus stored in the liver, largely as long-chain saturated retinyl esters. Other tissues including the kidneys, lungs, adrenals, retina and intraperitoneal fat, contain about 9% of the total and the serum contains approximately 1% of the body's reserves.

When required, vitamin A is released from the liver as the alcohol form, retinol. Hence, the mobilization and transportation of vitamin A from liver storage requires hydrolysis of the retinyl esters. The hepatic retinyl ester hydrolase is an important enzyme since it regulates the release of retinol from its storage. The free retinol is subsequently conjugated with a specific 20,000 molecular weight transport protein, retinol-binding protein (RBP), which is produced and secreted by hepatic parenchymal cells (Fig.11.3). RBP is α–globulin, a single polypeptide chain of 182 amino acid residues with a single binding site for retinol. The retinol–RBP complex is finally released to the circulation. In serum, RBP normally circulates bound to retinol, forming holo-RBP, and complexes with another protein, transthyretin (TTR), as a 1 : 1 : 1 molar ratio. The TTR (previously called prealbumin) is a tetrameric protein with

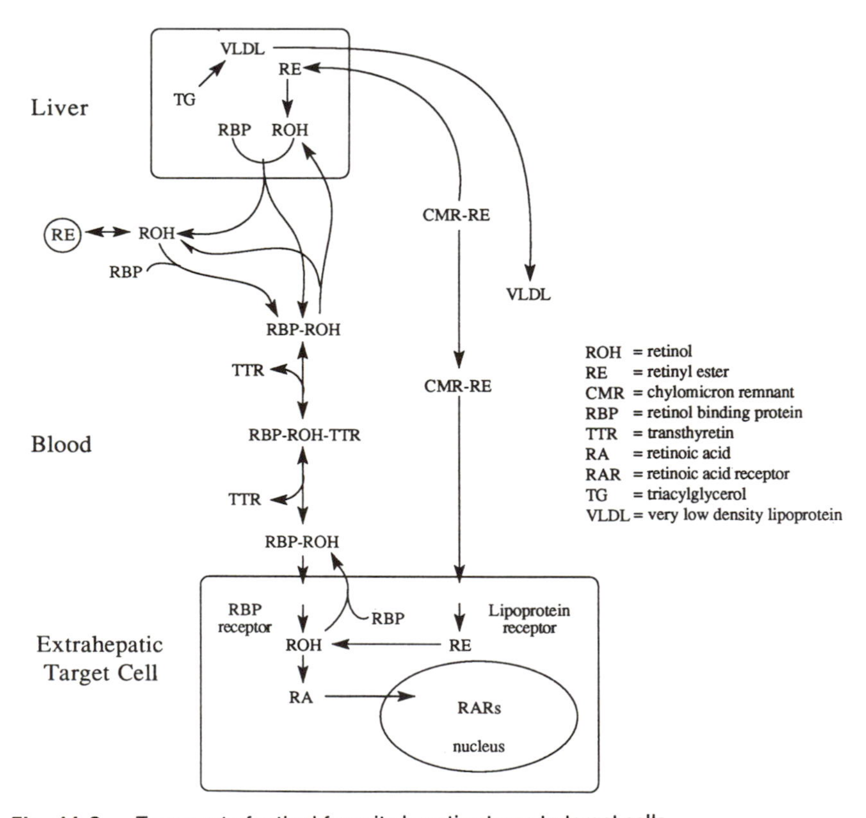

Fig. 11.3. Transport of retinol from its hepatic stores to target cells.

a molecular weight of 54,900. Although TTR is thought to have four binding sites for RBP, the two proteins form a 1 : 1 complex.

The interaction of retinol with RBP serves to solubilize the vitamin in serum and protect the retinol molecule against its oxidative damage. Formation of the protein–protein complex appears to stabilize the binding of retinol to RBP and prevent glomerular filtration of the relatively small RBP molecule with a very short half-life. Normally in the human, the plasma half-life of RBP is 4 hours, but its half-life is increased to 12 hours when bound to TTR (Goodman, 1984). It is estimated that approximately 95.5% of plasma retinol is present as TTR–RBP–retinol complex, 4.4% as RBP–retinol complex, and only a trace (0.1%) as unbound retinol (Blomhoff *et al.*, 1991).

Distribution within the cell

Circulating retinol that is bound to RBP and TTR is taken up by non-hepatic cells without a concomitant uptake of either RBP or TTR. The cellular uptake process of retinol is believed to be mediated by a specific cell membrane receptor that recognizes the proteins but not retinol (Goodman, 1984). The RBP molecule is thus involved in interacting with retinol, TTR and cell surface receptors. When retinol is delivered to target cells, RBP loses its affinity for TTR, returns in the blood as apo-RBP (lacking retinol) and is then eliminated via the glomerular filtration (Sporn *et al.*, 1984).

Once retinol enters a target cell, it is quickly bound by the cellular retinol-binding protein (CRBP) located in the cell cytosol. The CRBP is different from RBP in that it does not bind with TTR, does not cross-react immunologically with RBP, and its molecular weight is 14,600 (Chytil and Ong, 1979).

Other intracellular binding proteins with specificity for retinoic acid (cellular retinoic acid-binding protein, CRABP) and retinaldehyde (cellular retinal-binding protein, CRALBP) have also been identified in various tissues (Table 11.1). CRABP is found in most tissues other than in the liver, kidneys, lungs, spleen and muscle, whereas CRALBP is found only in the retina. An inter-photoreceptor retinol-binding protein (IRBP) has been identified in the extracellular space between the retinol pigment cells and the photoreceptor cells, which bind all-*trans*- and *cis*-retinol.

Our understanding of the vitamin A-binding proteins is far from being clear. It is, however, believed that these proteins serve as intracellular receptors, which facilitate the transfer of specific retinoids to the nucleus of the cell, where they can participate in the regulation of gene expression for controlling cell differentiation and growth (Chytil and Ong, 1987). Recently, a group of nuclear receptors that bind retinoic acid has been identified (Wolfe, 1993). Retinoic acid receptors (PAR α, β, γ) bind to retinoic acid to interact with responsive elements in various genes involved in development and differentiation. Additionally, there exists a class of structurally distinct nuclear receptors referred to as

Table 11.1. A list of identified tissue-specific vitamin A carrier proteins.

Proteins	Mol. wt	Endogenous ligand	Principal site
Retinol-binding protein (RBP)	21,000	all-*trans* retinol	Blood
Cellular retinol-binding protein (CRBP)	14,600	all-*trans* retinol	Vitamin A - sensitive tissues
Cellular retinoic acid-binding protein (CRABP)	14,600	all-*trans* retinoic acid	Vitamin A - sensitive tissues
Cellular retinal-binding protein (CRALBP)	33,000	11-*cis* retinal	Retina of the eye
Inter-photoreceptor retinol-binding protein (IRBP)	144,000	all-*trans* retinol 11-*cis* retinol	Extracellular space of the retina
Cellular retinol-binding protein (CRBP II)	16,000	all-*trans* retinol	Absorptive cells of the small instestine

Adapted from Ong (1985).

retinoid x receptors (P x R α, β, γ) which bind 9-*cis*-retinoic acid as ligand and may also function in the regulation of genes.

Bio-transformation and excretion

The biologically active form of vitamin A in mammalian tissue is all-*trans* retinol which can be oxidized to the aldehyde, retinal, a reversible reaction (Fig. 11.4). This reaction takes place in the retina of the eye as a means of providing retinal for the visual pigment rhodopsin (Ganguly, 1989). The oxidation of retinol to retinal is catalysed by alcohol dehydrogenase, which is a non-specific enzyme requiring NADP as a cofactor. In addition to alcohol dehydrogenase there exist specific enzymes, such as retinol dehydrogenase and retinal reductase, which may also be involved in catalysing the reversible reaction of dehydrogenation of retinol. These enzymes have also been found in the liver as well as in the intestine. However, the equilibrium of the reversible reaction in these sites of the body appears to be far towards reduction of retinal to retinol. A further oxidation of retinal, through an irreversible step, produces retinoic acid; the reaction is thought to be catalysed by aldehyde dehydrogenase or xanthine oxidase (Ganguly, 1989).

Retinoic acid is rapidly metabolized in the liver to more polar metabolites (Fig. 11.4), which have no known biological activity (Madani, 1986). Most oxidized products are either excreted in urine or conjugated in the liver with glucuronic acid, the principal metabolite being retinoyl-β-glucuronide. The resulting glucuronides are excreted in the bile, reabsorbed from the intestine, and then transported back to the liver, thus establishing an enterohepatic circulation of vitamin A metabolites. The majority of these biliary glucuronides, however, are excreted in the faeces. The overall dynamics of vitamin A recycling and excretion are given in Fig. 11.5. This recycling mechanism helps conserve

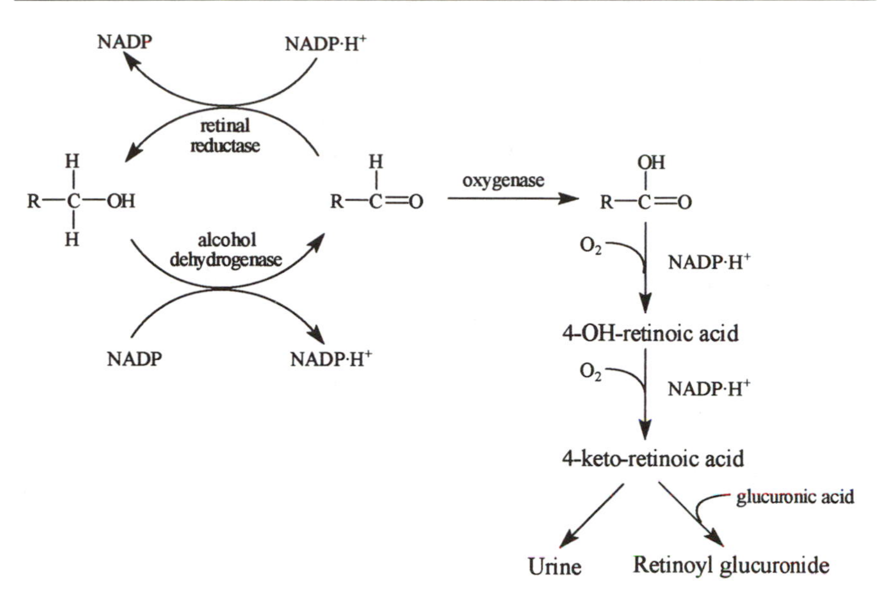

Fig. 11.4. Biotransformation of retinol.

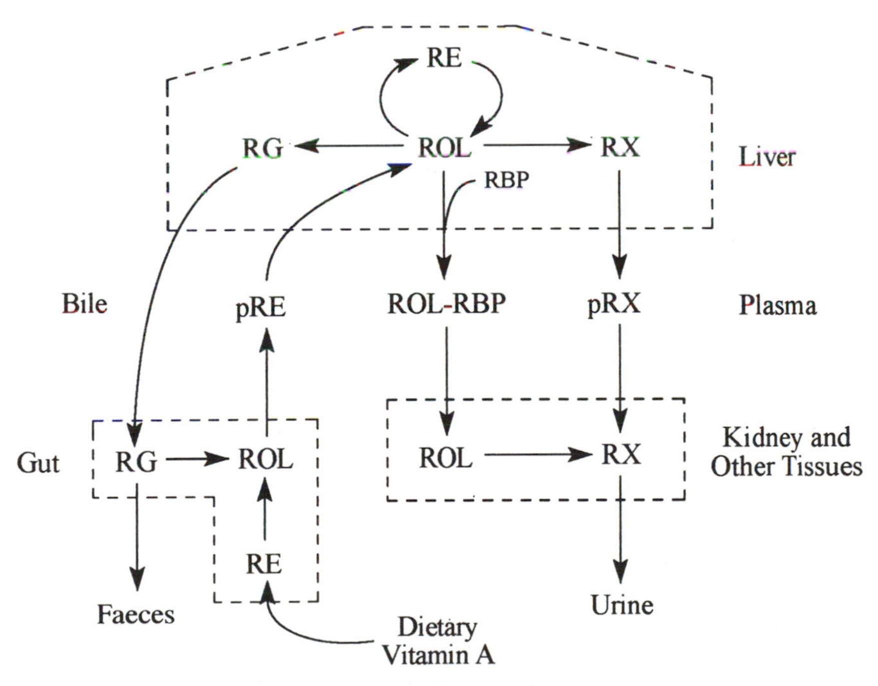

Fig. 11.5. Dynamics of vitamin A recycling and excretion (ROL, retinol; RE, retinylester; RG, retinyl glucuronide; RX, biologically inactive oxidation products; RBP, retinol-binding protein; p, plasma).

the body's supply of vitamin A as well as contributing toward its toxicity when taken in excess. Overall, an average of 10–20% of the dietary intake of vitamin A is not absorbed. Of the remaining, a further 20% appears in the faeces through the bile, 17% is excreted in the urine, 3% is expired as CO_2, and 40–50% is stored in the body, primarily in the liver (Olson, 1994).

Factors Affecting Circulatory Vitamin A and Holo-RBP (Retinol–RBP Complex)

Unlike carotenoids, serum concentrations of vitamin A and its transport protein remain constant over a wide range of dietary intakes and liver stores. The homeostatic level of circulatory vitamin A, however, is known to be affected by various factors. Essentially, these include dietary factors, such as vitamin A itself, lipid, protein, and zinc; age; stress; disease; and a variety of pharmacological agents, including alcohol (see Chapter 18).

Dietary Factors

Vitamin A

Serum levels of retinol are homeostatically regulated and so unaffected by its intake, except in a state of severe deficiency or toxicity. In the presence of an inadequate intake of vitamin A, its hepatic stores are drawn upon in order to maintain a relatively constant serum level until the stores are nearly exhausted. When the liver is depleted of its reserve (< 20 $\mu g\ g^{-1}$ liver), serum retinol levels fall rapidly (Olson, 1984). When the storage concentrations of retinol exceed the limit of 300 $\mu g\ g^{-1}$ liver, the homeostatic control over serum retinol concentrations is lost. A hypothetical relationship between mean serum vitamin A levels and hepatic concentrations of vitamin A is shown in Fig. 11.6.

Availability of vitamin A in the liver has been found to be an important factor for the release of RBP from the liver (Muto *et al.*, 1972). Thus, vitamin A deficiency reduces the serum concentrations of the protein; this effect does not appear to be primarily caused by its synthesis, since the RBP concentration in the hepatic tissue is considerably increased. Furthermore, the RBP concentration in the liver decreases concominantly with an increase of the serum concentration, when vitamin A is administered to rats deficient in vitamin A (Fig. 11.7). The underlying mechanism for homeostasis of serum retinol is essentially mediated through synthesis and secretion of RBP. However, when there is an excess intake of the vitamin for a long period of time, the hepatic storage capacity and the retinol-binding capacity of RBP are exceeded to their limits. In this situation, retinyl ester but not retinol can be mobilized from the liver by lipo-

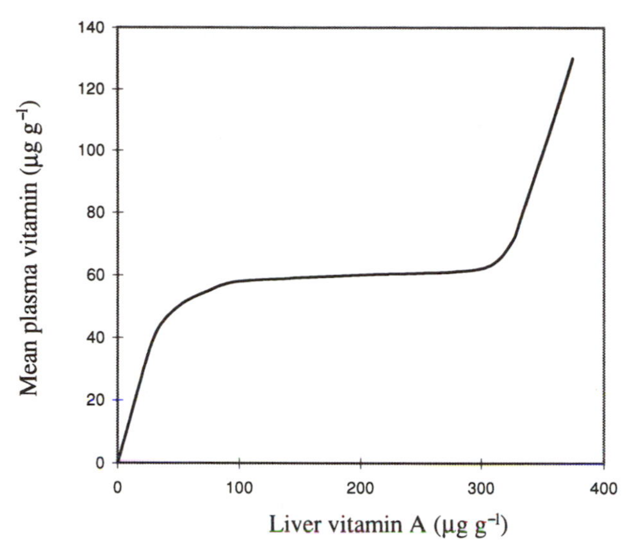

Fig. 11.6. Hypothetical relationship between mean plasma vitamin A levels and liver vitamin A concentrations. Source: Olson (1987).

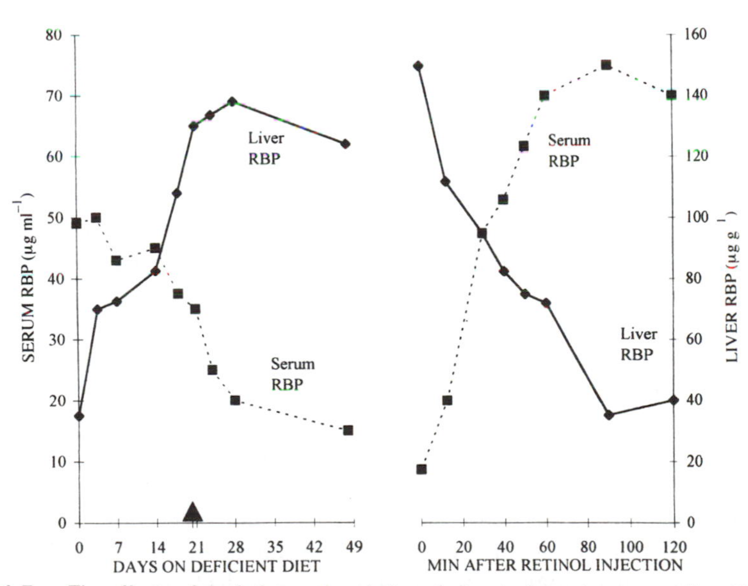

Fig. 11.7. The effects of depletion and repletion of vitamin A on plasma and liver RBP levels in rats. The hepatic stores of vitamin A were depleted on day 20, as indicated by the arrow. In the repletion study, each vitamin A deficient animal was given by i.v. injection 150 µg of retinol dispersed in 1 ml of aqueous solution of Tween 40. Source: Smith and Goodman (1979) (with permission).

proteins. This explains why the administration of large doses of vitamin A results in increased serum concentrations of retinyl ester while retinol levels remain unchanged (Smith and Goodman, 1976). The manifestations of hypervitaminosis A have been ascribed to excessive levels of serum retinyl esters, which normally contribute less than 5% of blood vitamin A.

It is thought that retinyl ester being carried by lipoproteins is more available to cell membranes. This pool of retinyl esters may be responsible for damaging cell membranes, leading to the clinical signs of vitamin A toxicity (see Chapter 21). It is not known whether the retinyl esters so transported to cell membranes are first hydrolysed to retinol at the membrane site, or the ester itself can be toxic. None the less, in contrast with retinyl ester the retinol is sequestered for biological membranes when bound to RBP, thus making this protein have a protective role against hypervitaminosis A.

Lipid

In order to be absorbed from the luminal phase to the mucosal phase of the small intestine, dietary vitamin A requires solubilization into a micellar solution, which is formed from bile salts. When a diet is very low in fat, or when there is an obstruction of the bile duct, the mucosal uptake of vitamin A along with other fat-soluble nutrients is seriously impaired. Since retinol has some polar characteristics, the impairment is less marked for retinol than for carotenes.

Protein

As discussed previously, absorption, transport and metabolic availability of vitamin A are all dependent upon several enzymes and specific proteins, which are synthesized in the body. Protein nutritional status, thus, becomes an important determining factor for vitamin A status. Children with protein–energy malnutrition (PEM) typically have low serum concentrations of retinol as well as its carrier protein. Interestingly, their vitamin A status appears to be reversed by giving protein without the vitamin A in a diet. It seems probable that the low serum concentrations of retinol are the reflection of an impairment of hepatic release of vitamin A due to defective synthesis of RBP in the liver.

The RBP molecule is rich in aromatic and many essential amino acids (Rask *et al.*, 1980). It has a short half-life (< 4 hours), a small body pool size (2 mg kg^{-1} body weight), and a rapid turnover rate. These characteristics make RBP highly sensitive to an inadequate intake of protein (Ingenbleek *et al.*, 1975). It can be anticipated therefore that in protein–energy malnourished children, there may be an adequate store size of vitamin A in the liver but due to a diminished protein synthesis an insufficient amount of RBP molecules are synthesized, making the vitamin metabolically unavailable to the target tissues.

Zinc

Zinc represents another nutrient whose status may influence serum levels of vitamin A and RBP (Smith, 1982). According to many experimental studies, deficiency of this trace element appears to be accompanied by a depression in serum retinol levels. The hepatic concentrations of RBP in zinc-deficient rats have been shown to be less than 60% of that in control animals, and that repletion of the zinc-deficient animals restores serum vitamin A values within the normal range. An association between low serum levels of retinol, zinc and RBP has also been noted in humans (Solomons and Russell, 1980). In a study carried out in the Philippines involving vitamin A-deficient pre-school children, zinc supplements (100 mg week^{-1} zinc sulphate) without concomitant intakes of vitamin A have been shown to improve vitamin A status considerably (Villanuera *et al.*, 1981).

The exact mechanism by which zinc influences vitamin A status is not at present clear. Zinc plays an important role in protein synthesis. Its deficiency may, therefore, limit the synthesis of RBP, TTR, as well as the enzymes involved in vitamin A metabolism, such as retinylester hydrolase and alcohol dehydrogenases (Smith *et al.*, 1973; Mejia, 1986). It has been shown that the activity of hepatic alcohol dehydrogenase (ADH), an enzyme involved in the conversion of retinol to retinaldehyde (Fig. 11.4), is significantly reduced in zinc deficiency (Boron *et al.*, 1988). Hepatic ADH is a metaloenzyme in which zinc acts as a prosthetic group and a stabilizer of the protein structure. A link between zinc and vitamin A has been further supported by the fact that the typical signs of vitamin A deficiency, such as impaired growth and development, taste acuity, dark adaptation and cell mediated immunity, are also known to occur as a result of zinc deficiency (Solomons and Russell, 1980).

Age

The fetus does not appear to synthesize its own RBP until the latter part of pregnancy (Takahashi *et al.*, 1977). Vitamin A is generally transported across the placenta bound to its transport protein, which is derived from the mother. Consequently, hepatic stores of the vitamin at birth are very limited, and serum concentrations of retinol and RBP of newborn babies are approximately 50% of their mothers' levels (Basu *et al.*, 1994). Whether these low values at birth are counterbalanced by an increased turnover rate of RBP, or whether they reflect a physiologically diminished demand for vitamin A, cannot be ascertained at this point. Using mice, Sundboom and Olson (1984), however, have shown that there is a steady increase in total hepatic concentrations of vitamin A from birth to a maximum at about 200 days of age. The ratio of the amount of vitamin stored in the liver to that metabolized and excreted was generally greater in young mice than in older mice. In rats, the age-related increases in hepatic store size of vitamin A and the serum levels of RBP and retinol, have been shown to be correlated well with the increased growth rates (Glover, 1983).

Stress

According to many studies (Underwood, 1984; Cynober *et al.*, 1985), stress of either physical or pathological origin may influence circulating retinol, causing a transient reduction in blood concentrations. This effect has been demonstrated in postoperative patients, and subjects with severe burns and infections. It is thought that stress-induced secretion of corticosteroids reduces tissue vitamin A by favouring its elimination from the body. Thus, administration of corticosterone to rats results in a rapid loss of vitamin A from the plasma, liver, adrenals and thymus (Atukorala *et al.*, 1981). The steroid-mediated depression of plasma and tissue contents of vitamin A is reversed when animals are treated with corticosterone in combination with vitamin A.

Disease

Serum concentrations of vitamin A are generally depressed in conditions associated with fat malabsorption, such as cystic fibrosis, obstructive jaundice, and sprue. Cirrhosis of the liver results in very low circulatory levels of vitamin A, probably because of a combination of decreased synthesis and secretion of RBP.

In kidney disease, there may be either increased or decreased serum concentrations of vitamin A and RBP. Hypervitaminosis A is generally seen in both acute and chronic renal disease (Smith and Goodman, 1971). Since free RBP normally is catabolized in the proximal convoluted tubules after glomerular filtration, the elevated levels of RBP in renal disease may reflect impaired renal clearance. A considerable amount of low molecular weight protein, including RBP, is lost in urinary excretion in nephrotic patients with tubular proteinuria.

Serum concentrations of vitamin A and RBP have been found to be decreased in diabetic patients (Basu *et al.*, 1989). Rats made diabetic with streptozotocin have also been shown to have reduced levels of vitamin A in the serum while hepatic concentrations were elevated (Basu *et al.*, 1990). Both human and animal studies point to the possibility of the vitamin A transport mechanism being defective in diabetes. Furthermore, patients with type 2 diabetes have been found to have reduced plasma levels of retinol while the levels of retinyl esters are elevated, suggesting an increased hepatic storage of vitamin A (Wako *et al.*, 1986). The underlying mechanism for the subnormal plasma retinol and its increased hepatic concentration in diabetics is not properly understood. There is, however, accumulating evidence that diabetes mellitus may lead to zinc deficiency. Since zinc is an important factor for the hepatic synthesis of RBP, it is possible that the reduced serum retinol levels in diabetics reflect reduced mobilization of the vitamin from the liver.

Pathophysiology of Vitamin A

During the last two decades, our knowledge of the roles and functions of vitamin A have broadened considerably from the concept of this vitamin being essential in the visual process to its role regulating gene expression and controlling cell differentiation and growth. Overall, vitamin A is necessary for normal differentiation of epithelial tissues, visual processes, and reproduction. These functions are mediated by the different forms of the molecule. Retinol and retinal are both capable of maintaining normal vision and reproductive functions, while retinoic acid can substitute for either of these vitamin A forms for normal growth and development, but it is not active in vision and reproduction.

Somatic (systemic) function

Vitamin A is well known for its importance in the general growth and differentiation and maintenance of epithelial tissue (DeLuca *et al.*, 1972). If vitamin A is deficient, differentiation switches in the pathway leading to keratinization of squamous cells which replace normal epithelium, a process known as squamous metaplasia (Fig. 11.8). During the process of squamous metaplasia, mucous membranes change from a single layer of mucin-secreting and ciliated

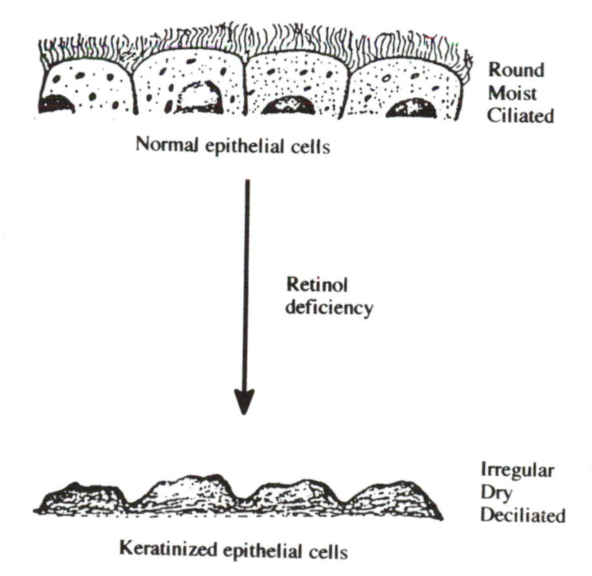

Fig. 11.8. Effect of retinol deficiency on the morphology of the epithelial cells.

epithelium to multiple layers of deciliated non-secretory epithelial cells, with overlying keratin resembling those of the skin.

Our understanding of the mechanism(s) underlying the abnormal epithelial cell differentiation due to vitamin A deficiency is far from clear. Much of the evidence points to cell membranes as the most likely biological structures for the site of action of vitamin A. This vitamin has long been known to be involved in regulating the stability of cell membranes (Wolfe, 1980). Both deficiency and excess of vitamin A appear to result in disruptions of lysosomal and other cellular membranes at the subcellular level.

A hypothesis has been suggested to explain the role of vitamin A on the cell membrane and differentiation relating to its involvement in the biosynthesis of glycoproteins (DeLuca, 1977). These sugar-containing proteins are the constituents of membranes and may serve as antigenic determinants, as virus receptors, and as 'markers' of cellular identity. In the process of synthesis of these proteins, vitamin A is thought to act like a coenzyme, carrying a sugar (mannose) molecule to a protein (acceptor of mannose). The mechanism of action for this process involves (i) phosphorylation of retinol, (ii) glycosylation of retinyl phosphate, and (iii) glycosyl transfer to a glycoprotein acceptor (Fig. 11.9). The importance of glycoprotein processing in cellular control mechanisms makes this hypothesis attractive. There is, however, a need to identify the specific protein that serves as an acceptor for the mannose of mannosyl retinyl phosphate to substantiate the hypothesis.

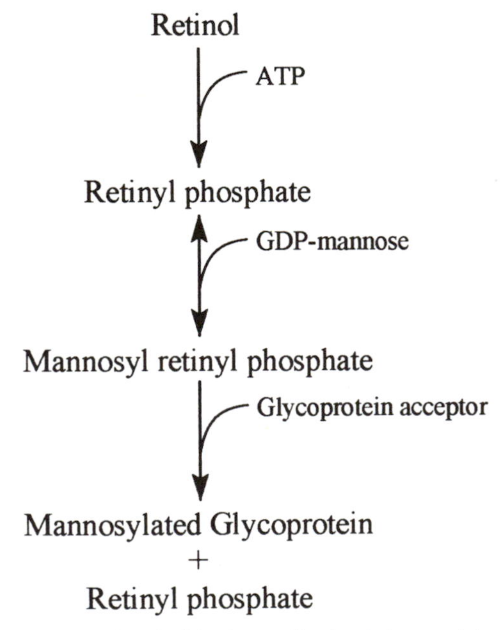

Fig. 11.9. A possible role of vitamin A in the synthesis of glycoproteins.

In vitamin A deficiency different epithelia are affected differently. In epithelial tissues such as intestinal mucosa, where there are normally no keratinizing cells, one observes a declining in mucus secreting cells. In other tissues such as epidermis, where there are no mucus secreting cells, one finds hyperkeratosis (cornification) as a result of vitamin deficiency.

Vitamin A and the immune system

Studies using experimental animals have suggested that vitamin A deficiency impairs both humoral and cell mediated immunity (Ross, 1992). Lack of vitamin A is accompanied by profoundly depressed serum immunoglobulins (e.g. IgG and IgA) responses, mitogen responses and delayed-type hypersensitivity reactions. Vitamin A deficiency has also been shown to depress *in vivo* antibody titre as well as *in vitro* antigen-specific T-lymphocyte proliferative response in chicks (Friedman and Sklan, 1989). Recently, the significant role of vitamin A in the immune system has been evident in humans. Thus, children treated with vitamin A for measles have displayed a more profound immune response than children with measles receiving conventional therapy (Coutsoudis *et al.*, 1991). Furthermore, IgG and T-cell subset responses to tetanus immunization in children are dramatically increased by vitamin A supplementation (Semba *et al.*, 1992). The role of vitamin A supplementation in enhancing immunity has received much attention in recent years; this aspect is discussed further in Chapter 19.

Vitamin A and the visual cycle

The visual cycle is the system in which the role of vitamin A is thoroughly understood. The specific role of vitamin A in the physiological mechanisms of vision has been elucidated largely by George Wald, who received a Nobel Prize in 1967 for his discovery of the role of vitamin A in vision. It is firmly established that the vitamin is required for vision in the dark and also for colour perception. The active form of vitamin A for this function is the aldehyde (retinal), derived from retinyl esters and retinol.

In the visual process the photoreceptors of the eye transmit the initial photostimulation to the optic nerves. These receptors are located in the retina and are classified as rods and cones. The rod cells are responsible for vision in dim light, and they contain the photosensitive pigment called rhodopsin. The cone cells, on the other hand, are responsible for colour and bright light vision; they contain the light sensitive pigment called iodopsin. Both pigments contain 11-*cis* retinal, attached through amide linkages to a lysyl residue of the protein called opsin (Fig. 11. 10). In the presence of bright light, rhodopsin is bleached resulting in the conversion of 11-*cis* retinal to all-*trans* retinal (Figs 11.10 and 11.11). Since the all-*trans* retinal shape is altered, it is separated from the opsin, resulting in an interruption of optic nerve electric impulses by Ca momentarily.

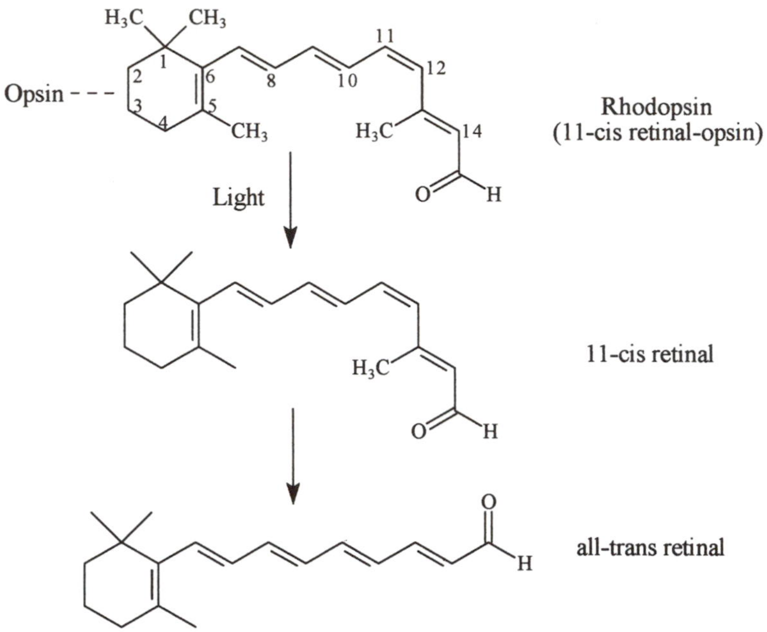

H3C
CH3
1
2
3
4
5
6
8
10
11
12
14
Opsin - - -
CH3
H3C
O
H
Rhodopsin
(11-cis retinal-opsin)
Light
H3C
O
H
11-cis retinal
O
H
all-trans retinal

Fig. 11.10. The bleaching effect of bright light on rhodopsin.

O H
11-cis retinal-C—N—(CH2)4—opsin
(Rhodopsin)
DARK
LIGHT
Opsin
11-cis retinal
all-trans retinal
NADH·H+
NAD
retinol
dehydrogenase
retinal
reductase
NADH·H+
NAD
11-cis retinol
all-trans retinol
isomerization
retinol (from circulation)

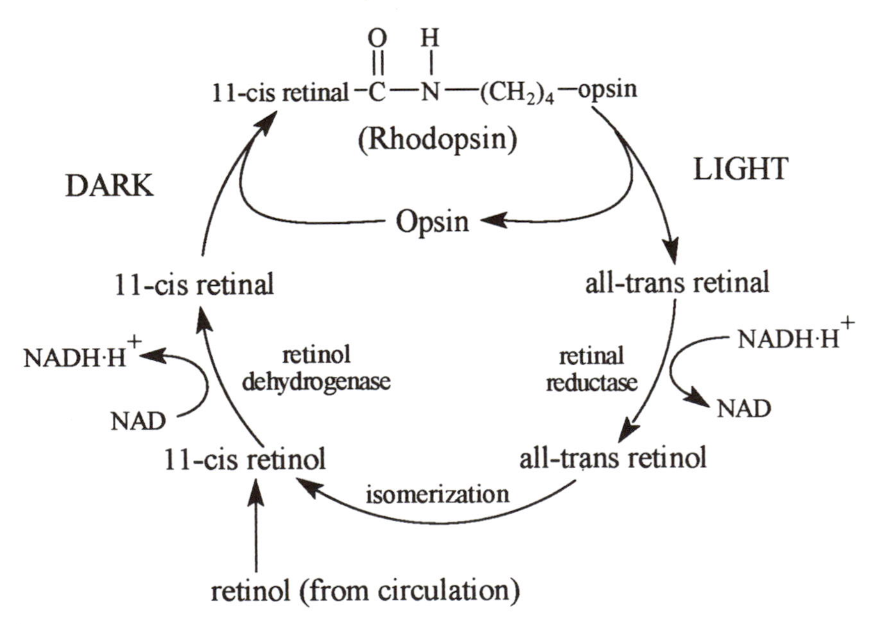

Fig. 11.11. Rhodopsin – vitamin A cycle.

The all-*trans* isomer of retinal is then reduced to the corresponding isomer of retinol through the action of NADH.H (Fig. 11.11). The all-*trans* retinol is subsequently isomerized to form 11-*cis* retinol. Through the action of retinol dehydrogenase (similar to alcohol dehydrogenase) and NAD the 11-*cis* isomer of retinol oxidizes to the corresponding isomer of retinal. In the dark, the 11-*cis* retinal combines with opsin to regenerate rhodopsin.

During the rhodopsin–vitamin A cycle (Fig. 11.11), some of the vitamin A is lost, and consequently there is a shortage of 11-*cis* retinal to form an adequate amount of rhodopsin. Normally, the retinal pigment selectively absorbs all-*trans* retinol from the circulation to make up the shortage. When there is a deficiency of vitamin A, however, it is not readily available in the circulation, and the situation consequently results in an inability to see in dim or dark light. This condition, commonly known as night blindness (nyctalopia), is generally the first specific clinical manifestation of vitamin A deficiency in humans. In severe cases, the early sign through various stages (see later) can ultimately progress to a total blindness (Oomen, 1974).

Vitamin A and reproduction

It has long been recognized that vitamin A, due to its role in cell differentiation, is an important nutritional factor, required for spermatogenesis and embryonic development (Underwood, 1984). Vitamin A deficiency in experimental animals thus appears to result in both desquamation of germinal cells in seminiferous tubules in males and resorption of fetuses in females. In males, the deficiency causes changes in accessory organs, such as the prostate and seminal vesicles. In addition to this somatic role of vitamin A in epithelial cell maintenance, it seems to have a specific coenzymatic role in the biosynthesis of testosterone (Fig. 11.12). For this function, however, retinoic acid cannot be substituted for retinol.

Occurrence of Vitamin A Deficiency

Involving eyes

In the eye there are essentially three sites that are affected by vitamin A deficiency: these are the retina, conjunctiva, and cornea (Fig. 11.13). The retina, which contains the rod cells, is very sensitive to vitamin A deficiency, and is thus responsible for the early occular sign of night blindness. If the deficiency state persists, the conjunctiva followed by the cornea undergo the changes of xerosis (i.e. dryness). The conjunctival xerosis is often accompanied by Bitot's spots, which consist of a heaping up of keratinized epithelial cells. They usually take the form of a small plaque of a silver-grey hue with a foamy surface, occurring

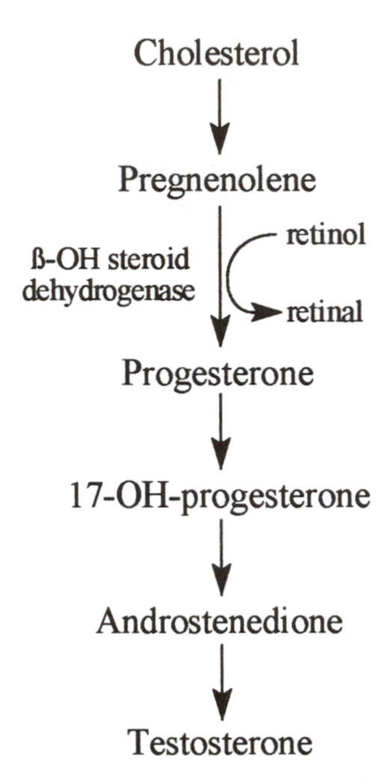

Fig. 11.12. Possible role of vitamin A in testosterone synthesis.

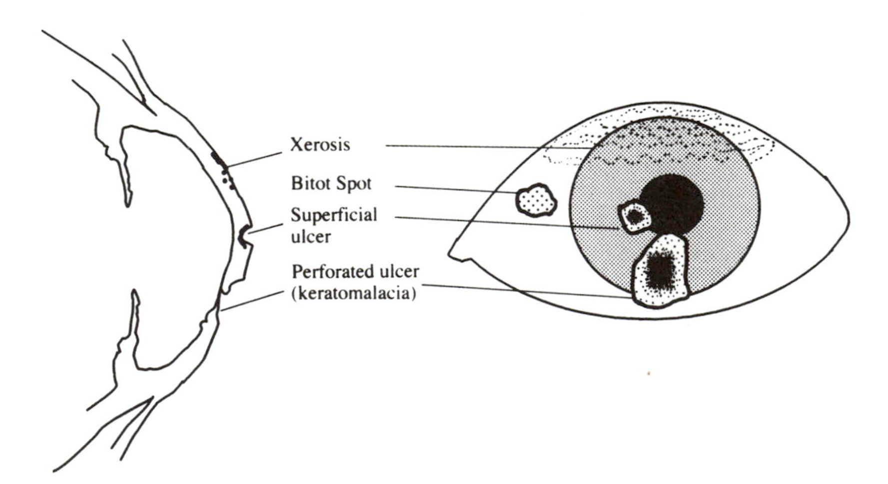

Fig. 11.13. The eye sites that are affected by vitamin A deficiency.

most commonly on the temporal aspect of the bulbar conjunctiva in the interpalpebral fissure near the limbus. The concurrent appearance of conjunctival xerosis and Bitot's spots, especially in young children, is usually a good indication of vitamin A deficiency. However, these signs can also be unrelated to vitamin A deficiency. In these circumstances, local irritation and trauma appear to be involved, the conditions are unresponsive to vitamin A, and the subjects are usually older children or adults.

The damage to the cornea is more serious than the damage to the conjunctiva. The earliest corneal change is xerosis of the epithelium, giving this structure a hazy appearance. At this stage the condition is reversible if promptly treated. In young children the corneal xerosis may rapidly progress to the final stage involving a destructive process of softening and melting of the corneal substance, and ulceration and perforation of the cornea. Corneal ulceration associated with partial or complete loss of sight is generally referred to as keratomalacia. This condition is frequently accompanied by severe protein–energy malnutrition. The codes used by the World Health Organization to describe various stages of xerophthalmia are listed in Table 11.2. The term 'xerophthalmia,' meaning dryness of the eye, is generally used to describe vitamin A deficiency-related eye problems, which include night blindness, conjunctival lesions, Bitot's spot, corneal xerosis, corneal damage, and keratomalacia. In industrialized countries, xerophthalmia due to primary vitamin A deficiency is rarely seen because of the availability of adequate fruits, vegetables and animal food combined with fortified food such as margarine. Deficiency of this vitamin, however, remains the second major nutritional problem among children in many countries, such as Bangladesh, India, Indonesia, Brazil, as well as other developing countries. It is estimated that 20–40 million children worldwide have at least mild vitamin A deficiency, and nearly half are said to reside in India (United Nations Administrative Committee, 1987). The most vulnerable group to vitamin A deficiency is children from birth to six years of age because of their low liver reserves, low protein, lipid, and vegetable intakes, and increased requirements due to rapid growth rate. In addition, in the areas where vitamin A deficiency is prevalent, food such as margarine is not generally fortified with the vitamin.

Table 11.2. The codes used by World Health Organization to classify the stages of xerophthalmia.

Code	Symptom
XN	Night-blindness
X1A	Conjunctival xerosis
X1B	Bitot's spot
X2	Corneal xerosis
X3A	Corneal ulceration/keratomalacia (< 1/3 corneal surface)
X3B	Corneal ulceration/keratomalacia (≥ 1/3 corneal surface)
XS	Corneal scar
XF	Xerophthalmic fundus

Adapted from Sommer (1994).

Involving epithelial cells

Many studies have shown that even a mild deficiency of vitamin A leads to patchy keratinization of the epithelial lining of the gastrointestinal tract, respiratory tract, urogenital tract as well as the skin. The mucous linings of the mouth, oesophagus, and all other moist surfaces lose their water content as the cells become filled with keratin, a tough fibrous protein that is insoluble in water. Indeed, one of the earliest signs of vitamin A deficiency is a loss of appetite associated with changes in gut cell morphology.

The epithelial barrier is the first step of the protective mechanism against infection, and it has been suggested that keratinization leads to bacterial colonization and infection. Indeed, the circulating levels of vitamin A are usually found to be depressed in children with active infectious diseases (Reddy *et al.*, 1986; Feachem, 1987). The cause and effect of this relationship is, however, far from being understood. The disease may deplete vitamin A reserves by increasing requirements at a time when the intake and absorption of nutrients are generally decreased because of the disease. The low serum vitamin A during active infections could also represent the redistribution of vitamin A from the circulation as an effect of the acute phase response (Beisel, 1977). The possibility that this internal redistribution during infection might reduce the vitamin A content of peripheral tissues such as lungs, gut, and skin has been the underlying rationale for justifying the administration of vitamin A supplements to children with severe measles, irrespective of their prior vitamin A status (Hussey and Klein, 1990). Vitamin A deficiency and infections can establish a vicious cycle that induces and perpetuates systemic disease. The severity of the synergism between vitamin A deficiency and infection is responsible for excessive childhood mortality in many developing regions of the world.

In areas where vitamin A deficiency and undernutrition are public health problems, studies have shown that children with mild vitamin A deficiency are at an increased risk of respiratory disease and diarrhoea (Feachem, 1987). Furthermore, children who have recently suffered from measles are many times more likely to develop vitamin A deficiency leading to corneal xerophthalmia than those without the history of measles (Sommer *et al.*, 1984).In recent years, the mortality rates from infectious diseases have been reported to be markedly reduced by vitamin A supplementation (see Chapter 19).These studies point to the fact that vitamin A may have both preventive as well as prophylactic effects against infectious disease.

Requirements

Our knowledge of vitamin A requirements for humans has been derived essentially from two kinds of studies. These are, firstly, epidemiological studies where serum levels or dietary intakes of vitamin A have been correlated with clinical evidence of deficiency of the vitamin. Secondly, there are controlled depletion/repletion experiments involving both humans and animals, in which clinical manifestations or changes in serum levels of vitamin A are first produced through restricting intake, and then an estimate of the requirement of vitamin A is derived from specific increases in intake.

Epidemiological studies have indicated that in countries such as North America and Europe, where the daily vitamin A intake is in the range of 900–2700 μg RE per person, vitamin A deficiency is rarely seen. On the other hand, deficiency is known to occur more frequently in countries such as parts of Asia, the Far East, and South America, where the reported daily intake of vitamin A is in the range of 300–750 μg RE. The findings of these demographic studies are important, but do not provide enough information to define the amount of vitamin A required to maintain health.

Many deprivation and recovery studies involving adult volunteers have been conducted in an attempt to determine the requirement for vitamin A (Rodriguez and Irwin, 1972). Of these, the Sheffield experiment in England during World War II (Hume and Krebs, 1949) and the more recent study conducted in the United States (Sauberlich *et al.*, 1974) have influenced the recommended requirements of vitamin A in North America, Japan, and the United Kingdom (Table 11.3). According to these two studies, an intake of 500–600 μg of retinol per day is sufficient to prevent all deficiency signs and to restore consistent serum retinol levels in adults. The appropriate authorities in Japan (National Institute of Health and Nutrition, 1985) and in the United Kingdom (Dietary Reference Values for Food, Energy and Nutrients, 1991) have adopted a recommended intake of approximately 600 μg retinol per day for the normal adult (Table 11.3). However, in the United States (Food and Nutrition Board, 1989) as well as in Canada (Health and Welfare Canada, 1990), an intake above the minimum requirement is considered necessary in order to produce liver storage. In these countries the daily recommended intake of retinol thus arrived at is 1000 μg for an adult male; the intake for the adult female is

Table 11.3. Recommended intakes of vitamin A (μg RE) in various countries.

Category	Age (yrs)	USA[1]	Canada[2]	UK[3]	Japan[4]
Infants	< 1	375	400	350	400
Children	1–10	400–700	400–700	400–500	400–500
Males	> 10	1000	1000	600–700	600
Females	> 10	800	800	600	540
Pregnancy		+ none	+ none	+ 100	+ 60
Lactation		+ 350	+ none	+ 350	+ 400

[1]Food and Nutrition Board, USA (1989).
[2]Report of the Scientific Review Committee, Canada (1990).
[3]Dietary Reference Values for Food, Energy and Nutrients, UK (1991).
[4]National Institute of Health and Nutrition, Japan (1985).

800 μg of retinol or 80% of the male's recommended intake because of the usually smaller body size of women. In most countries, there have been some adjustments for pregnancy and lactation to this recommended intake (Table 11.3). However, the latest recommendation in Canada does not include any additional allowances during either pregnancy or lactation (Health and Welfare Canada, 1990), except that the consumption of foods that are rich in vitamin A has been strongly encouraged during these phases of life.

It should be pointed out that retinol is teratogenic (see Chapter 21). A relationship has been suggested between incidence of birth defects in infants and high maternal vitamin A intakes (more than 3300 μg day^{-1}) during pregnancy. As a precautionary measure women who are, or might become, pregnant have therefore been advised not to take supplements containing vitamin A unless advised to do so by a physician or antenatal clinic (Department of Health, 1991). Furthermore, recent analyses of animal livers commonly consumed in the United Kingdom have shown them to contain on average 13,000–40,000 μg/100 g depending on species. It is therefore advisable that liver or liver products are also avoided during pregnancy.

Recommended intakes of vitamin A for infants under one year of age are generally based on the amount of retinol in human milk. According to the WHO Expert Committee on Nutrition in Pregnancy and Lactation (1965), an average well-nourished lactating mother produces about 850 ml of milk each day, with a retinol content of 49 μg/100 ml. A breast-fed infant would, therefore, be expected to receive approximately 420 μg of retinol per day. The recommendation for infants is, indeed, set arround this amount in most countries (Table 11.3).

Assessment of Vitamin A Status

The overall methods that are used to determine vitamin A status in humans include dietary assessment, clinical evaluation, retinal function tests, and biochemical assessment. These methods of evaluation have been discussed in a recent review article (Underwood, 1990). This text will deal with only the retinal function and biochemical tests, which are used more frequently in nutrition surveys to diagnose mild deficiencies of vitamin A.

Retinal Function Tests

Rapid dark adaptation (RDA)

This is a measurement of the functional changes in visual acuity, called night blindness, an early sign of vitamin A deficiency. The procedure of this test involves first an exposure of the eyes to a bright light; after this the light is extinguished and the subject is shown small illuminated test objects. A rapid dark adaptation test has been developed (Thornton, 1977) and tested for its applications in the field (Vinton and Russell, 1981). There appear to be some limitations to this method of vitamin A assessment. Young children are generally unfit for this test. In addition, there are many conditions other than vitamin A deficiency where dark adaptation may be impaired; these conditions include detachment of the retina, retinitis pigmentosa or congenital night blindness.

Conjunctival impression cytology (CIC)

This is one of the new methods for assessment of subclinical signs of vitamin A deficiency (Wittpenn *et al.*, 1987). It detects early physiological changes such as the progressive loss of goblet cells in the conjunctiva and the appearance of enlarged, partially keratinized epithelial cells. These cytological changes precede clinical signs of vitamin A deficiency. Criteria for interpretation of CIC are given in Table 11.4.

Table 11.4. An interpretive guideline for conjunctival impression cytology.[1]

Interpretation	Mucin spots	Goblet cells	Epithelia
Normal	Densely covering ≥ 25% of sample	±	Normal > abnormal
Borderline normal	Covering ≥ 25% of sample	±	Abnormal > normal
Borderline abnormal	Diffusely covering < 25% of sample	Rare	Abnormal
Abnormal	Diffusely covering < 25% of sample	None	Abnormal

[1]Modified from Kjolhede *et al.* (1989)

CIC is a simple, rapid, and inexpensive method, which is suitable for a field survey. The procedure involves touching a strip of cellulose ester filter paper (0.45 μm pore size) to the lower, temporal (outer) portion of the conjunctiva for 3–5 s. The cells are transferred to the filter paper strip and this strip is then placed in a tightly closed vial until it can be stained and examined. Manuals describing the procedure in detail are available (ICEPO, 1988).

Rose bengal staining test (RBST)

This is another simple and inexpensive method, which can conveniently be used in a field survey for an early detection of conjunctival xerosis (Vijauaraghavan *et al.*, 1978). In this method, a drop of 1% solution of rose bengal stain is instilled carefully in the eye and signs of positive or negative staining are observed. The test is generally considered positive when there is a pink spot on the conjunctiva irrespective of its size and location on close examination.

Biochemical Tests

Serum vitamin A and its carrier proteins

Serum retinol is the most commonly used biochemical index for the vitamin status. The serum values, however, do not reflect body stores of vitamin A because (i) serum contains only 1% of the total body reserve, and (ii) the serum value is homeostatically controlled. Serum vitamin A is changed only when the hepatic store is severely depleted ($< 20\ \mu g\ g^{-1}$) or excessively high ($> 300\ \mu g\ g^{-1}$). In addition, factors such as low protein status, acute catabolic states, and decreased hepatic function are all known to affect the RBP production and therefore the metabolic availability of vitamin A from its hepatic stores. Serum vitamin A levels alone do not reveal if there is an impairment in its metabolic availability. These values can be interpreted better, however, if they are measured along with RBP and TTR (Tyler and Dickerson, 1984). An interpretive guideline for these parameters is given in Table 11.5.

Table 11.5. Values of serum retinol and its carrier proteins as indicative of an inadequate vitamin A status.

	Serum values	References
Retinol ($\mu g\ dl^{-1}$)	< 20[1]	Grant *et al.*, 1981
RBP ($mg\ dl^{-1}$)	< 2.6	Grant *et al.*, 1981
TTR ($mg\ dl^{-1}$)	< 15.0	Pilch, 1987
RDR[2] (%)	> 20	Flores *et al.*, 1984

[1]Conversion factor to SI units ($\mu mol\ l$) = × 0.035.
[2]Relative dose response test.

Relative dose response test (RDR)

This is a modification of the serum retinol index for vitamin A nutriture. RDR is the rise in serum retinol concentration that occurs 5 hours after a loading oral dose of retinyl ester (A_5), expressed as a percentage of the serum retinol concentration before dosing (A_0). It is calculated according to the following formula (Eq. 11.1):

$$\text{RDR}\ (\%) = \frac{A_5 - A_0}{A_5} \times 100 \qquad (11.1)$$

The RDR test is based on the principle that in the presence of vitamin A deficiency, RBP accumulates in the liver as apo-RBP (free retinol). Following a test dose of vitamin A, the latter binds to this relative excess of apo-RBP in the liver and holo-RBP (retinol–RBP complex) is subsequently released from the liver. It is expected, therefore, that in vitamin A deficiency there will be a rapid increase in serum retinol after ingestion of a small dose of the vitamin (Loerch *et al.*, 1979). The use of this test is thought to be a better index for the hepatic stores of vitamin A as has been evident from animal studies. An RDR greater than 20% has been suggested to be an indication of inadequate hepatic reserves (Flores *et al.*, 1984).

Methods for analysing serum vitamin A

Several methods are available for vitamin A analysis in the serum. These include colorimetry, fluorometry, and high-performance liquid chromatography (HPLC) with a UV detector. HPLC appears to offer many advantages over the other assays. It provides high specificity and makes the assay very sensitive. Only 100 μl of serum is needed to obtain a detection limit of 50 μg of retinol l^{-1}. Only HPLC can distinguish retinol from retinyl ester in the serum (Chaudrey and Nelson, 1984).

References

Atukorala, T.M.S., Basu, T.K. and Dickerson, J.W.T. (1981) Effect of carticosterone on the plasma and tissue concentrations of vitamin A in rats. *Annals of Nutrition Metabolism* 25, 234–238.

Basu, T.K., Tze, W.J. and Leichler, J. (1989) Serum vitamin A and RBP in type 1 diabetic patients. *American Journal of Clinical Nutrition* 50, 329–331.

Basu, T.K., Leichter, J. and McNiell, J.H. (1990) Plasma and liver vitamin A concentrations in STX induced diabetic rats. *Nutrition Research* 10, 421–427.

Basu, T.K., Wein, E.E., Gaggopadhyay, K.C., Wolever, T.M. and Godel, J.C. (1994) Plasma vitamin A (retinol) and retinol-binding protein in newborns and their mothers. *Nutrition Research* 14, 1297–1303.

Beisel, W.R. (1977) Magnitude of the host nutritional response to infection. *American Journal of Clinical Nutrition* 30, 1236–1247.

Blomhoff, R. (1994) Transport and metabolism of vitamin A. *Nutrition Reviews* 52, 513–523.

Blomhoff, R., Green, M.–H., Green, J.–B., Berg, T. and Norum, K.–R. (1991) Vitamin A metabolism: New perspectives on absorption, transport and storage. *Physiological Reviews* 71, 951–990.

Boron, B., Hupert, J., Barch, D.H., Fox, C.C., Friedman, H., Layden, T.J. and Mobaghan, S. (1988) Effect of zinc deficiency on hepatic enzymes regulating vitamin A status. *Journal of Nutrition* 118, 995–1001.

Chaudrey, L.R. and Nelson, E.C. (1984) Separation of vitamin A and retinyl esters by reversed phase high performance liquid chromatography. *Journal of Chromatography* 294, 466–470.

Chytil, F. and Ong, D.E. (1979) Cellular retinol and retinoic acid binding proteins in vitamin A action. *Federation Proceedings* 38, 2510–2514.

Chytil, R. and Ong, D.E. (1987) Intercellular vitamin A-binding proteins. *Annual Review of Nutrition* 7, 321–335.

Coutsoudis, A., Broughton, M. and Coovadia, H.M. (1991) Vitamin A supplementation reduces measles morbidity in young African children: a randomized, placebo-controlled, double-blind trial. *American Journal of Clinical Nutrition* 54, 890–895.

Cynober, L., Desmoulins, D., Lioret, N., Aussell, C., Hirsch-Marie, H. and Saizy, R. (1985) Significance of vitamin A and retinol-binding protein levels after burn injury. *Clinica Chimica Acta* 148, 247–253.

Dietary Reference Values for Food Energy and Nutrients for the United Kingdom, London (1991) Her Majesty's Stationery Office, London.

DeLuca, L.M. (1977) The direct involvement of vitamin A in glycosyl transfer reactions in mammalian membranes. In: *Vitamins & Hormones.* Vol. 35. Academic Press, New York, pp. 1–57.

DeLuca, L., Maestri, N., Bonanni, F. and Nelson, D. (1972) Maintenance of the cell differentiation: The mode of action of vitamin A. *Cancer* 30, 1326.

Department of Health (1991) *Dietary Reference Values for Energy and Nutrient for the United Kingdom.* Report on Health and Social Subjects, No. 41. HMSO, London.

Feachem, R.G. (1987) Vitamin A deficiency and diarrhoea: A review of interrelationships and their implications for the control of xerophthalmia and diarrhoea. *Tropical Disease Bulletin* 84, 2–16.

Flores, H., Campos, F., Araujo, C.R.C. and Underwood, B.A. (1984) Assessment of marginal vitamin A deficiency in Brazilian children using the relative dose response procedure. *American Journal of Clinical Nutrition* 40, 1281–1289.

Food and Nutrition Board (1980) *Recommended Dietary Allowances.* National Academy of Sciences, Washington, D.C.

Food and Nutrition Board (1989). National Research Council: *Recommended Dietary Allowances,* 10th edn. National Academy Press, Washington, D.C.

Friedman, A. and Sklan, D. (1989) Antigen-specific immune response impairment in the chick as influenced by dietary vitamin. *American Journal of Nutrition* 119, 790–795.

Ganguly, J. (1989) In: *Biochemistry of Vitamin A.* CRC Press, Inc. Boca Raton, Florida, pp. 19–40.

Glover, J. (1983) Factors affecting vitamin A transport in animals and man. *Proceedings of the Nutrition Society* 42, 19–24.

Goodman, D.S. (1984) Plasma retinol-binding protein. In: Sporn, M.B., Roberts, A.B. and Goodman, D.S. (eds.) *The Retinoids,* Vol 2. Academic Press, Inc., New York, pp. 41–88.

Grant, J.P., Custer, P.B. and Thurlow, J. (1981) Current techniques of nutritional assessment. *Surgical Clinics of North America* 61, 437–463.

Health and Welfare Canada (1990) In: *Nutrition Recommendations.* Canadian Government Publishing Centre, Ottawa.

Hume, E.M. and Krebs, H.A. (compilers) (1949) *Vitamin A Requirement of Human Adults. Reports of the Vitamin A Subcommittee of the Accessory Food Factors Committee.* Medical Research Council (Great Britian), Special Report Ser. No. 264. H.M. Stationery Office, London.

Hussey, G.D. and Klein, M.A. (1990) A randomized controlled trial of vitamin A in children with severe measles. *New England Journal of Medicine* 323, 160–164.

ICEPO (1988) *Training Manual: Assessment of Vitamin A Status by Impression Cytology.* Data centre for preventive ophthalmology, The Johns Hopkins University, Baltimore, MD.

Ingenbleek, T., Van den Schrieck, H.G., De Nayer, P. and De Visscher, M.L. (1975) (TBPA-RBP) complex in assessment of malnutrition. *Clinica Chimica Acta* 63, 61–67.

IUPAC-IUB Joint Commission on Biochemical Nomenclature (1982) Nomenclature of retinoids. Recommendations 1981. *European Journal of Biochemistry* 129, 1–5.

Kjolhede, C.L., Gadomski, A.M., Wittpenn, J., Bulux, J., Rosas, A.R., Solomans, N.W., Brown, K.H. and Forman, M.R. (1989) Conjunctival impression cytology: feasibility of a field trial to detect subclinical vitamin A deficiency. *American Journal of Clinical Nutrition* 49, 490–494.

Loerch, J.D., Underwood, B.A. and Lewis, K.C. (1979) Response of plasma levels of vitamin A to a dose of vitamin A as an indicator of hepatic vitamin A reserves in rats. *Journal of Nutrition* 109, 778–786.

Madani, K.A. (1986) Retinoic acid: A general overview. *Nutrition Research* 6, 107–123.

Mejia, L.A. (1986) Vitamin A – nutrition interrelationships. In: J.C. Bauernfeind (ed.), *Vitamin A Deficiency and Its Control.* Academic Press, Orlando, FL, pp. 69–100.

Muto, Y., Smith, J.E., Milch, P.O. and Goodman, D.S. (1972) Regulation of retinol-binding protein metabolism by vitamin A status in the rat. *Journal of Biological Chemistry* 247, 2542–2550.

National Institute of Health and Nutrition (1985) *Recommended Intake of Nutrients and Energy.* Ministry of Health Press, Tokyo, 1985.

Norum, K.K. and Blomhoff, R. (1992) McCollum award lecture, 1992: Vitamin A absorption, transport, cellular uptake and storage. *American Journal of Clinical Nutrition* 56, 735–744.

Olson, J.A. (1984) Serum levels of vitamin A and carotenoids as reflectors of nutrition status. *Journal of the National Cancer Institute* 73, 1439–1444.

Olson, J.A. (1987) Recommended dietary intakes (RDI) of vitamin A in humans. *American Journal of Clinical Nutrition* 45, 704–716.

Olson, J.A. (1994) Vitamin A, retinoids and carotenoids. In: Shils, M.E., Olson, J.A., Shike, M. (eds.), *Modern Nutrition in Health and Disease,* 8th edn. Lea & Febiger, Philadelphia, pp. 287–307.

Ong, D.E. (1985) Vitamin A binding protein. *Nutrition Review* 43, 225–232.

Oomen, H.A.P.C. (1974) Vitamin A deficiency, xerophthalmia and blindness. *Nutrition Review* 32, 161–170.

Pilch, S.M. (1987) Analysis of vitamin A data for the health and nutrition examination surveys. *Journal of Nutrition* 117, 636–640.

Rask, L., Anundi, H., Bohme, J., Eriksson, U., Fredriksson, A., Nilsson, S.F., Ronne, H., Vahlquist, A. and Peterson, P. (1980) The retinol-binding protein. *Scandinavian Journal of Clinical Laboratory Investigation* 154 (supplement 40), 45–61.

Reddy, V., Bhaskaran, P. and Raghuramulu, N. (1986) Relationship between measles, malnutrition and blindness: A prospective study in Indian children. *American Journal of Clinical Nutrition* 44, 924–930.

Rigtrup, K.M. and Ong, D.E. (1992) A retinyl ester hydrolase activity intrinsic to the brush border membrane of rat small intestine. *Biochemistry* 31, 2920–2926.

Rigtrup, K.M., McEwen, L.R., Said, H.M. and Ong, D.E. (1994) Retinyl ester hydrolytic activity associated with human intestinal brush border membranes. *American Journal of Clinical Nutrition* 60, 111–116.

Rodriguez, M.S. and Irwin, M.I. (1972) A conspectus of research on vitamin A requirements of man. *Journal of Nutrition* 102, 909–968.

Ross, A.C. (1992) Vitamin A status: relationship to immunity and the antibody responses. *Proceedings of the Society of Experimental Biological Medicine* 200, 303–320.

Sauberlich, H.E., Hodges, R.E., Wallace, D.L., Kolder, H., Canham, J.E., Hood, J., Raica, N. and Lowry, L.K. (1974) Vitamin A metabolism and requirements in the human studied with the use of labelled retinol. *Vitamin Hormones* 32, 251–275.

Semba, R.D., Scott, A.L., Natadisastra, G., Wirasmita, S., Mele, L., Ridwan, E., West, K.P. Jr. and Sommer, A. (1992) Depressed immune response to tetanus in children with vitamin A deficiency. *Journal of Nutrition* 122, 101–107.

Smith, F.R. and Goodman, D.S. (1971) The effect of diseases of liver, thyroid, and kidneys on the transport of vitamin A in human plasma. *Journal of Clinical Investigation* 50, 2426–2436.

Smith, F.R. and Goodman, D.S. (1976) Vitamin A transport in human vitamin A toxicity. *New England Journal of Medicine* 284, 805–808.

Smith, F.R. and Goodman, D.S. (1979) Retinol-binding protein and regulation of vitamin A transport. *Federation Proceedings*, 38, 2504–2509.

Smith, J.C. (1982) Interrelationship of zinc and vitamin A metabolism in animal and human nutrition: a review. In: *Clinical, Biochemical and Nutritional Aspects of Trace Elements*. Alan R. Liss, New York, pp. 239–258.

Smith, JC. Jr., McDaniel, E.G., Fan, F.F. and Halsted, J.A. (1973) Zinc: A trace element essential in vitamin A metabolism. *Science* 181, 954–955.

Solomons, N.W. and Russell, R.M. (1980) The interaction of vitamin A and zinc: implications for human nutrition. *American Journal of Clinical Nutrition* 33, 2031–2040.

Sommer, A. (1994) Vitamin A: Its effect on childhood sight and life. *Nutrition Reviews* 52, S60–S66.

Sommer, A., Katz, J. and Tarwotjo I. (1984) Increased risk of respiratory disease and diarrhoea in children with preceding mild vitamin A deficiency. *American Journal of Clinical Nutrition* 40, 1090–1095.

Sporn, M.S., Roberts, A.B. and Goodman, D.S. (eds) (1984) Plasma retinol protein. In: *The Retinoids*, Vol. 2. Academic Press, New York, Chapter 8.

Sundboom, J. and Olson, J.A. (1984) Effects of ageing on the storage and catabolism of vitamin A in mice. *Experimental Gerontology* 19, 257–265.

Takahashi, Y.I., Smith, J.E. and Goodman, D.S. (1977) Vitamin A and retinol-binding protein metabolism during fetal development in the rat. *American Journal of Physiology* 233, 263.

Thornton, S.P. (1977) A rapid test for dark adaptation. *Annals of Ophthalmology* 9, 731–735.

Tyler, H. and Dickerson, J.W.T. (1984) Determination of serum retinol in cancer studies. *European Journal of Cancer and Clinical Oncology* 20, 1205–1206.

Underwood, B.A. (1984) Vitamin A in animal and human nutrition. In: Sporn, M.B., Roberts, A.B. and Goodman, D.S. (eds) *The Retinoids*, Vol. 1, Academic Press, New York, pp. 281–392.

Underwood, B.A. (1990) Methods for assessment of vitamin A status. *Journal of Nutrition* 120, 1459–1463.

United Nations Administrative Committee on Coordination – Subcommittee on Nutrition (1987) *First Report on the World Nutrition Situation*. United Nations, Geneva, 1987: 36.

Vijayuaraghavan, K., Sharma, K.V.R., Reddy, V. and Bhaskaram, P. (1978) Rose bengal staining for detection of conjunctival xerosis in nutrition surveys. *American Journal of Clinical Nutrition* 31, 892–894.

Villanuera, I.E. Santos, L.S., Martin, J.S. and Roxas, B.V. (1981) The effects of zinc supplementation on serum vitamin A levels of pre-school children. *Philippine Journal of Nutrition* 34, 5–11.

Vinton, N.E. and Russell, R.M. (1981) Evaluation of a rapid test of dark adaptation. *American Journal of Clinical Nutrition* 34, 1961–1966.

Wako, Y., Suzuki, K., Goto, Y. and Kimura, S. (1986) Vitamin A transport in plasma of diabetic patients. *Tohoku Journal of Experimental Medicine* 149, 133–143.

Wittpenn, J., Tseng, S. and Sommer, A. (1987) Detection of early xerophthalmila by impression cytology. *Archives of Ophthalmology* 104, 237–239.

Wolfe, G. (1980) Vitamin A. In: Alfin-Slater, R.B. and Kritchevsky, D. (eds) *Human Nutrition: A Comprehensive Treatise*. Plenum Press, New York, Vol. 3B, pp. 97–203.

Wolfe, G. (1993) The newly discovered retinoic acid-X receptors (PXRs). *Nutrition Reviews* 51, 81–84.

World Health Organization (1965) Nutrition in pregnancy and lactation: Report of a WHO Expert Committee. *WHO Technical Report Series No. 302*. Geneva.

β-Carotene and Related Substances 12

Carotenoids are members of the terpenoid family of compounds which are characterized by their polyunsaturated nature. They form an important group of natural photosynthetic pigments, providing colour tones from yellow to red for many foods such as tomatoes, carrots and oranges. These substances are fat soluble; hence they are suitable as a colouring agent in foods such as cheese and margarine. More than 500 carotenoids are known to exist in nature. The most common carotenoids in nature are α-, β-, γ-, and δ-carotenes, and some alcohol derivatives, such as xanthophyll, zeaxanthine, lycophyll and cryproxanthine. All carotenoids are synthesized by plants but not by humans, whose sources of carotenes are dietary. Some of the carotenoids with provitamin A activity include α-, β-, γ-carotenes, and cryptoxanthine (3-hydroxy-β-carotene).

Biosynthesis of Carotenes

Most naturally occurring carotenoids have all-*trans* configuration of double bonds in their chains with 40 carbon atoms and eight isoprene residues. In plants, they are constructed symmetrically arising by head to head condensation of four C_{10} precursors (gerenyl pyrophosphate), which originate from mevalonic acid (Fig. 12.1).

In the process of carotenoid biosynthesis, mevalonate is first phosphorylated to form 5-pyrophosphomevalonate with the consumption of 2 moles of ATP (Fig. 12.1). Subsequently, an active isoprene unit with C_5, isopentenyl pyrophosphate, is formed. This reaction proceeds with condensation of two molecules of the isoprene unit forming a C_{10} unit, geranyl pyrophosphate. A tetraisoprenoid geranyl pyrophosphate is eventually formed to yield phytoene, which is a colourless product with 40 hydrocarbons; of its nine double bonds only three are conjugated, i.e. double bonds alternate between single chemical bonds (Fig. 12.2). The conjugated system grows by gradual dehydrogenation producing successively carotenes and lycopene.

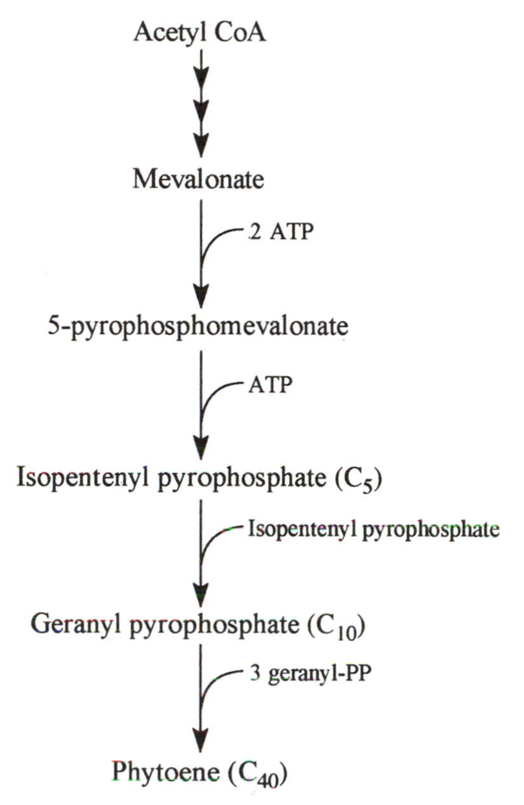

Fig. 12.1. Biosynthesis of phytoene.

Structure of Carotenes

Lycopene is a long-chain hydrocarbon with 13 double bonds, eleven of which are in conjugation. Because of the very many conjugated bonds, lycopene unlike the phytoene is a coloured substance. It is the main pigment of the tomato, pepsicum and many fruits.

The structural difference between lycopene and other carotenes lies only in their end rings (Fig. 12.2). Thus in lycopene, both rings are open, while ring closure on only one end results in γ-carotene, and with ring closure on both ends, α- and β-carotene are produced. α- and β-carotene differ in the position of the double bonds in the rings as shown in Fig. 12.2. In β-carotene, both annular double bonds are in conjugation with the system of double bonds to the long chain (β-ionone structure) and in α-carotene one of the annular double bonds is removed from the system of conjugation by one position (α-ionone structure).

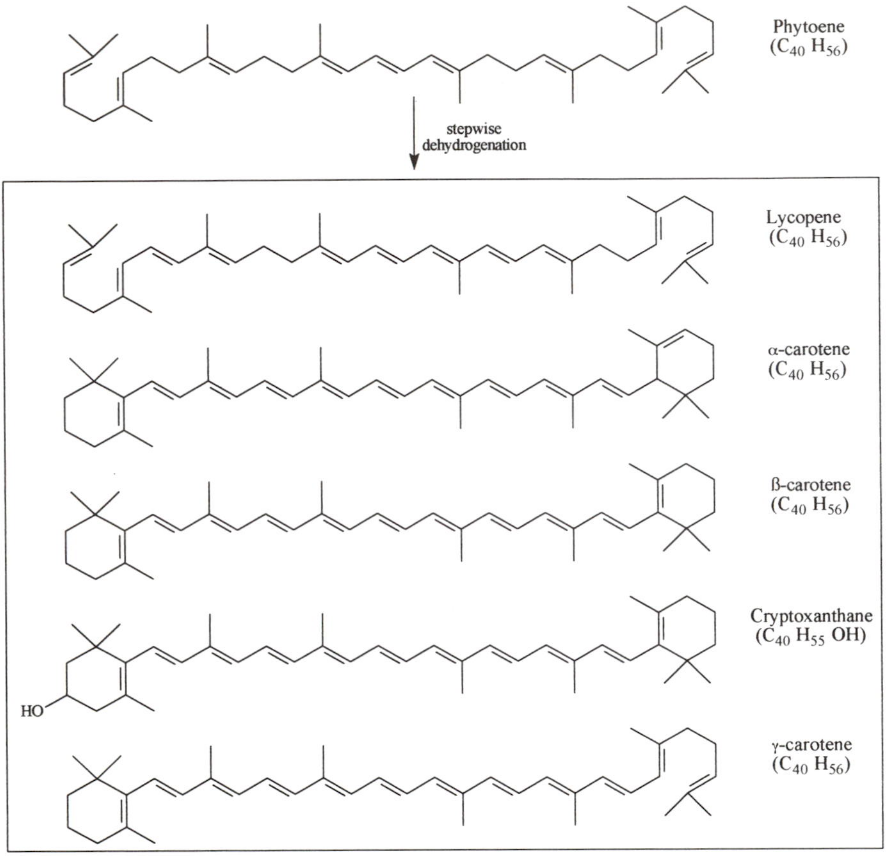

Fig. 12.2. Major carotenoids.

The structural details of the carotenes have physiological significance because the β-ionone ring (conjugated trimethyl cyclohexane) is essential for vitamin A activity. Thus, unlike α-, β-, and γ-carotene, the lycopene does not have any provitamin A activity because of its having both rings not closed (Fig. 12.2).

Of all carotenoids, β-carotene, because of its appropriate structural configuration, has the highest vitamin A activity. In recent years, its actions independent of its provitamin A activity have attracted many investigators. Evidence to date indicates while the importance of vitamins for human health seems to have been largely established, this is not yet the case for β-carotene. The indications that β-carotene is a very important substance in human nutrition have prompted this nutrient to be dealt with separately in this chapter rather than together with vitamin A in Chapter 11.

Properties of β-Carotene

The empirical formula for β-carotene is $C_{40} H_{56}$ with a molecular weight of 536·9. It is a reddish-brown to deep-violet crystalline substance; insoluble in water; very sparingly soluble in alcohol, fats and oils, ether, and acetone; slightly soluble in chloroform and benzene. Its melting point is 176–182°C and its solution in cyclohexane exhibits a characteristic absorption spectrum at 456–484 nm.

Generally, β-carotene is very stable during cooking, but it may lose some activity during storage due to destruction by light and air. Food processing such as hydrogenation of vegetable fats and dehydration of vegetables and fruits reduces its activity. Freezing foods helps retain stability of β-carotene.

Dietary Sources of β-Carotene

Generally, all vegetations with chlorophyll are rich in β-carotene. The main sources of β-carotene for human beings are fruits and vegetables. The food items rich in β-carotene include carrots, spinach, sweet potatoes, and winter squash, with contents ranging from 6–8 mg/100 g (Block *et al.*, 1985; LaChance, 1988). Produce such as broccoli, cantaloupe, collards, pink grapefruit, mangos, papayas, vegetable juice (V-8), tomato juice and peaches are also significant sources of β-carotene in our diets. Food such as tomatoes, although not very concentrated in β-carotene, are important sources because of the large volume eaten.

Fruits and vegetables, such as carrots, mangos and papayas, almost exclusively contain carotenoids with vitamin A activity, whereas others such as tomatoes, spinach and yellow corn contain only a small amount of provitamin A carotenoids (Mangels *et al.*, 1993). The latter sources contain mainly xanthophylls, which have no provitamin A activity. In countries where red palm oil is used, its carotene content represents a major source of vitamin A. It should be pointed out, however, that all carotenes may have biological potentials other than their provitamin A activities. Most commonly eaten carotene sources in various countries include carrots in North America, yellow–green vegetables in Japan, red palm oil in West Africa and dark green leafy vegetables among the Chinese population.

Absorption, Transport and Storage of β-Carotene

β-Carotene often exists in combination with protein in foods from which it must be released through the action of pepsin in the stomach and proteolytic enzymes in the proximal small intestine (Simpson and Chichester, 1981). The released β-carotene is subsequently solubilized into micellar solutions along with the other fat-soluble components. Hence, bile salts and dietary fat are needed for the absorption of carotene in the upper small intestine. The micellar solutions diffuse into the glycoprotein layer surrounding the microvilli of the duodenum and jejunum. Only about 5–50% of the total β-carotene consumed is absorbed (Blomhoff *et al.*, 1991). Normally, the absorption of β-carotene decreases with increasing dosage. The bioavailability of the carotene is thought to be greater from cooked or mashed vegetables than from raw ones (Erdman *et al.*, 1988).

β-Carotene is converted into retinol, primarily in the intestinal mucosa (Laksman *et al.*, 1989). The biosynthetic process for retinol from carotene involves two soluble enzymes, which are predominantly located in the cytosol of the intestine; these are β-carotene-15, 15′-dioxygenase and retinaldehyde reductase. The first enzyme is responsible for the oxidative split of the β-carotene molecule at the plane of symmetry between C-15 and C-15′ (Fig. 12.3). This initial cleavage product is retinal (vitamin A aldehyde), which is then reduced

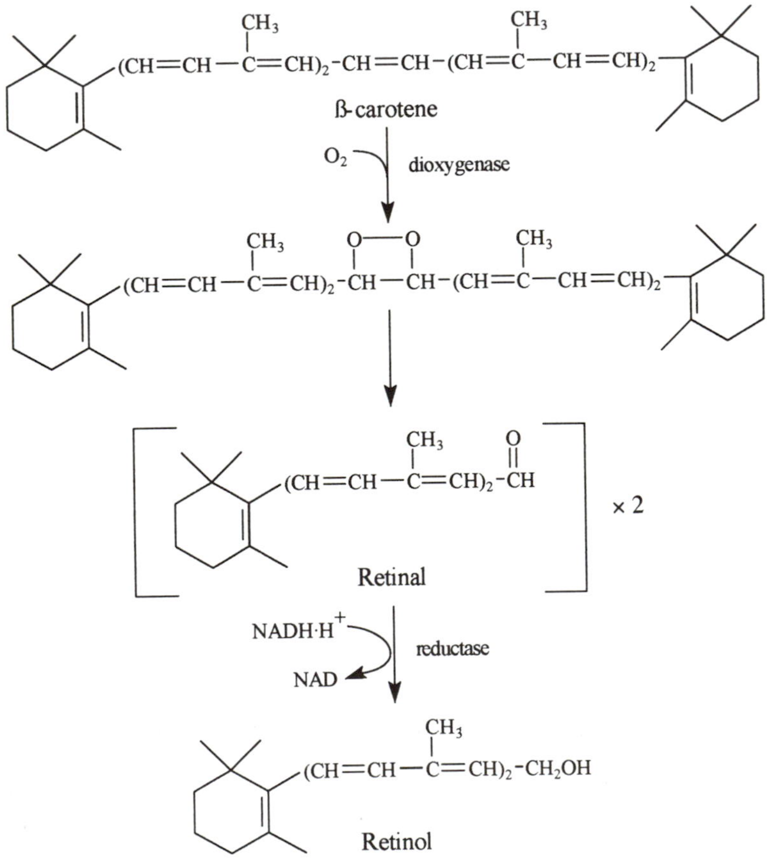

Fig. 12.3. Conversion of β-carotene to retinol.

to retinol (vitamin A alcohol) by the second enzyme which is NADH dependent. The cleavage enzyme is also present in the liver but the rate of conversion of β-carotene to retinol in this site is markedly less than that in the intestine. The rat intestinal dioxygenase enzyme has been shown to be twice as active as the enzyme in the liver on a per total organ basis, and 4–7 times as active on a per gram basis (Zachman and Olson, 1963). Among all species studied, the cat is one species that has been found to be devoid of the dioxygenase enzyme, making cats unable to obtain vitamin A from β-carotene. Hence, a cat's diet must have the preformed vitamin A. Retinol derived from β-carotene or other carotenoids or the hydrolysis of dietary retinyl esters is esterified in the mucosal cell to form retinyl palmitate. These newly formed esters along with any carotenoids that have been absorbed unchanged are incorporated into chylomicrons. Retinyl esters and unchanged carotenoids in chylomicrons are carried to the circulation and then to the liver via the lymphatic system. The transport of retinyl ester and its fate in the body are discussed in more detail in Chapter 11.

Unchanged carotenoids reaching the liver are generally subject to three possible fates. As mentioned earlier, a small portion may be converted into retinol; some may be incorporated into VLDL synthesized in the liver and then transported to fatty tissues via the circulatory VLDL and LDL; and some may also be stored in the liver.

Unlike vitamin A, serum carotene levels are variable and directly reflect recent dietary intake. The half-life of carotene is not clearly established, essentially due to the erratic pattern of serum levels and absence of uniform declining slope after the carotene is discontinued. The fluctuating pattern of the declining slope cannot be explained at this time. However, an approximate estimation of the half-life of β-carotene is believed to be of the order of about 5 days (Dimitrov *et al.*, 1986).

Bio-potency of β-Carotene

Provitamin A activity

As discussed earlier, the cleavage can yield retinol only if at least one of the two end rings of a carotenoid molecule possesses the unsubstituted conjugated trimethyl cyclohexane structure (β-ionone ring). Thus, only the α-, β-, γ-carotene, and cryproxanthine have the potential to be the precursors of vitamin A (Fig. 12.2). Of these, β-carotene, because of its two β-ionone rings, is best converted to vitamin A. There are many factors that are believed to influence the conversion of β-carotene to retinol:

1. Dietary fat. Carotene absorption appears to be very poor (5%) when substituting on a low-fat diet (7% of total energy intake). A small lipid supplementation to this diet appears to raise absorption by 50%.

2. Dietary protein. The oxygenase activities of the intestinal mucosa have been shown in experimental animals to be proportional to protein intake.
3. Thyroid status. Administration of thyroxine increases while antithyroid drugs, such as thiouracil, decrease the cleavage of β-carotene in rats.
4. Availability of pre-formed vitamin A. The amount of β-carotene converted to vitamin A is controlled by vitamin A status. This explains why β-carotene does not contribute to plasma vitamin A levels unless the levels fall below homeostatic levels. β-Carotene, thus, becomes a more significant source of vitamin A in developing countries where pre-formed vitamin A intake is limited than in developed countries where animal foods and hence the pre-formed vitamin A intake is adequate (Peto *et al.*, 1981).
5. Vitamin E status. The cleavage of β-carotene into retinal requires the presence of vitamin E, which is probably necessary to protect the substrate and product from their oxidation.

The β-carotene molecule is essentially a double retinol structure (Fig. 12.3). Theoretically, therefore, one molecule of β-carotene can be cleaved into two molecules of retinol. Biologically, however, β-carotene usually yields only about one-sixth of its weight of retinol, and the rate of conversion decreases as the amount of β-carotene in the diet increases. The disparity between the chemical structure and true provitamin A activity of β-carotene is primarily due to its variable absorption, and conversion to vitamin A. It is assumed that one-sixth (on a weight basis) of dietary β-carotene and one-twelfth of other provitamin carotenoids is converted to retinol in the enterocytes, and absorbed (Blomhoff *et al.*, 1991).

One international unit (IU) of vitamin A is defined as 0.3 μg of all-*trans* retinol (see Chapter 11). The term 'retinol equivalents' (RE) is more appropriately used to describe the amount of vitamin A intake from all dietary sources of vitamin A and carotenoids. Thus, 1 μg of all-*trans* retinol equals 1 RE, and 1 μg of retinol is assumed to be biologically equivalent to either 6 μg of β-carotene or 12 μg of mixed dietary carotenoids.

Antioxidant property

β-Carotene was once thought to have no value, except as a precursor of vitamin A. Several lines of evidence in recent years have indicated that carotenoids (with or without vitamin A activity) have many functions, which are independent of their provitamin A roles. One of these functions is their antioxidant properties that help trap free radicals, which are highly reactive. This concept first emerged from the fact that carotenoids are located in all photosynthetic tissues, usually adjacent to chloroplasts. Their conjugated double-bond structure is capable of trapping singlet oxygen generated during the process of photosynthesis when light energy enters and is trapped by the cell.

Energized molecules are formed during normal biochemical reactions in the body, such as in immune or inflammatory responses, or through external sources, such as cigarette smoke or air pollution. These molecules can damage cell membranes and the genetic material of cells. Carotenoids, because of their antioxidant properties, are thought to be involved in inhibiting the oxidations of LDL and arachidonic acid as well as in promoting the immune system.

LDL oxidation

An emerging theory is that oxidatively modified LDL tends to trigger foam cell formation (Fig. 12.4), which is responsible for stimulating chemotactic responses in macrophages and promoting smooth muscle cell proliferation and differentiation (Parthasarathy *et al.*, 1992; Gletsu and Basu, 1993). Arterial cells are believed to be affected by this oxidative process and the changes induced in these cells may contribute to the progression of atherosclerosis. The likely candidate responsible for the oxidative modification of LDL and arterial cells is the oxygen derived free radical (Kukreja *et al.*, 1992). They are thought to interact with the lipid component of the LDL particle and cause this lipid to become peroxidatively modified (Oen *et al.*, 1992). Polyunsaturated fatty acids contained within the LDL particle are especially susceptible to attack by free-radical species. These reactive compounds are capable of abstracting one of the double allylic hydrogen atoms on the carbon atom between the double bonds of the fatty acid. The reaction then proceeds as follows:

$$LH + R^{\cdot} \rightarrow L^{\cdot} \quad (12.1)$$

$$L^{\cdot} + O_2 \rightarrow LOO^{\cdot} \quad (12.2)$$

$$LOO^{\cdot} + LH \rightarrow LOOH + L^{\cdot} \quad (12.3)$$

$$LOOH + Fe^{3+} \rightarrow LOO^{\cdot} + H^{+} + Fe^{2+} \quad (12.4)$$

$$LOOH + 2\ GSH \rightarrow LOH + GSSG + H_2O \quad (12.5)$$

The product of a reaction between a free radical ($R^{\cdot}$) and a polyunsaturated fatty acid, LH (Eq. 12.1), is the pentadienyl radical ($L^{\cdot}$), which can immediately react with oxygen to form the peroxyl radical, $LOO^{\cdot}$ (Eq. 12.2). This process may be propagated if the peroxyl radical reacts with another polyunsaturated fatty acid to form another lipid free-radical and a lipid peroxide, LOOH (Eq. 12.3). Lipid peroxides may reinitiate this chain reaction if converted to a free radical (Eq. 12.4). Instead, they may be converted to an alcohol (LOH) by enzymes such as selenium-dependent glutathione peroxidase (GSH), and thus become inactivated (Eq. 12.5) (Krinsky, 1992). Hence, the propagation phase of the chain reaction generating free radicals can be inhibited by the addition of a chain-breaking antioxidant, such as vitamin E (see Chapter 14).

In 1984, Burton and Ingold postulated that β-carotene is a novel type of antioxidant, particularly at the low oxygen pressures (15–160 mmHg) found

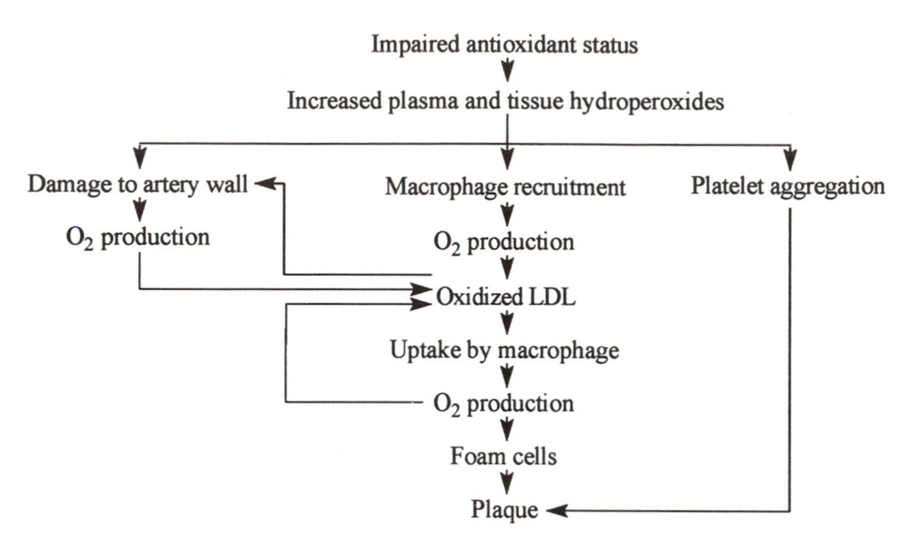

Fig. 12.4. Flow diagram of possible oxidation reactions contributing to plaque development.

within cells. It inhibits lipid peroxidation by acting as a scavenger of singlet oxygen ($^1O_2$) and by quenching peroxyl radicals (Cranfield *et al.*, 1992). The reaction of β-carotene (CAR) with lipid peroxyl radicals (LOO$^\cdot$) to form multiple carbon-centred radicals [LOO-CAR$^\cdot$, LOO-CAR-OOL, (LOO)$_2$-CAR-OOL$^\cdot$, and (LOO)$_2$-CAR-(OO L)$_2$], is shown below (Krinsky, 1992):

$$\text{CAR} + \text{LOO}^\cdot \longrightarrow \text{LOO–CAR}^\cdot \tag{12.6}$$

$$\text{LOO–CAR}^\cdot + \text{LOO}^\cdot \longrightarrow \text{LOO–CAR–OOL} \tag{12.7}$$

$$\text{LOO–CAR–OOL} + \text{LOO}^\cdot \longrightarrow (\text{LOO})_2\text{–CAR–OOL}^\cdot \tag{12.8}$$

$$(\text{LOO})_2\text{–CAR–OOL}^\cdot + \text{LOO}^\cdot \longrightarrow (\text{LOO})_2\text{–CAR–(OOL)}_2 \tag{12.9}$$

Because carotenoids feature a conjugated double-bond system, they are highly efficient quenchers of singlet oxygen. A single β-carotene molecule is believed to eliminate up to 1000 singlet oxygens by physical mechanisms involving energy transfer, before it is oxidized and loses its antioxidant properties. The rate of oxidation of β-carotene is dependent on the oxygen partial pressure. The carbon-centred radicals are resonance-stabilized when the oxygen pressure is lowered (as found *in vivo*). The equilibrium reactions (Eq. 12.7–12.9) shift sufficiently to the left to effectively lower the concentrations of peroxy radicals (LOO$^\cdot$); the rate of auto-oxidation of β-carotene is thus reduced. The reactivity of β-carotene towards peroxy radicals and the stability of the resulting carbon-centred radicals are two important features that give the carotene molecule antioxidant capabilities (Burton, 1989).

Specific oxidation products of β-carotene, particularly certain epoxides, have been detected and will in future be useful as markers of singlet oxygen reactions (Kennedy and Liebler, 1991). As carotenoids trap singlet oxygen, they release energy in the form of heat, thus a regeneration system, such as that required for vitamin E, is not needed (Sies *et al.*, 1992). *In vitro* studies have revealed that 2 μmol of β-carotene is more potent than 40 μmol of α-tocopherol in inhibiting the oxidative modification of LDL (Jialal *et al.*, 1991). A large case-controlled study involving physicians reported preliminary findings of reduced cardiac myopathy in a β-carotene-supplemented group (Hennekens and Everlein, 1985). In this study, subjects with a history of cardiac myopathy were randomized into a β-carotene-supplemented group (50 mg every two days), and a placebo group. A statistically significant reduction in cardiovascular events of almost 50% was observed in the β-carotene group compared with the placebo group. At the dosage used in this study, no unwanted side-effects were noted. These findings, along with evidence that β-carotene may reduce the incidence of certain cancers suspected to have free radical involvement, lend support to the concept that β-carotene may be involved in the protection against atherosclerosis, possibly due to its antioxidant action. Further investigation is warranted to define β-carotene's role in atherosclerosis prevention and therapy.

Arachidonic acid oxidation

Another proposed function of β-carotene is its inhibition of arachidonic acid oxidation. This oxidation follows two major pathways: the lipoxygenase and the cycloxygenase pathway. In the lipoxygenase oxidation, arachidonic acid is oxidized to hydroxyicosatetraenoic acids (HETE) which subsequently form the leukotrienes. Leukotrienes are involved in the inflammatory and anaphylactic allergic responses of the body. Similarly, in the cycloxygenase pathway, the oxidation of arachidonic acid leads to the formation of prostaglandins which are involved in inflammation, inducing labour at term as well as vascular, bronchial and platelet clotting control in the body. In a study carried out by Halevy and Sklan (1987), it was shown that β-carotene can inhibit the production of prostaglandins and, to a lesser extent, HETE in bovine seminal vesicles. It should be noted from this study that retinol had a similar effect but at concentrations much greater than an equivalent amount of β-carotene to produce the same effect. This would indicate that β-carotene has an intrinsic inhibitory arachidonic acid oxidation action separate from its provitamin A activity. These findings also suggest that β-carotene may inhibit the inflammatory, oxytocic, vascular, bronchial, platelet and allergic effects of prostaglandins and leukotrienes.

Immunomodulatory effect

There is strong evidence that β-carotene enhances the capability of the immune functions including T and B lymphocyte proliferation and induction of cells capable of killing tumour cells (Bendich, 1989). Alexander *et al.* (1985) showed that 180 mg of β-carotene daily taken orally over 14 days could increase the number of T-4 cells *in vivo* by approximately 30%. Other T cells did not appear to be affected. A further study (Rhodes *et al.*, 1984) suggested that interferon inhibition caused by tumours could be reversed by β-carotene *in vitro*.

Canthaxanthine, which does not have any provitamin A activity, has been shown to have the same ability to enhance immune responses as β-carotene in rats (Fig. 12.5). These results suggest that the immunomodulatory effect of β-carotene is independent of its provitamin A activity. It has been suggested that the immuno-enhancement may be due to carotene's ability to prevent the loss of antigen activity in macrophages.

Requirements

To date, there is no recommended dietary intake (RNI) for β-carotene, because its only nutritional function has been considered to be the exclusive role as provitamin A. Current dietary intakes of β-carotene are traditionally expressed as part of the RNI for vitamin A. However, based upon recent studies indicating that some of β-carotene's (and other carotenoids) functions are independent of its provitamin A role, a guideline involving 4–5 servings of a combination of fruits and yellow vegetables each day, amounting to 6–10 mg of β-carotene, has been suggested; this amount is normally found in half a cup of cooked carrots. The suggestion comes from at least two sources including the National Research Council (1989) and National Cancer Institute (1991).

Factors that Influence β-Carotene Status

Carotene does not readily cross the placental barrier and, in newborn infants, the level of carotene in the blood is one-fifth to one-tenth of that in the maternal blood. Generally, the rate of absorption of carotene and its conversion to retinol are not as efficient as in adults. Hence, in infants, carotene is believed to be an unsatisfactory substitute for the pre-formed vitamin A (Barker, 1982).

According to many studies (Davis *et al.*, 1983; Chow *et al.*, 1986; Stryker *et al.*, 1988), low serum carotene levels appear to be associated with smoking. Smokers ingesting the same amount of β-carotene as non-smokers achieve lower serum β-carotene levels, suggesting that the effect is smoking specific. A study investigating the effect of smoking on the dietary components of antioxidants, such as vitamins E and C, β-carotene, and selenium, revealed that it was only the serum levels of β-carotene and vitamin C that were depressed in smokers (Chow *et al.*, 1986). There appears to be a strong dose–response relationship between number of cigarettes smoked and serum carotene levels (Davis *et al.*, 1983). Other groups at risk of low carotene status include subjects suffering from conditions in which lipid absorption is impaired, such as sprue,

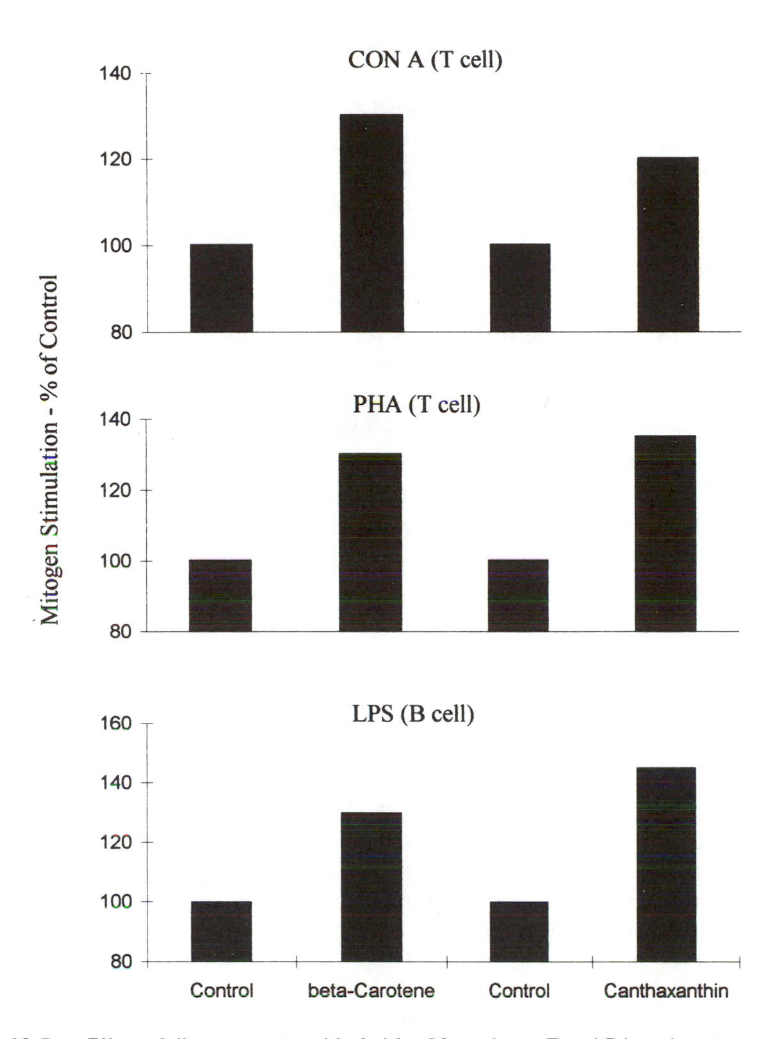

Fig. 12.5. Effect of dietary carotenoids fed for 20 weeks on T and B lymphocyte responses in rats. Rats were fed diets containing β-carotene, canthaxanthine or no carotenoid (control) for 20 weeks. T lymphocyte responses to the mitogens Con A and PHA and B lymphocyte responses to the mitogen LPS were determined and compared to the level of mitogenic activity seen in the control group. Source: Bendich (1989) (with permission).

jaundice, cirrhosis of the liver, and cystic fibrosis. Hypercarotenaemia, associated with elevated serum carotene levels and yellowish tint of the skin, has been reported to occur in hypothyroidism, diabetes mellitus, and an inborn error of metabolism with failure to convert carotene to vitamin A (Vakil *et al.*, 1985).

Assessment of Carotene Status

In the past, many investigations involved determinations of the intensity of the colour of a crude extract as a measure of β-carotene concentration in serum. In humans, the serum carotenoids reflect dietary intakes of complex mixtures of carotenes, of which β-carotene constitutes less than 20% (Brown *et al.*, 1989). Recently, the use of high performance liquid chromatography (HPLC) has been recommended in the analysis of carotenoids in serum. The development of this method has made possible the efficient separation and detection of a diverse class of carotenoid compounds. Procedures for microdetermination of the major human blood carotenoids by HPLC separation and quantitation are detailed in many papers (Broich *et al.*, 1983; Katrangi *et al.*, 1984; Milne and Botnen, 1986). There are, however, a number of factors that should be taken into account in the interpretation of the serum values. These include: (i) the serum carotenoid levels are correlated with recent carotenoid consumption and are not maintained within a narrow physiological range like serum vitamin A levels (Willett *et al.*, 1983); (ii) serum concentrations of carotenoids are subject to seasonal variations (Rantalahti *et al.*, 1993); and (iii) β-carotene constitutes a higher percentage of serum carotenoids for females than for males (Stacewicz-Sanpuntzakis *et al.*, 1987).

References

Alexander, M., Newmark, H. and Miller, R.G. (1985) Oral β-carotene can increase the number of OKT4+ cells in human blood. *Immunology Letters* 9, 221–224.

Barker, B.M. (1982) Vitamin A. In: Barker, B.M.R and Bender, D.A. (eds) *Vitamins in Medicine*, 4th edn, Vol. 1. William Heinemann, London, pp. 211–290.

Bendich, A. (1989) Carotenoids and the immune response. *Journal of Nutrition* 119, 112–115.

Block, G., Dresser, C.M., Hartman, A.M. and Caroll, M.D. (1985) Nutrient sources in the American diet: Quantitative data from the NHANES II survey – I. Micronutrients. *American Journal of Epidemiology* 122, 13–26.

Blomhoff, R., Green, M.H., Green, J.B., Berg, T. and Norum, K.R. (1991) Vitamin A metabolism: new perspectives on absorption, transport and storage. *Physiological Reviews* 71, 951–990.

Broich, C.R., Gerber, L.E., Erdman, Jr., J.W. (1983) Determination of lycopene, α- and β-carotene and retinyl esters in human serum by reversed-phase high performance liquid chromatography. *Lipids* 18, 253–258.

Brown, E.D., Micozzi, M.S., Craft, N.E., Bieri, J.G., Beecher, G., Edwards, B.K., Rose, A., Taylor, P.R. and Smith, J.C. (1989) Plasma carotenoids in normal men after a single ingestion of vegetable or purified β-carotene. *American Journal of Clinical Nutrition* 49, 1258–1265.

Burton, G.W. (1989) Antioxidant action of carotenoids. *Journal of Nutrition* 1991, 109–111.

Burton, G.W. and Ingold, K.V. (1984) β-carotene: an unusual type of lipid antioxidant. *Science* 224, 569–573.

Chow, C.K., Thacker, R.R., Changchit, C., Bridges, R.B., Rehm, S.R., Humble, J. and Turbek, J. (1986) Lower levels of vitamin C and carotene in plasma of cigarette smokers. *Journal of the American College of Nutrition* 5, 305–312.

Cranfield, L.M., Forage, J.W. and Valenzuela, J.G. (1992) Carotenoids as cellular antioxidants. *Proceedings of the Society of Experimental Biology and Medicine* 200, 260–266.

Davis, C., Brittain, E., Hunninghake, D., Graves, K., Buzzard, M. and Tyroler H. (1983) Relations between cigarette smoking and serum vitamin A and carotene in candidates for the lipid research clinics coronary prevention. *Trials of the American Journal of Epidemiology* 118, 445.

Dimitrov, N.V., Boone, C.W., Hay, M.B., Whetter, P., Pins M., Kelloff, G.J. and Malone, W. (1986) Plasma β-carotene levels – kinetic patterns during administration of various doses of β-carotene. *Journal of Nutrition, Growth and Cancer* 3, 227–237.

Erdman, J.W., Poor, C.L. and Dietz, J.M. (1988) Factors affecting the bioavailability of vitamin A, carotenoids and vitamin E. *Food Technology*, October, pp. 214–221.

Halevy, O. and Sklan, D. (1987) Inhibition of arachidonic acid oxidation by β-carotene, retinol and α-tocopherol. *Biochimica et Biophysica Acta* 918, 304–307.

Hennekens, C.H. and Everlein, K.A. (1985) A randomized trial of aspirin and β-carotene among U.S. physicians. *Preventive Medicine* 14, 165–168.

Jialal, I., Norkus, E.P., Cristol, L. and Grundy, S.M. (1991) β-carotene inhibits the oxidative modification of low-density lipoprotein. *Biochimica et Biophysica Acta* 1086, 134–138.

Katrangi, N., Kaplan, L.A. and Stein, E.A. (1984) Separation and quantitation of serum β-carotene and other carotenoids by high performance liquid chromatography. *Journal of Lipid Research* 25, 400–406.

Kennedy, T.A. and Liebler, D.C. (1991) Peroxyl radical oxidation of a-carotene: formation of carotene epoxides. *Chemical Research in Toxicology* 4, 290–295.

Krinsky, N.I. (1992) Mechanism of action of biological antioxidants. *Proceedings of the Society for Experimental Biology and Medicine* 200, 248–256.

Kukreja, R.C., Jesse, R.L. and Hess, M.L. (1992) Singlet oxygen: a potential culprit in myocardial injury. *Molecular and Cellular Biochemistry* 111, 17–24.

LaChance, P. (1988) Dietary intake of carotenes and the carotene gap. *Clinical Science* 7, 118–122.

Laksman, M.R., Mychrovsky, I. and Attlesey, M. (1989) Enzymatic conversion of all-trans β-carotene to retinol by a cytosolic enzyme from rabbit and rat intestinal mucosa. *Proceedings of the National Academy of Sciences, USA* 86, 9124–9128.

Mangels, A.R., Holden, J.M., Beecher, G.R., Forman, M.R. and Lanza, E. (1993) Carotenoid content of fruits and vegetables: an evaluation of analytical data. *Journal of American Dietetic Association* 93, 284–296.

Milne, D.B. and Botnen, J. (1986) Retinol, α-tocopherol, lycopene, and α- and β-carotene simultaneously determined in plasma by isocratic liquid chromatography. *Clinical Chemistry* 32, 874–876.

National Cancer Institute (1991) Eat more fruits and vegetables: five a day for better health. October, *NIH Publication* 92, 32–48.

National Research Council (1989) *Diet and Health: Implications for Reducing Chronic Disease Risk*. National Academy Press, Washington, D.C., p. 15.

Oen, L.H., Utomo, H., Suyanta, F., Hanafiah, A. and Asikin, N. (1992) Plasma lipid peroxides in coronary heart disease. *International Journal of Clinical Pharmacology Therapy and Toxicology* 30, 77–80.

Parthasarathy, S., Steinberg, D. and Witztum, J.L. (1992) The role of oxidized low-density lipoproteins in the pathogenesis of atherosclerosis. *Annual Review in Medicine* 43, 219–225.

Peto, R., Doll, R., Buckley, J.D. and Sporn, M.B. (1981) Can dietary β-carotene materially reduce human cancer rates? *Nature* 290, 201–208.

Rantalahti, M., Albanes, D., Hankka, J., Roos, E., Gref, C. and Virtamo, J. (1993) Seasonal variation of serum concentrations of β-carotene and α-tocopherol. *American Journal of Clinical Nutrition* 57, 551–556.

Rhodes, J., Stokes, P. and Abrams, P. (1984) Human tumour-induced inhibition of interferon action *in vitro*: Reversal of inhibition by β-carotene (provitamin A). *Cancer Immunology and Immunotherapy* 16, 189–192.

Sies, H., Stahl, W. and Sundquist, A.R. (1992) Antioxidant functions of vitamins: vitamins E and C, β-carotene, and other carotenoids. *Annals of the New York Academy of Sciences* 669, 7–19.

Simpson, K.L. and Chichester, C.O. (1981) Metabolism and nutritional significance of carotenoids. *Annual Review of Nutrition* 1, 351–374.

Stacewicz-Sanpuntzakis, M., Bower, P.E., Kikendall, J.W. and Burgess, M. (1987) Simultaneous determination of serum retinol and various carotenoids: their distribution in middle-aged men and women. *Journal of Micronutrient Analysis* 3, 27–45.

Stryker, W.S., Kaplan, L.A. Stein, E.A., Stampfer, M.J., Sober, A. and Willett, W.C. (1988) The relation of diet, cigarette smoking and alcohol consumption to plasma β-carotene and α-tocopherol levels. *American Journal of Epidemiology* 127, 283–296.

Vakil, D.V., Ayiomamitis, A., Nizami, N. and Nizami, R.M. (1985) Hypercarotenaemia: A case report and review of the literature. *Nutrition Research* 5, 911–915.

Willett, W.C., Stampfer, M.J., Underwood, B.A., Speizer, F.E. Rosner, B. and Hennekens, C.H. (1983) Validation of a dietary questionnaire with plasma carotenoid and α-tocopherol levels. *American Journal of Clinical Nutrition* 38, 631–639.

Zachman, R.D. and Olson, J.A. (1965) Uptake of ^{14}C β-carotene and its conversion to retinol ester by the isolated perfused rat liver. *Journal of Biological Chemistry* 238, 541.

Vitamin D

13

The term 'vitamin D' refers to a group of seco-steroid compounds with anti-rick-etic properties. These compounds each possess a conjugated triene system of double bonds. The two forms of the vitamin that are of nutritional significance include ergocalciferol, vitamin D_2 and cholecalciferol, vitamin D_3 (Fig. 13.1). Ergocalciferol is commercially more important than cholecalciferol, being used as a food additive for man and livestock and derived from the ergosterol of yeast. Ergosterol was discovered almost accidentally when it was found that antirick-etic activity could be generated in many foodstuffs by exposing them to ultravio-let light. Cholecalciferol, on the other hand, is synthesized in mammals by the action of sunlight on its precursor, 7-dehydrocholesterol, in the skin (Fig. 13.2). It is also naturally present in a greater variety of foods than ergocalciferol.

Irradiation of ergosterol produces a number of molecular species in addition to ergocalciferol. These are produced by further photoisomerism of the primary

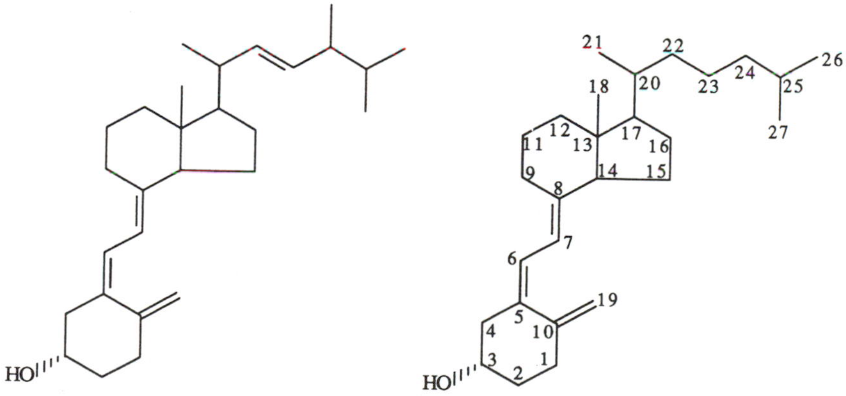

A

B

Fig. 13.1. Molecular structure of vitamin D: (A) ergocalciferol; (B) cholecalciferol.

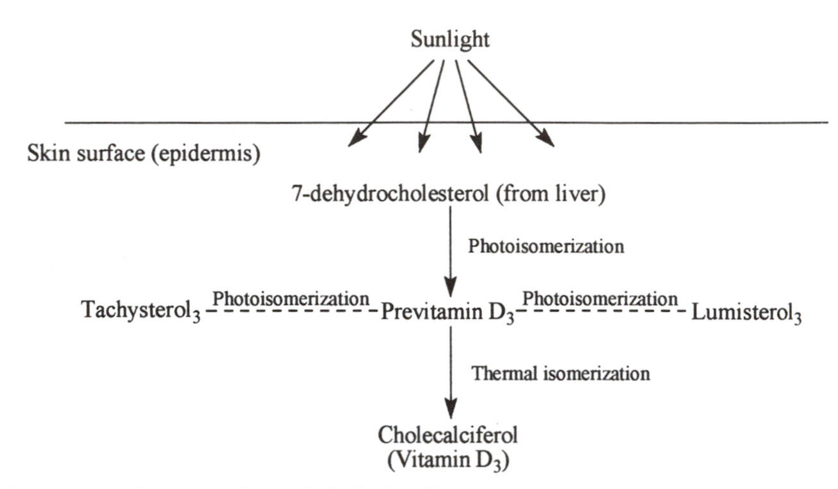

Fig. 13.2. Synthesis of vitamin D_3 in the skin.

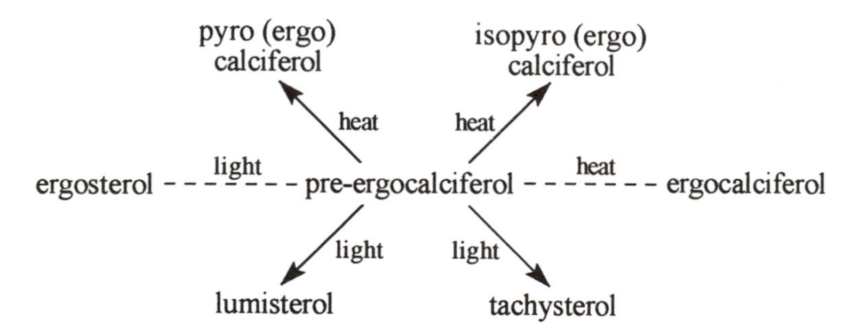

Fig. 13.3. Synthesis of ergocalciferol (vitamin D_2).

product, pre-ergocalciferol (Fig. 13.3). This photochemical isomerization of ergosterol involves the ring system only and applies equally to the irradiation of 7-dehydrocholesterol to form cholecalciferol (Fig. 13.3).

Of the many possible radiation products of the provitamin, only the vitamin itself has significant activity in preventing rickets in man, and this explains the importance of its exact molecular configuration for biological activity and function. It is, however, true that a number of compounds exist which have different levels of antiricketic activity in different species. Biological activity depends on three readily identifiable features of the molecule: (i) a free hydroxyl group at carbon 3; (ii) an intact side chain preferably of the cholesterol type; and (iii) three conjugated double bonds at carbons 5–6, 7–8 and 10–19.

Although vitamins D_2 and D_3 both appear to be equally active and have been shown to undergo similar metabolic conversions, the discussion that follows refers mainly to vitamin D_3. This is because vitamin D_3 is the predominant form available to humans, and it has been used in most of the studies that have resulted in our current knowledge of vitamin D metabolism and its roles.

Sources

Food sources

Vitamin D, being fat soluble, is found in worthwhile amounts in only relatively few foods. No vegetables, fruits or nuts contain any detectable vitamin D, and meats and non-fatty fish contain only trace amounts. Herring and mackerel contain the highest concentrations (Table 13.1) and canned fish somewhat less than half. Vitamin D is to some extent stored in the liver and therefore mammalian liver contains worthwhile amounts. Eggs are a good source. In addition to the natural sources, all skimmed milk, some powdered milk, margarine, breakfast cereals, bread and chocolate bars have varying amounts of added vitamin D, especially in most industrialized countries.

Cutaneous synthesis

Vitamin D is also synthesized in the body by the action of ultraviolet light on 7-dehydrocholesterol (provitamin D_3), which is present in the skin. The solar ultraviolet B photons with energies between 290 and 315 nm penetrate the skin

Table 13.1. Vitamin D content of some foods*.

	μg/100 g	μg/average portion[3]
Butter[1]	0.76	0.06 (medium layer on bread)
Milk		
winter[1]	0.03	0.06 (1/3 pint)
summer[1]	0.03	
Cheese – cheddar[1]	0.26	0.10
Egg – chicken, boiled[1]	1.75	1.0
Herring – grilled[2]	25	27.5
Mackerel – fried[2]	21	23
Pilchards – canned[2]	8	8.4
Sardines – canned[2]	7.5	5.2
Tuna – canned[2]	5.8	5.5
Liver – lambs[2]	0.5	0.45

[1]Holland *et al.* (1989).
[2]Paul and Southgate (1978).
[3]Davies and Dickerson (1991) for portion sizes.
*Vitamin D content of food and its daily requirement are also expressed as International Units (IU). 1 μg of vitamin D_3 equivalent is equal to 40 IU.

photolysing provitamin D_3 which results in opening up the B ring of the sterol nucleus and forming previtamin D_3 (Fig. 13.2). This previtamin D_3 is thermodynamically unstable, and undergoes an isomerization to form cholecalciferol (vitamin D_3). The isomerization involves a sigmatropic shift of a proton from C-9 to C-10.

The naturally occurring vitamin D precursor, 7-dehydrocholesterol, is primarily synthesized in the sebaceous glands. At first, it is secreted on to the surface where it is reabsorbed into the various layers of the epidermis. In rats, the sterol compound is distributed superficially in the stratum corneum, while in humans it is concentrated in the deeper malpighian layer of the epidermis.

In humans, the production of vitamin D_3 in the skin varies with the geographical location, season, atmospheric conditions and, of course, the time spent out of doors, clothing and skin pigmentation (Holick, 1995). Dark skin containing significant amounts of melanin markedly reduces the transformation of 7-dehydrocholesterol to vitamin D_3 (Clemens *et al.*, 1982). Only about 1% of the sun's radiant energy is in the ultraviolet wavelength at the earth's surface. The shortest wavelength is about 290 nm. Studies on isolated human epidermis (Fraser, 1980b) have shown that 51% of light in the wavelength range 290–340 nm penetrates white Caucasian stratum corneum compared with only 18% in black Africans. The penetration of total epidermis by light of 320 nm wavelength was 44% for white Caucasian and only 3% for black African. The stratum corneum of black human skin is generally thicker than that of white human skin. A considerable amount of UV is thought to be lost due to its absorption by the thick stratum corneum in a black complexioned individual. It has been difficult to obtain reliable estimates of the amount of vitamin D produced in the skin of man exposed to ultraviolet light. Latitude is another determinant factor for the transmission of vitamin D-producing UV irradiation. The UV light declines with the distance from the earth's equator. Thus, in winter, there is almost no UV light at latitudes above 50°. Applications of sunscreens have been shown to prevent the production of vitamin D in human skin, both *in vitro* and *in vivo* (Webb and Holick, 1988). *p*-Aminobenzoic acid (PABA), which is the protective component in sunscreen, appears to be particularly responsible for this effect (Matsuoka *et al.*, 1987). A single topical application of PABA can interfere significantly with the cutaneous synthesis of vitamin D.

Absorption

Dietary vitamin D is absorbed through the small intestine in association with lipid and with the aid of bile salts. The vitamin is then taken up in lymph along with chylomicrons and lipoproteins. Vitamin D in plasma, coming from either diet or the skin, is picked up for transport to the liver by a plasma protein called vitamin D-binding protein (DBP), which is synthesized in the liver. Orally administered vitamin D appears to be less effective than biosynthesized vitamin D in raising circulatory levels of 25-OH-D_3. A portion of all vitamin D reaching the liver is 25-hydroxylated and released for circulation in the plasma. Thus, plasma levels of 25-OH-D_3 are related to the size of the liver stores. In the plasma, the 25-OH-D_3 form circulates on another DBP (α_1-globulin), also known as the group-specific component, GC (Schoentgen *et al.*, 1986).

Metabolism

Research on vitamin D has been one of the most fruitful areas of nutritional biochemistry in the last 50–60 years. The elucidation of the metabolism of cholecalciferol has led to increased knowledge of the role of this substance in calcium and phosphate metabolism, its interrelationships with the hormones, parathormone and calcitonin, which affect calcium metabolism, and has helped to identify the molecular basis of some of the genetic disorders affecting bone calcification. At the same time it has led to developments in the treatment of other diseases, such as chronic renal failure (CRF) in which calcium/phosphate metabolism is disturbed with important skeletal consequences.

The liver and kidney are the main sites for the metabolic activation of vitamin D_3. A number of metabolites have been characterized and the sequence of production of the most important of them is shown in Fig. 13.4 (Haussler and McCain, 1977). Vitamin D_3 is first hydroxylated in the liver at the 25-carbon atom by a vitamin D_3-25-hydroxylase enzyme (25-OHase) present in the hepatic microsomal fraction of all species studied (Bhattacharyya and DeLuca, 1974). In the chicken, this enzyme exists in the extrahepatic tissues including the intestine and kidney, but in mammals the liver is the predominant site. The product of this hydroxylation, 25-hydroxyvitamin D_3(25-OH-D_3), is the principal circulating metabolite. The conversion of vitamin D_3 to 25-OH-D_3 is carried out by a reaction requiring NADPH and molecular oxygen. The resultant 25-OH-vitamin D_3 is bound to an α_2-globulin in plasma and transported to the kidney where it undergoes a second obligatory hydroxylation before it can function. The second hydroxylation is catalysed by the enzyme 25-hydroxy-vitamin D_3-1-hydroxylase (1-OHase), to produce 1,25-$(OH)_2D_3$ (calcitrol). The

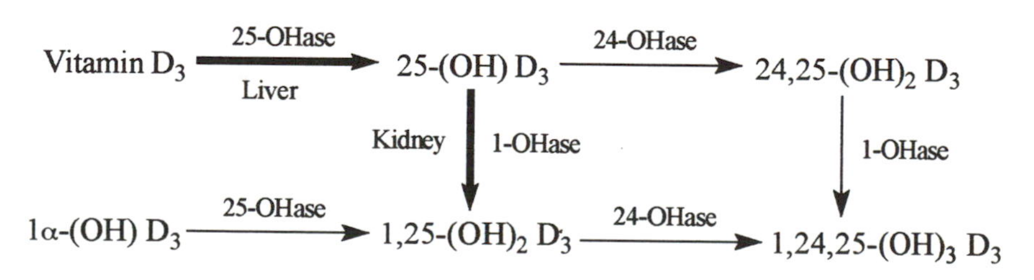

Fig. 13.4. Metabolism of vitamin D. Bold arrows indicate the most important pathways to the functional hormone. Adapted from Haussler and McCain (1977).

renal rate-limiting enzyme is found exclusively in mitochondria. It is this dihydroxy metabolite of vitamin D that is believed to stimulate intestinal calcium transport, intestinal phosphate transport, bone calcium mobilization and other functions attributed to the function of vitamin D. It prevents rickets, is at least five times as biologically active as vitamin D_3 or the 25-hydroxy derivative, and functions three times faster than either of its precursors in promoting calcium absorption (Raisz *et al.*, 1972).

Kidney mitochondria contain another enzyme, 25-hydroxy-vitamin-D_3-24-hydroxylase(24-OHase) which is able to hydroxylate 25-$(OH)D_3$ to form 24,25-$(OH)_2D_3$. This metabolite has limited biological activity and no certain function seems yet to have been found for it.

Solubility and transport of the vitamin D_3 metabolites is achieved by their binding to a specific binding protein, and it is the relative binding affinities of the protein which account for the relative concentrations of the metabolites in human plasma (Table 13.2). Formation of 1,25-$(OH)_2D_3$ is under homeostatic control regulated by the supply of calcium and the prevailing 1,25-$(OH)_2D_3$ concentrations. This is the prime reason why 1,25-$(OH)_2D_3$ has come to be considered a steroid hormone rather than a vitamin and the kidney as the endocrine secretory gland (Fraser, 1980a).

Vitamin D does not meet the classical description of vitamin on a number of counts. It can be synthesized in mammals by the action of sunlight on its precursor, 7-dehydrocholesterol, naturally present in the skin. Only when irradiation is inadequate is there a need for a dietary source. Furthermore, vitamin D_3 is metabolized at sites distant from its origin. Its principal mode of action is mediated through stimulation of the synthesis of specific proteins requiring receptor proteins. The production of 1,25-$(OH)_2D_3$ is strongly feedback regulated at the physiological level (see below). All these characteristics are more akin to the action of a hormone.

Regulation of Metabolism

It seems that there is no direct regulation of 25-hydroxylase activity and the formation of the hepatic derivative is determined largely by the vitamin D supply

Table 13.2. Concentration of vitamin D metabolites in human plasma.

	Concentration	
Metabolite	($mol\ l^{-1}$)	($ng\ ml^{-1}$)
25-OH-D_3	$0.025\text{-}0.125 \times 10^{-6}$	10–50
24,25-$(OH)_2D_3$	$0.020\text{-}0.02 \times 10^{-7}$	0.8–8
1,25-$(OH)_2D_3$	$0.072\text{-}0.123 \times 10^{-9}$	0.03–0.05

Source: Fraser (1981a).

(Garabedian *et al.*, 1972). In contrast, there is considerable evidence that both the synthesis and secretion of 1,25-$(OH)_2D_3$ are under fine regulatory control (Henry and Norman, 1984). The secretion of this derivative is increased during growth, pregnancy, and lactation, presumably due to the hormonal influences that operate during these periods.

Direct modulation of renal 1-hydroxylase activity is most likely influenced by 1,25-$(OH)_2D_3$ itself and parathyroid hormone (PTH). PTH is probably the physiological mediator of altered calcium homeostasis on 1,25-$(OH)_2D_3$ production (Fraser, 1980a). A fall in plasma calcium concentration increases PTH secretion and *in vitro* PTH has been shown to stimulate 1,25-$(OH)_2D_3$ production in kidney homogenates (Henry, 1979). PTH also causes a rise in the serum concentration of 1,25-$(OH)_2D_3$ in human subjects (Sorensen *et al.*, 1982). Furthermore, calcitonin, calcium and phosphate have been shown to affect renal 1,25-$(OH)_2D_3$, but it is unlikely that these effects are due to a direct action on 1-hydroxylase activity. Oestrogens elevate serum 1,25-$(OH)_2D_3$ concentrations. This physiologically important effect may reflect increased production and/or affinity of renal PTH receptors. Growth hormone and thyroid hormone may have permissive roles in 1,25-$(OH)_2D_3$ regulation.

Functions

The hormonal form of vitamin D regulates calcium and phosphate metabolism by its action on three target tissues – small intestine, bone and kidney. In the intestine, 1,25-$(OH)_2D_3$ regulates the absorption of calcium from the diet and in the kidney it regulates the reabsorption of calcium and phosphate from the tubules. The action of vitamin D on bone is less well defined (see below). It is possible that the 1,25-dihydroxy derivative may have a role in bone resorption (Fraser, 1981), and it has been proposed that another metabolite, 24,25-$(OH)_2D_3$, plays a significant role in normal bone formation (Ornoy *et al.*, 1978).

However, it seems that there is no direct effect of 24,25-$(OH)_2D_3$ (or any other vitamin D metabolite) on mineralization and an earlier report (Boyle *et al.*, 1973) had suggested that the 24,25-$(OH)_2D_3$ metabolite must undergo 1-hydroxylation in order to be biologically active. Later it was found that the trihydroxy derivative can be formed from either 24,25-$(OH)_2D_3$ or 1,25-$(OH)_2D_3$ and it would seem possible that 24-hydroxylation may be an inactivation, rather than an activation step in the metabolism of vitamin D (Castillo *et al.*, 1978).

More recent work has increased our understanding of the mechanism of vitamin D function (Lawson and Muir, 1991). As with other steroid hormones, 1,25-$(OH)_2D_3$ interacts with a specific receptor protein in its target tissues and the complex is taken up and retained within the nucleus. Although mRNA in a number of tissues responds to 1,25-$(OH)_2D_3$, it is only the transcription of the calbindin gene in the intestine which is clearly dependent on the hormone (Emtage *et al.*, 1973). Thus it has been suggested that the major biochemical response of the intestine to vitamin D is an increase in calbindin synthesis. The larger form of this protein (mol. wt 30,500) is more widely distributed, with a smaller form (mol. wt 9000) being present in mammalian intestine, placenta and kidney. It seems that only mammalian kidney has both forms of this protein.

The function of calbindin is not yet clear. There is some evidence to suggest that it is related to the intestinal transport of calcium because (i) both are vitamin D dependent in the intestine, (ii) both respond in a similar way to low dietary levels, and (iii) the order of affinity of divalent ions for calbindin is the same as the order of their rate of intestinal absorption. These ions are absorbed by the same route as calcium.

Since the initial discovery of a vitamin D dependent calcium-binding protein, CaBP (Wasserman and Taylor, 1969), a considerable amount of work has been done to determine the role of this protein in calcium absorption. That synthesis of CaBP is vitamin D dependent has been shown for several species. Moreover, CaBP is present along the entire length of the intestine. The use of immunohistochemical and, more recently, sensitive radioimmunoassay, has made possible the determination of the tissue distribution of CaBP. High concentrations have been found in pig duodenum (Table 13.3) with lower but still appreciable concentrations in kidney, thyroid, liver, pancreas and plasma. Using a fluorescent antibody procedure in studies on chick intestine, it was found that the protein was present in goblet cells and was also associated with the absorptive surface of all the intestinal cells. CaBP at this latter location would be at the point of control for calcium passing across the microvilli and thus it was described as 'functional CaBP'. This protein has been reported in a similar situation in the intestine of a number of other species, including man. Experiments in ricketic chicks have shown that the concentration of CaBP is directly proportional to the amount of vitamin D given. Further information about CaBPs will be found in a review by Wasserman *et al.* (1978).

The calbindin proteins isolated by Lawson and his colleagues are synthesized more quickly than CaBP in response to vitamin D. At the molecular level,

Table 13.3. The concentration of intestinal CaBP in various organs and tissues of the pig as determined by radioimmunoassay.

Tissue or organ	CaBP concentration ($ng\ g^{-1}$ tissue, wet wt)
Duodenum	400,000–2,000,000
Kidney	2,000–11,000
Thyroid	610–300
Liver	400–1100
Pancreas	120–160
Plasma	30–80[1]

[1]Plasma CaBP expressed as $ng\ ml^{-1}$.
Source: Wasserman *et al.* (1978).

$1,25\text{-}(OH)_2D_3$ stimulates the mRNA for CaBP. The dihydroxy vitamin D metabolite also stimulates the synthesis of alkaline phosphatase. The relationship of this protein to CaBP is unclear. A Ca-stimulated ATPase has also been described in the intestinal brush border which responds to vitamin D.

At least two steps are involved in the transport of calcium from the lumen to the bloodstream. The mineral must first be moved across the brush border into the cell. This process takes place by diffusion. The ions then move across the cell bound to CaBP to the basal-lateral membrane. Movement across this membrane involves an active pump to move the ions against a concentration gradient. Some workers have considered that vitamin D affects the first of these processes and others either or both of the other processes. Norman and his colleagues, as the result of studies with the polyene antibiotic, filipin, have suggested that the microvillous membrane is the most probable region for vitamin D action. This would involve a carrier mechanism with saturation kinetics (similar to that for active absorption of sodium) through which vitamin D enhances calcium transport (Norman, 1978).

Earlier studies suggested that an effect of vitamin D on phosphate absorption is secondary to that on calcium absorption. However, more recent work suggests that vitamin D has differing effects in differing regions of the small intestine. In the jejunum and ileum, vitamin D stimulates phosphate uptake independently of calcium transport. This is in contrast to what occurs in the duodenum where the presence of calcium greatly increases the response of phosphate transport to vitamin D. We can conclude, though, that regardless of site or mechanism, the transport of both calcium and phosphate across the intestinal mucosa is stimulated by the vitamin D dihydroxy metabolite, $1,25\text{-}(OH)_2D_3$ (Chen *et al.*, 1974). Figure 13.5 summarizes the actions of this metabolite on the intestinal mucosal cell.

When considering the effect of vitamin D on bone (Braidman, 1990) it must be borne in mind that bone contains several types of cells, each of which may respond to the sterol in different ways. It is also necessary to consider if these bone-cell targets for the sterol are part of the mechanism for the systemic control of calcium. Receptors for $1,25\text{-}(OH)_2D_3$ are present in osteoblasts and the hormone is thought to increase mineralization and osteoblast differentiation. In this way bone formation, hence growth, is promoted. In tissue culture studies of fetal calvariae it has been shown that $1,25\text{-}(OH)_2D_3$ stimulates the maturation of osteoclasts from osteoclast precursor cells. In summary, it seems that it is unlikely that bone is part of the vitamin D_3 endocrine system for mammals. The overall effect of $1,25\text{-}(OH)_2D_3$ on bone depends on its stage of remodelling. Thus, the hormone is one of the factors controlling the balance between bone formation and resorption. This is in contrast to the situation in birds in which bone acts as a calcium reserve and provides 40% of egg shell calcium.

We have referred thus far only to the major sites of vitamin D action, but many other body tissues and cells have receptors for and responses to $1,25\text{-}(OH)_2D_3$. These include pancreas, and pituitary gland, lymphocytes, and

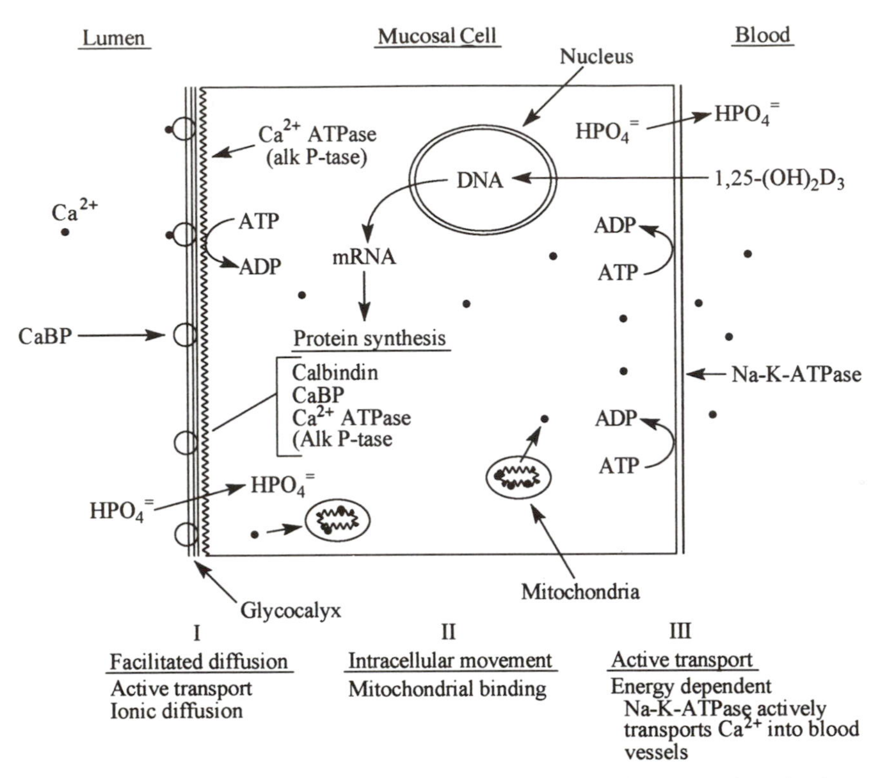

Fig. 13.5. Diagrammatic representation of essential features of a proposed mechanism of action of 1,25-$(OH)_2D_3$ in the absorption of calcium. Adapted from Norman (1978).

monocytes (Braidman, 1990). The biological importance of some of the apparent contradictory effects on these tissues and cells awaits further elucidation.

Recommended Intake

From what has been said already, it is clear that whether or not there is a dietary requirement for vitamin D depends on the amount of exposure of the skin to UV radiation, and the opportunity for endogenous synthesis of cholecalciferol. The recommended intake or allowance will therefore vary for different countries and for different groups of people within those countries. For the United States and Canada, the recommended intake for vitamin D ranges from 10 μg day^{-1} for infants (0–1 year) to 5 μg day^{-1} for adults (National Research Council, 1989; Health and Welfare, Canada, 1990). For the UK, Dietary Reference Values are given only for children (0–6 months): 8.5 μg day^{-1} (1 μg = 40 IU); children (7 months–3 years): 7 μg day^{-1}; and for men and women (aged 65 years and over): 10 μg day^{-1} (Department of Health, 1991). It is also recommended that pregnant and lactating women should receive 10 μg day^{-1}. In addition to these recommendations, it is recognized that without casual exposure to sunlight, the dietary intake for vitamin D should be at least 2–3 times more than the recommended intake (Holick, 1995). It is noteworthy that exposure to sunlight through plastic or glass will not produce vitamin D in the skin.

It is likely that elderly persons are at an increased risk of vitamin D deficiency because of their insufficient intake of the vitamin, limited exposure to sunlight, reduced cutaneous synthesis and reduced absorption of vitamin D, and possibly its impaired activation in the kidney (Holick, 1986). Because of these factors the dietary requirement for vitamin D in the elderly has been suggested to be greater than that of younger adults. Indeed, in recent years, the requirement of the vitamin for the elderly population has been increased in countries such as the United Kingdom (Department of Health, 1991) and Canada (Health and Welfare, Canada, 1990).

Vitamin D and Diseases of Man

Rickets

This disease used to be widespread in city children. In Britain and North America the prevalence was substantially reduced by fortification of National Dried Milk at the end of the Second World War and by fortification of milk in 1934, respectively. In the UK, vitamin D deficiency had disappeared as a public health problem until the 1960s when reports of rickets began to appear in the medical literature. There was a particular problem amongst the Asian immigrant population. However, reports then began to appear from a number of other countries. Between 1972 and 1984, 48 cases of vitamin D-deficient rickets were documented at Winnipeg Children's Hospital, Canada; 40 of these children were Canadian natives (Hayworth and Dilling, 1986). Most affected children are generally between 6 and 24 months of age, live in poor socioeconomic conditions and have little exposure to sunlight even though they live in places with sunny climates. The disease is common in children of Islamic mothers, in tropical countries and particularly where the mother is subject to 'purdah'.

Clinical features

The earliest sign of rickets is craniotabes: late closure of the fontanelle. This is detected mainly before 12 months of age as round unossified areas in the skull which yield like parchment under finger pressure. These lesions occur on the upper occipital, posterior parietal bones, and sometimes in the upper temporal lobe. These are all regions on which the weight of the head may rest.

Beading of the ribs, referred to as the 'ricketic rosary', is almost a constant sign after the age of 6 months, and is caused by the swollen cartilaginous ends of the ribs. The chest may be narrow and rather funnel-shaped, and described as a 'pigeon' chest. In severe cases this may interfere with breathing and necessitate surgical intervention. The arms become deformed when the child begins to crawl or sit upright. The already enlarged radial and ulnar epiphyses broaden still further.

When the child begins to toddle and puts weight on the legs, the femur becomes bowed resulting in separation of the knees ('genu varum'). Alternatively there is abnormal incurving of the leg with the knees in contact ('knock knee' or 'genu valgum') and there is a gap between the feet. Greenstick fractures are common and may be overlooked, being often painless and giving no more deformity than would have occurred from twisted limbs. The pelvis may be sufficiently distorted as to interfere with normal childbirth in later life. Gluteal muscle weakness may produce a waddling gait by the second year of life. Severe vitamin D deficiency reduces the growth rate and causes microcephaly with reduction in brain growth; eruption of teeth is delayed.

On radiological examination the expanded epiphyseal cartilage can be seen and the metaphyseal boundary is uneven and ragged. Epiphyseal secondary centres of ossification appear late and are often indistinct due to poor mineralization. The shaft of the bone appears rarefied with poorly defined trabeculae; the cortex is thin and translucent and its boundary blurred due to excess subperiosteal osteoid. Overall in rickets, the skull, the ribs, the pelvis, the arms and legs can all become softened and bent and widened at the ends (Fig. 13.6).

Biochemically, the plasma calcium concentration is often near to normal. Similarly, the plasma inorganic phosphate concentration, though often lowered, may be normal. Plasma alkaline phosphatase is usually increased, but this finding is non-specific and therefore not diagnostic of rickets. A low concentration of 25-OH-D_3 is often associated with clinical rickets, but may be found in individuals without clinical symptoms.

Simple vitamin D-deficient rickets can be completely cured with oral administration of 50–125 $\mu g\ day^{-1}$ of cholecalciferol or ergocalciferol. Recurrence of the deficiency should be prevented by exposure to solar irradiation.

Osteomalacia

The term 'rickets' is used only to describe the disorder due to vitamin D deficiency in children. The term 'osteomalacia' on the other hand, applies to defective mineralization of bone at any age which may be due to either vitamin D deficiency or impaired vitamin D function. In the UK osteomalacia due to vitamin D deficiency, and described as 'privational' (Table 13.4), is found mainly in Asian immigrants and the elderly. Secondary deficiency and impaired vitamin D function may occur in gastrointestinal, hepatic, and renal disorders.

Vitamin D-deficient osteomalacia in elderly women often coexists with osteoporosis and has been attributed to dietary lack of the vitamin. However,

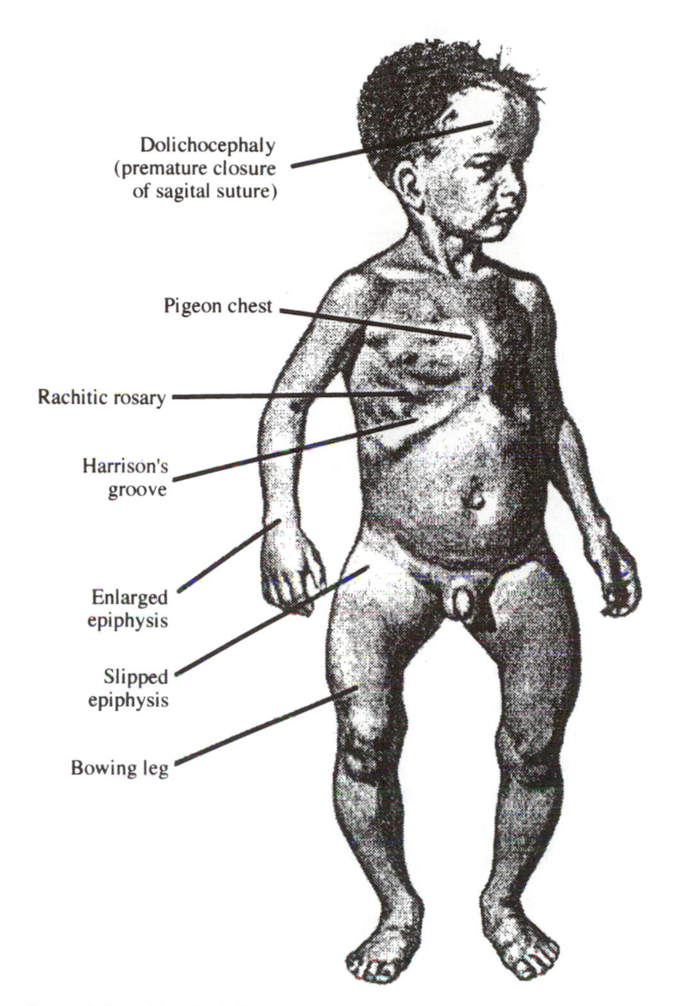

Fig. 13.6. Bone deformities in rickets.

skin synthesis of cholecalciferol during the summer is probably more important than diet as a source (Lawson *et al.*, 1979).

The main clinical features of osteomalacia are skeletal pain and muscle weakness. As the disease progresses severe pain and body tenderness occur in the thorax, shoulder, hips, thighs, forearms and feet. In contrast to osteoporosis, pain is less common in the axial skeleton. Radiologically, bone density is reduced as with osteoporosis, but in contrast to osteoporosis, poor mineralization is seen more commonly in the peripheral bones than in the vertebrae. The most characteristic radiological signs of osteomalacia are pseudofractures or 'Looser's

Table 13.4. Classification of rickets and osteomalacia (Lawson and Muir, 1991).

A. *Nutritional and metabolic*
Privational
Elderly, immigrants in Europe
Gastrointestinal
Small bowel disease, partial gastrectomy
Liver
Biliary cirrhosis, alcoholism, chronic hepatitis
B. *Endocrinological*
Renal osteodystrophy
Other acquired renal disease
Tumours
Heavy metal poisoning
Hereditary renal disease
Vitamin D resistance
Vitamin D dependence
Acidosis
Cystinosis

zones', appearing as translucent bands of decalcification at right angles to the bone surfaces. Fractures are common. Iliac crest biopsies show accumulation of uncalcified osteoid.

Plasma calcium and inorganic phosphate levels are reduced and those of alkaline phosphatase increased. Urinary calcium is usually reduced and the calcium balance often negative. The plasma concentration of 25-OH-D_3 is reduced.

Treatment with vitamin D reverses the clinical signs and results in healing of fractures.

Vitamin D deficiency was first reported amongst Asian immigrants in Glasgow in the UK in 1962. The affected children were aged between 5 and 15 years and severe deformities were confined to girls who presented with leg pains, walking difficulty, genu valgum, broadened wrists and narrowed pelvises. In contrast to the cases of infant rickets, the children had no evidence of malnutrition. Treatment with vitamin D (75 μg day^{-1} for 4 months) resulted in clinical improvement and normalized the biochemical changes. A number of factors have been suggested as contributing to the vulnerability of Asians to vitamin D deficiency. These include inadequate exposure to sunlight, a low dietary intake and a genetically determined higher requirement. There have been suggestions that the consumption of unleavened bread ('chappaties') with its high phytate content might reduce the availability of dietary calcium. The balance of evidence at present suggests that simple lack of sunshine is the main cause of vitamin D deficiency in Asians living in Britain (Fraser, 1980b).

Drug-induced osteomalacia

Prolonged treatment of epileptics with anticonvulsants, particularly phenytoin, may induce osteomalacia. Since drugs induce the mixed function oxidase, drug-metabolizing enzyme system in the liver which acts on steroids, it has been suggested that cholecalciferol is converted to polar inactive metabolites (Stamp, 1974). It has also been suggested that the drugs may, in some way, induce resistance to the vitamin, and this may account for the rather large dose of vitamin D, 250 $\mu g\ m^{-2}$ body area per week in divided doses together with 500 mg calcium per day which it has been suggested should be given for the treatment of anticonvulsive osteomalacia (Hahn and Alvioli, 1975).

Osteomalacia after gastrectomy

A substantial proportion of individuals, particularly women, develop osteomalacia after partial or total gastrectomy. The disease responds well to quite low doses of vitamin D, although there is some evidence that women may not respond as well as men. Dietary deficiency and/or failure to absorb the vitamin may be a cause. Certainly in older patients a restricted diet may well be deficient in vitamin D.

Osteomalacia and intestinal malabsorption

Patients with long-term malabsorption due to cystic fibrosis or gluten-induced enteropathy are at risk of developing osteomalacia. It seems that in these conditions, and also after gastrectomy, there may be enhanced destruction of vitamin D as a result of enhanced activity of hepatic hydroxylating enzymes (Shefer *et al.*, 1969).

Osteomalacia and liver disease

Bone disease, commonly osteomalacia and osteoporosis, may occur in patients with biliary obstruction or hepatocellular damage. There may be decreased absorption of vitamin D and defective production of 25-OH-D_3 as a result of decreased 25-dehydroxylase activity.

Risk of hip fracture may be greater in alcoholics as a group than in the general population. The cause of this increased risk is not clear. Bone weakness could result from poor nutrition, reduced physical activity or as a result of concomitant hepatic or other chronic disease. Abnormal vitamin D metabolism could also be a factor since the liver plays such an important role in the conversion of cholecalciferol to the 25-hydroxy derivative. However, in considering these various factors it is necessary to bear in mind that the diet is a relatively insignificant source of vitamin D and therefore it seems unlikely that the part played by poor nutrition is a major factor. Despite evidence of defective vitamin D metabolism, osteomalacia occurs less frequently than osteoporosis in alcoholics with liver disease (Stanbury and Mawer, 1990).

Osteomalacia and renal disease

In acquired renal disease, the 25-OH-D_3-1-hydroxylase enzyme activity is severely reduced leading to inadequate production of 1,25-$(OH)_2D_3$. Longitudinal studies have shown that plasma concentrations of this metabolite remain normal until the renal function is so reduced that the creatinine clearance has fallen to about 40 ml min^{-1} (normal values: men, 85–125; women 75–115 ml min^{-1}). Renal osteodystrophy can be treated effectively with pharmacological doses of vitamin D or with small doses of 1,25-$(OH)_2D_3$(0.5–1 μg day^{-1}) (Stanbury and Mawer, 1990).

Osteoporosis and ageing

Osteoporosis is characterized by too little calcified bone and is apparently the end result of a number of diseases that affect the skeleton. It is caused when bone mineral resorption exceeds new bone formation. The disease is rare in young adults, and its incidence rises in the 60s in women and 70s in men (Cummings *et al.*, 1985). Especially, it is prevalent in postmenopausal women. Osteoporosis can be classified into two categories, primary and secondary. Causes of the condition such as intestinal malabsorption, liver disease, renal failure, and hyperparathyroidism explain secondary osteoporosis, but less is known about primary osteoporosis. Among the factors proposed as having a significant role in the development of primary osteoporosis are oestrogen deficiency (Richelson *et al.*, 1984) and ageing (Lindeman *et al.*, 1985).

A causal role for oestrogen deficiency in the pathogenesis of osteoporosis is supported by accelerated loss of cortical and trabecular bone in the early postmenopausal period and by retardation of this bone loss with oestrogen administration (Cummings *et al.*, 1985). One hypothesis is that diminished oestrogen blood levels in the postmenopausal state cause hypersensitivity of the skeleton to the resorbing action of PTH (Riggs *et al.*, 1976). This increased sensitivity of the skeleton leads to transient hypercalcaemia, which in turn elicits suppression of PTH. The reduced PTH affects renal calcium conservation and is proposed to curtail 1,25-$(OH)_2D_3$ biosynthesis.

Treatment of osteoporosis patients with 1,25-$(OH)_2D_3$ has, however, produced mixed results. In some studies, this therapeutic regime has increased calcium absorption and improved calcium balance, accompanied by a reduction in bone loss, while in other trials the treatment was without effect (Brautbar, 1986).

In a recent intervention study (Tilyard *et al.*, 1992), patients with osteoporosis were divided by severity into those with five or less and those with five or more fractures at baseline. Following three years of treatment of those patients with daily oral administration of 0.5 μg 1,25-$(OH)_2D_3$, there appeared to be no differences in outcome in those with severe disease, but in those with fewer fractures the treatment was found to be very effective in reducing fracture rate.

The importance of vitamin D nutrition in the older person is best exemplified by its possible relationship to the increased incidence of osteoporosis and osteomalacia. A decrease in bone mass occurs with increasing age and appears to be a normal concomitant of the ageing process. In some individuals this 'physiological osteoporosis' becomes excessive (i.e. pathological) and is characterized clinically by the collapse of vertebral bodies and a tendency for long bones to fracture.

Hormone imbalance may lead to a negative calcium balance especially in postmenopausal women (Tsai *et al.*, 1984). The absorption of calcium falls in men and women after 70 years of age (Spencer *et al.*, 1982) and this may in part be caused by vitamin D deficiency. In many cases the vitamin D deficiency is not detected until clinical osteomalacia or osteoporosis has developed. A poor intake of vitamin D is often associated with a persistence in thinking that butter is better than margarine, when the latter would give them more vitamin D. In people who are housebound there is a further factor, for they are not exposed to sunlight and therefore do not make the vitamin in the skin. A further reason for vitamin D deficiency in the elderly could well be a decreased ability to convert vitamin D to its active metabolite, $1,25\text{-}(OH)_2D_3$ in the kidney. This may result in impairment of calcium absorption, which in turn may account for the fact that calcium intake necessary to establish equilibrium in the elderly exceeds the RDI of 800 mg day^{-1} (Spencer *et al.*, 1982).

Hereditary rickets

Elucidation of the metabolism of vitamin D, combined with the application of the techniques of molecular biology, has led to developments in our knowledge of the specific defects which cause the different forms of hereditary rickets (Table 13.5). The commonest form is vitamin D resistant, or hypophosphataemic rickets, in which there is no real change in vitamin D metabolism. This condition is generally characterized by a phosphate leak in the kidney and accompanied by a blunted adaptive increase in $1,25\text{-}(OH)_2D_3$ in response to hypophosphataemia. It is treated with oral phosphate and $1,25\text{-}(OH)_2D_3$. Another type of renal rickets is due to an inability by the kidney to convert $25\text{-}OH\text{-}D_3$ to

Table 13.5. Genetic forms of rickets.

Description	Transmission	Molecular defect
Vitamin D-resistant hypophosphataemic	Dominant X-linked	Defective, phosphate or transporter gene (kidney)
Vitamin D dependency (Type I)	Autosomal recessive	Defective $1,25\text{-}(OH)_2D_3$ synthesis (kidney)
Vitamin D dependency (Type II)	Autosomal recessive	Target-tissue defective in $1,25\text{-}(OH)_2D_3$ receptor

1,25-$(OH)_2D_3$; presumably because of genetic error in 1-OHase. This is called Type 1 vitamin D-dependent rickets. Successful treatment can be effected with up to 4 mg of vitamin D daily, or 1–2 μg 1,25-$(OH)_2D_3$. In the type II form of the disease, there may be a defective receptor protein in the intestinal cell causing impaired absorption of calcium (Lawson and Muir, 1991). This condition is characterized by hypocalcaemia, secondary hyperthyroidism and rickets. Patients have normally high levels of 1,25-$(OH)_2D_3$ in the blood.

Biochemical Assessment of Vitamin D Status

Vitamin D status is usually assessed from circulating levels of 25-OH-D_3, the major serum metabolite of vitamin D. In general, serum total 25-OH-D_3 reflects the amount of vitamin D in the liver as well as the supply of vitamin D from both synthesis and dietary sources (Gibson, 1990). It has also the longest half-life of all vitamin D metabolites. Serum levels of 25-OH-D_3, however, vary seasonally, with serum concentrations being higher in the summer than in the winter (Table 13.6). This is essentially because of high solar ultraviolet B radiation in the summer months. Serum level of 25-OH-D_3 appears to be generally lower in females than males irrespective of season.

Table 13.6. Effect of season on serum 25-OH-D_3 concentrations (mg ml^{-1}) in elderly persons living in the United Kingdom.

	Summer	Winter
Male	16.2 ± 8.6 (258)	11.1 ± 5.7[1] (250)
Female	13.1 ± 7.0 (199)	9.9 ± 4.1[1] (209)

Source: Dattani *et al.*, 1984, with permission.
[1]$P < 0.005$; figures in parentheses indicate the total number of subjects involved.

Serum 25-OH-D_3 is measured by a competitive protein-binding assay (Haddad and Chyu, 1971). In this assay the vitamin D metabolite is first extracted with an organic solvent (silicic acid). Its recovery is monitored by adding tracer amounts of methyl-^{3}H-cholecalciferol to the serum prior to extraction and chromatography (Duncan and Haddad, 1981).

The normal values of serum 25-OH-D_3 indicative of normal vitamin D status are not yet clearly established. According to some studies, however, concentrations below 3 μg ml^{-1} (7.5 nmol l^{-1}) have been associated with clinical signs of vitamin D deficiency in both children and adults (Gibson, 1990). Concentrations in the range of 3–10 μg ml^{-1} could be interpreted as borderline normal.

References

Bhattacharyya, M.H. and DeLuca, H.F. (1974) Subcellular location of rat liver calciferol-25-hydroxylase. *Archives of Biochemistry and Biophysics* 160, 58–62.

Boyle, I.T., Omdahl, J.L., Gray, R.W. and DeLuca, H.F. (1973) The biological activity and metabolism of 24,25-dihydroxy-vitamin D_3. *Journal of Biological Chemistry* 248, 4174–4180.

Braidman, I.P. (1990) Vitamin D and other extracellular factors in the control of growth. *Proceedings of the Nutrition Society* 49, 91–101.

Brautbar, N. (1986) Osteoporosis: Is $1,25(OH)_2D_3$ of value in treatment? *Nephron* 44, 161–166.

Castillo, L., Tanaka, Y., DeLuca, H.F. and IkeKawa, N. (1978) On the physiological role of 1,24,25-trihydroxyvitamin D_3. *Mineral and Electrolyte Metabolism* 1, 198–207.

Chen, T.C., Castillo, L., Korycka-Dahl, M. and DeLuca, H.F. (1974) Role of vitamin D metabolites in phosphate transport of rat intestine. *Journal of Nutrition* 104, 1056–1060.

Clemens, R.L., Adams, J.S., Henderson, S.L. and Holick, M.F. (1982) Increased skin pigment reduces the capacity of skin to synthesize vitamin D_3. *Lancet* 1, 74–76.

Cummings, S.R., Kelsey, J.L., Nevitt, M.C. and O'Dowd, K.J. (1985) Epidemiology of osteoporosis and osteoporotic fractures. *Epidemiologic Reviews* 7, 178–208.

Dattani, J.T., Exton-Smith, A.N. and Stephen, J.M. (1984) Vitamin D status of the elderly in relation to age and exposure to sunlight. *Human Nutrition: Clinical Nutrition* 38 c, 131–137.

Davies, J. and Dickerson, J.W.T. (1991) *Nutrient Content of Food Portions*. The Royal Society of Chemistry, Cambridge.

Department of Health (1991) Dietary reference values for food energy and nutrients for the United Kingdom. *Report on Health and Social Subjects, No. 41*. HMSO, London.

Duncan, W.E. and Haddad, J.G. (1981) Vitamin D assessment. The assays and their applications. In: Labbe, R.E. (ed.) *Laboratory Assessment of Nutritional Status*. Clinics in Laboratory Medicine 1, 713–727.

Emtage, J.S., Lawson, D.E.M. and Kodicek, E. (1973) Vitamin D-induced synthesis of m.RNA for calcium-binding protein. *Nature* 246, 100–101.

Fraser, D.R. (1980a) Regulation of the metabolism of vitamin D. *Physiological Reviews* 60, 551–613.

Fraser, D.R. (1980b) Vitamin D. In: Barker, B.M. and Bender, D.A. (eds) *Vitamins in Medicine*, 4th edn, Vol. 1. Heinemann Medical Books, London, pp. 42–146.

Fraser, D.R. (1981) Biochemical and clinical aspects of vitamin D function. *British Medical Bulletin* 37, 37–42.

Garabedian, M., Holick, M.F., DeLuca, H.F. and Boyle, I.T. (1972) Control of 25-hydroxy-cholecalciferol metabolism by parathyroid glands. *Proceedings of the National Academy of Sciences, USA* 69, 1673–1676.

Gibson, R.S. (1990) In: *Principles of Nutritional Assessment*. Oxford University Press, New York, pp. 391–397.

Haddad, J.C. and Chyu, K.J. (1971) Competitive protein-binding radio-assay for 25-hydroxycholecalciferol. *Journal of Clinical Endocrinology and Metabolism* 33, 992–995.

Hahn, T.J. and Alvioli, L.V. (1975) Anticonvulsant osteomalacia. *Archives of Internal Medicine* 135, 997–1000.

Haussler, M.R. and McCain, T.A. (1977) Basic and clinical concepts related to vitamin D metabolism and action. *New England Journal of Medicine* 297, 974–983.

Hayworth, J.C. and Dilling, L.A. (1986) Vitamin-D-deficient rickets in Manitoba, 1972–84. *The Canadian Medical Association Journal* 134, 237–241.

Health and Welfare Canada (1990) In: *Nutrition Recommendation.* Canadian Government Publishing Centre, Ottawa.

Henry, H. (1979) Regulation of the hydroxylation of 25-hydroxy-vitamin D_3 *in vivo* and in primary cultures of chick kidney cells. *Journal of Biological Chemistry* 254, 2722–2729.

Henry, H.L. and Norman, A.W. (1984) Vitamin D: metabolism and biological actions. *Annual Review of Nutrition* 4, 493–520.

Holick, M.F. (1986) Vitamin D requirements for the elderly. *Clinical Nutrition* 5, 121–129.

Holick, M.F. (1995) Environmental factors that influence the cutaneous production of vitamin D. *American Journal of Clinical Nutrition* 61, 638–645.

Holland, B., Unwin, I.D. and Buss, D.H. (1989) *Milk products and Eggs.* The fourth supplement to McCance and Widdowson's '*The Composition of Foods*' (4th edn.). The Royal Society of Chemistry, Cambridge; and Ministry of Agriculture, Fisheries and Food, London.

Lawson, D.E.M. and Muir, E. (1991) Molecular biology and vitamin D function. *Proceedings of the Nutrition Society* 50, 131–137.

Lawson, D.E.M., Paul, A.A., Black, A.E., Cole, T.J., Mandal, A.R. and Davie, M. (1979) Relative contributions of diet and sunlight to vitamin D state in the elderly. *British Medical Journal* 2, 303–305.

Lindeman, R.D., Tobin, J. and Shock, N.W. (1985) Longitudinal studies on the rate of decline in renal function with age. *Journal of American Geriatric Society*. 33, 278–285.

Matsuoka, L.Y., Ide, L., Wortsman, J., MacLaughlin, A. and Holick, M.F. (1987) Sunscreens suppress cutaneous vitamin D_3 synthesis. *Journal of Clinical Endocrinology and Metabolism* 64, 1165–1168.

National Research Council (1989) In: *Recommended Dietary Allowances.* National Academy Press, Washington, D.C.

Norman, A.W. (1978) Calcium and phosphate absorption. In: Lawson, D.E.M. (ed.) *Vitamin D.* Academic Press, London, pp. 93–132.

Ornoy, A., Goodwin, D., Noff, D. and Edelstein, S. (1978) 24,25 Dihydroxy vitamin D is a metabolite of vitamin D essential for bone formation. *Nature* 276, 517–519.

Paul, A.A. and Southgate, D.A.T. (1978) *McCance and Widdowson's The Composition of Foods*, 4th revised edition. Her Majesty's Stationery Office, London.

Raisz, L.G., Trummel, C.L., Holick, M.F. and DeLuca, H.F. (1972) 1,25-Dihydroxycholecalciferol: a potent stimulator of bone resorption in tissue culture. *Science* 176, 768–769.

Richelson, L.S., Wahner, H.W., Melton, L.J. and Riggs, B.L. (1982) Relative contributions of ageing and oestrogen deficiency to postmenopausal bone loss. *The New England Journal of Medicine* 311, 1273–1275.

Riggs, B.L., Jowsey, J. and Kelly, P.J. (1976) Role of hormonal factors in the pathogenesis of postmenopausal osteoporosis. *Israel Journal of Medical Science* 12, 616–620.

Riggs, B.L., Wahnen, H.W., Dunn, W.L., Mazess, R.B., Offord, K.P. and Melton, I.J. (1981) Differential changes in bone mineral density of the appendicular and axial skeleton with aging: relationship to spinal osteoporosis. *Journal of Clinical Investigation* 67, 328–335.

Schoentgen, F., Metz-Boutigue, M.J., Jolles, J., Constans, J. and Jolles, P. (1986) Complete amino acid sequence of human vitamin D-binding protein (group-specific component): evidence of a three-fold internal homology as in serum albumin and α-fetoprotein. *Biochimica et Biophysica Acta* 871, 189–193.

Shefer, S., Hauser, S., Bekersky, I. and Mosbach, E.H. (1969) Feedback regulation of bile acid biosynthesis in the rat. *Journal of Lipid Research* 10, 646–655.

Sorensen, O.H., Lumholtz, B., Lund, B.J. and Helmstrand, I.L. (1982) Acute effects of parathyroid hormone on vitamin D metabolism in patients with bone loss of ageing. *Journal of Clinical Endocrinology and Metabolism* 54, 1258–1261.

Spencer, H., Kramer, L. and Osis, D. (1982) Factors contributing to calcium loss in ageing. *American Journal of Clinical Nutrition* 36, 776–782.

Stamp, T.C. (1974) Effects of long-term anticonvulsant therapy on calcium and vitamin D metabolism. *Proceedings of the Royal Society of Medicine* 67, 64–68.

Stanbury, S.W. and Mawer, E.B. (1990) Metabolic disturbances in acquired osteomalacia. In: Cohen, R.D., Lewis, B., Alberti, K.G.M.M. and Denman, A.M. (eds) *The Metabolic and Molecular Basis of Acquired Disease*. Baillière Tindall, London, pp. 1717–1782.

Tsai, K., Heath, H., Kumar, R. and Riggs, B.L. (1984) Impaired vitamin D metabolism with ageing in women: Possible role in pathogenesis of senile osteoporosis. *Journal of Clinical Investigation* 73, 1668–1672.

Tilyard, M.W., Spears, G.F., Thomson, J. and Dovey, S. (1992) Treatment of postmenopausal osteoporosis with calcitrol or calcium. *New England Journal of Medicine* 326, 357–362

Wasserman, R.H. and Taylor, A.N. (1969) Vitamin D_3-induced calcium-binding protein in chick intestinal mucosa. *Science* 152, 791–793.

Wasserman, R.H., Fullmer, C.S. and Taylor, N. (1978) The vitamin D-dependent calcium-binding proteins. In: Lawson, D.E.M. (ed.) *Vitamin D*. Academic Press, London, pp. 133–166.

Webb, A.R. and Holick, M.F. (1988) The role of sunlight in cutaneous production of vitamin D_3. *Annual Review of Nutrition* 8, 325–399.

Vitamin E 14

Chemistry

The term 'vitamin E' is used as a generic designation for eight compounds synthesized by plants. These compounds fall into two classes: tocopherol and tocotrienol exhibiting quantitatively the biological activity of vitamin E. The basic structure consists of a hydroxylated ring system (chromanol ring) and an isoprenoid side chain (Fig. 14.1). The isoprenoid side chain is saturated in the tocols, while it is unsaturated in the trienols. Each class is comprised of four vitamers that differ in the number and position of the methyl groups on the chromanol ring. Vitamins in both classes are designated by the Greek letters α, β, γ, and δ. Biologically the most active is D-α-tocopherol. The higher biological activity of this isomer has been attributed to its optimal tissue storage. Both the isoprenoid side chain and the aromatic ring have been implicated in the antioxidant function of the vitamin. According to Jacob and Lux (1968) the benzene ring may trap free radicals such as exist in the presence of hydrogen peroxide

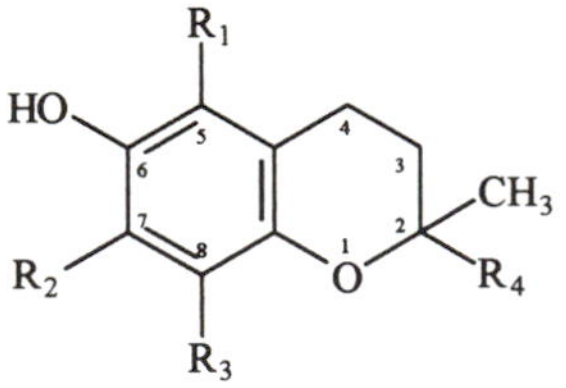

Tocopherols	Tocotrienols	CH_3-position
α	α	5, 7, 8
β	β	5, 8
γ	γ	7, 8
δ	δ	8

$$\text{Tocopherols}: R_4 = CH_2{-}(CH_2{-}CH_2{-}\underset{}{\overset{CH_3}{\overset{|}{C}H}}{-}CH_2)_3{-}H$$

$$\text{Tocotrienols}: R_4 = CH_2{-}(CH_2{-}CH{=}\overset{CH_3}{\overset{|}{C}H}{-}CH_2)_3{-}H$$

Fig. 14.1. Structure of tocopherol vitamers.

and its breakdown products and so prevent the initiation of peroxidation of the polyunsaturated fatty acid component of cellular membranes. Lucy (1972) suggested that the isoprenoid side chain, in addition to being responsible for the fat solubility, may cause stabilization of cellular membranes through physico-chemical interaction with fatty acyl chains of polyunsaturated phospholipids. Tocopherols are preferentially oxidized thus protecting the fats.

Sources

Vitamin E is synthesized only by plants so that plant oils are the main dietary sources with meat and dairy products providing only moderate amounts of the daily need. The amount of vitamin E in foods at the point of consumption is difficult to assess for it depends not only upon the amount originally present in the raw food but also upon the effects of processing, storage and preparation. Because it is fat soluble, vitamin E is not lost by leaching into processing water and is generally stable. However, being an antioxidant in many vegetable oils it is destroyed under oxidizing conditions such as exposure to air and light, accelerated by heat and the presence of copper. Such changes are relatively slow. Thus, potato crisps stored at room temperature lost 48% of their vitamin E in 2 weeks, 70% in 4 weeks and 77% in 8 weeks (Bunnell *et al.*, 1965). Similarly, French fried (chipped) potatoes lost 68% of the vitamin during frozen storage for 4 weeks and 74% when stored for 8 weeks. Free tocopherol is slowly oxidized in air while the esters are more stable. Tocopherol acetate is only 10–20% destroyed under conditions where free tocopherol is completely destroyed (Bunnell *et al.*, 1965). Boiling destroys 30% of the tocopherol in sprouts, cabbage, carrots and leeks and the losses on canning are considerable (Table 14.1; Bender, 1978). However, vegetables are not important sources of vitamin E in the human diet. Flour loses 50% of its vitamin E during bread making due to the use of chlorine dioxide as a flour-bleaching agent.

The approximate concentration of tocopherol in some common foodstuffs is shown in Table 14.2 (Nelson and Fischer, 1980). An assessment of the

Table 14.1. Vitamin E content of fresh and canned vegetables.

	Total tocopherol (mg 100 g^{-1})	α-tocopherol (mg 100 g^{-1})
Fresh green peas	1.73	0.55
Canned green peas	0.04	0.02
Frozen green beans	0.24	0.09
Canned green beans	0.05	0.03
Frozen kernel corn	0.49	0.19
Canned kernel corn	0.09	0.05

Sources: Bunnell *et al.* (1965); Bender (1978).

Table 14.2. Concentrations of total vitamin E and α-tocopherol in some common foods (mg kg^{-1}).

		Total vitamin E[1]	α-Tocopherol[2]
Animal foods	Lard	6–13	
	Butter	10–50	
Plant oils	Soybean	560–1600	
	Maize	530–1620	
	Peanut	200–320	
	Palm	330–730	
	Safflower	250–490	
	Olive	50–150	
Common foods	Meats		0.5–1.6
	Poultry		1.6–4.0
	Fish		6–10
	Fruit		2.3–7.2
	Vegetables		0.5–4.5
	Dairy products		0.4–10.0
	Margarine		280
	Peanuts (roasted)		77
	Potato chips		64
	Oatmeal		30
	Whole wheat bread		20

[1]Chow (1985).
[2]Nelson and Fischer (1980).

biological potency of a food should not only take account of the α-tocopherol content but also of its total tocopherol content and it is usual to express the vitamin E activity of foods in terms of α-tocopherol equivalents; the biological potency of α-tocopherol is therefore 1.0; that of γ-tocopherol is 0.08; α-tocotrienol is 0.21 and γ-tocotrienol is 0.01. Examples of the distribution of different tocopherols and tocotrienols contributing to the total vitamin E activity in grains are shown in Table 14.3.

Absorption and Transport

α-Tocopherol is absorbed unchanged from the small intestine by non-saturable passive diffusion. Tocopheryl esters are first hydrolysed by pancreatic esterase (Bjorneboe *et al.*, 1990). Thus, pancreatic juice together with bile are essential for the absorption of vitamin E. Absorption appears to occur mostly between the upper and middle thirds of the small intestine. The absorption efficiency of α-tocopherol and its esters is generally considered to be variable. In rats given a single bolus of α-tocopherol absorption was approximately 40% (Bjorneboe

Table 14.3. Compounds with vitamin E activity in some grains.

	Tocopherols (μg g^{-1})				Tocotrienols (μg g^{-1})	
Grain	α-	β-	γ-	δ-	α-	γ-
Maize	6–15		29–55		5–10	34–77
Soyabean	1–3		3–33	2–6		trace
Cotton	1–18		5–18			1–2
Oats	4–8	1			10–22	
Milo	4–7		14–17		1	
Barley	8–10	1–2	3–4		23–28	
Wheat	8–12	4–6			2–3	

Sources: Cort *et al.* (1983); Combs (1992).

et al., 1986), whereas when α-tocopheryl acetate was given as a slow continuous infusion (Traber *et al.*, 1986) absorption was 65%. In human studies over 24 hours absorption of α-tocopherol and its acetate ester was in the range of 21–86% (Gallo-Torres, 1980). Limited sample numbers and the variety of experimental approaches used by different investigators make the interpretation of the results of human studies somewhat complicated. Moreover, determination of absorption of vitamin E under experimental conditions may give little real indication of the efficiency of absorption of dietary vitamin E. So far it seems that information about the absorption of vitamin E from foods is lacking.

According to Combs (1992) the absorption of vitamin E shows biphasic kinetics which reflects the initial uptake of the vitamin by existing chylomicrons followed by a time lag due to the need to assemble new chylomicrons. Absorbed vitamin E, like other hydrophobic substances, enters the lymphatic circulation. Whilst the absorption of α- and γ-tocopherols is similar, γ-tocopherol is preferentially excreted in the bile and this accounts for its lower concentration in plasma despite its widespread distribution in the diet. The other tocopherols (β and δ) are poorly absorbed.

In the circulation chylomicrons are hydrolysed by lipoprotein lipase and the resulting remnants are taken up by the liver. α-Tocopherol is then secreted in very low density lipoproteins (VLDL) (Traber *et al.*, 1988). Subsequent breakdown of VLDL liberates α-tocopherol into LDL and HDL. Uptake of the vitamin by peripheral tissues may occur during the breakdown of chylomicrons and VLDL by lipoprotein lipase with or without the participation of the LDL receptor. In rats, it has been found that uptake is most rapid in lung, liver, small intestine, plasma kidney and red cells and slowest in brain, testes, adipose tissue and spinal cord. α-Tocopherol is mostly stored in adipose tissue, liver and muscle (Drevon, 1991). However, the mobilization of α-tocopherol from adipose tissue in response to a dietary deficiency is very slow.

Intracellular transport of the vitamin, at least in the liver, appears to involve specific tocopherol-binding proteins (TBPs). Several TBPs have been described with different binding affinities, for instance, in erythrocyte plasma membrane protein. TBPs have been identified also in liver, heart, brain and intestinal mucosa. Tocopherol has also been found to bind to retinol-binding protein from which it may be displaced by retinol.

The vitamin E in non-adipose cells is localized in their membranes and two pools of the vitamin, one 'labile' and the other 'fixed', differing in their rate of turnover, have been described. 'Labile' pools predominate in those tissues, such as plasma and liver, from which vitamin E is mobilized in conditions of deprivation of the vitamin. In contrast, as mentioned above, the vitamin E of adipose tissue appears to be a 'fixed' pool of only long-term physiological significance. So resistant to release is the vitamin E of adipose tissue that its content may be near 'normal' whilst animals may show clinical signs of vitamin E deficiency. The concentrations of α-tocopherol in human tissues vary greatly (Table 14.4) and are not clearly related to their lipid content.

Table 14.4. Concentrations of α-tocopherol in human tissues.

	α-Tocopherol	
Tissue	μg g^{-1} tissue	μg g^{-1} lipid
Plasma	9.5	1.4
Erythrocytes	2.3	0.5
Platelets	30	1.3
Adipose tissue	150	0.2
Kidney	7	0.3
Liver	13	0.3
Muscle	19	0.4
Heart	20	0.7
Uterus	9	0.7
Ovary	11	0.6
Testis	40	1.0
Adrenal	132	0.7
Hypophysis	40	1.2

Source: Machlin (1984).

Metabolism

Most absorbed tocopherols are transported unchanged to the tissues. The antioxidant function of the vitamin results in its oxidation to α-tocopherolquinone (Fig. 14.2) through a semi-stable intermediate α-tocopheroxyl radical. The first stage in this reaction is reversible whereas the conversion of the intermediate to the quinone is not. The conversion of α-tocopherol to its quinone is virtually a catabolism of the vitamin because the quinone does not possess any vitamin E activity. The quinone can be changed by another reversible reaction to the hydroquinone. This molecule can be conjugated with glucuronic acid and the

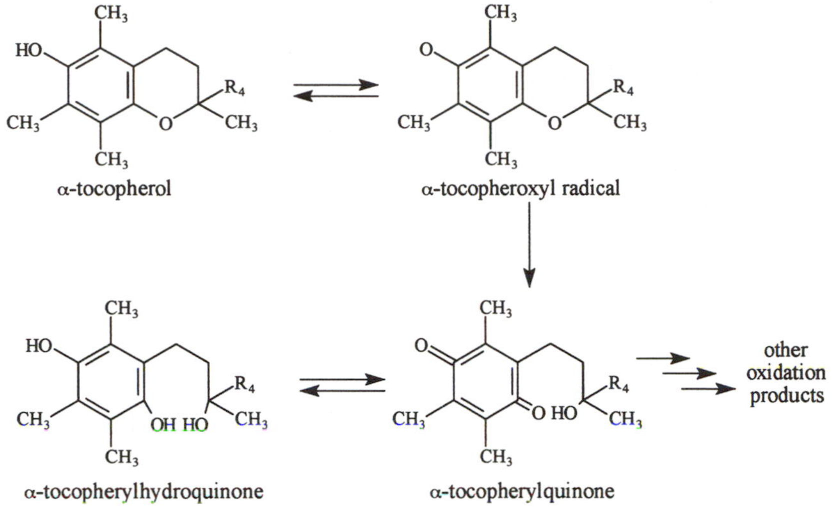

α-tocopherol

α-tocopheroxyl radical

α-tocopherylhydroquinone

α-tocopherylquinone

other oxidation products

Fig. 14.2. Metabolism of α-tocopherol.

product secreted in the bile and excreted in the faeces. This is the major route of excretion of the vitamin; at normal levels of intake of vitamin E less than 1% of the absorbed vitamin is excreted in the urine. The urinary metabolites that have been identified are α-tocopheronic acid and α-tocopheronolactone. These are side-chain oxidation products of tocopherol and may be present as conjugates with glucuronic acid. Some vitamin E may be eliminated via the skin.

The presence of the reversible reaction to the production of the tocopheryl radical has raised the possibility that vitamin E may be recycled and the proposition that a significant portion of the vitamin may be regenerated in this way. According to Combs (1992) two mechanisms have been proposed for this recycling, one with ascorbic acid and the other with thiol as the reducing agent. There are difficulties in accepting that ascorbic acid may be important and its role has not been supported by studies in experimental animals. In support of the second mechanism, it has been shown that the reduced form of glutathione (GSH) can reduce membrane-bound tocopheroxyl in the presence of a tocopheroxyl reductase activity found in the endoplasmic reticulum and mitochondria.

Function

The potent antioxidant properties of vitamin E were first demonstrated by Olcott and Mattill in 1931 and it was later proposed that the major function of the vitamin was the protection of polyunsaturated fatty acids (PUFAs) from oxidation *in vivo* to hydroperoxides. Other oxidation reactions prevented are the conversion of free or protein-bound sulfhydryls to disulphides. However, it was not until more recent years that the precise function of vitamin E was elucidated and its central role in protection against free-radical induced cellular damage recognized. Oxygen radicals are produced as a consequence of the normal process of reduction of oxygen to water and represent by-products of oxidative cellular metabolism. Figure 14.3 shows the series of reactions that take place in the conversion of oxygen to water (Diplock, 1985) and demonstrate the site of formation of the superoxide ($O^{\cdot}_{2}$) and the hydroxy ($OH^{\cdot}$) free radical series. These free radicals are highly reactive species and may attack the double bonds of PUFA chains of membrane phospholipids. The lipoperoxyl free radicals thus formed can attack adjacent PUFA residues and thereby initiate a chain of free radical reactions with widespread harmful consequences to membrane structure. Peroxidation chain reactions are characterized by initiation, propagation and termination stages (Fig. 14.4) with lipid hydroperoxides as the primary products (Halliwell and Gutteridge, 1989). The generally accepted mechanism of peroxide formation involves the formation of a free radical ($R^{\cdot}$) by a PUFA (RH) molecule, followed by the addition of oxygen to form peroxide ($ROO^{\cdot}$). The peroxide can react with another PUFA molecule to produce another free radical, thus the reaction is propagated. The chain reaction is thought to be inhibited

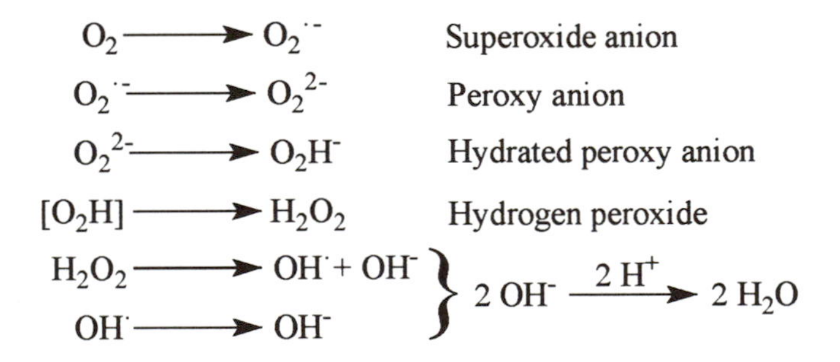

Fig. 14.3. Sites of formation of free radicals in the reduction of oxygen to water. Source: Dickerson and Williams, 1990.

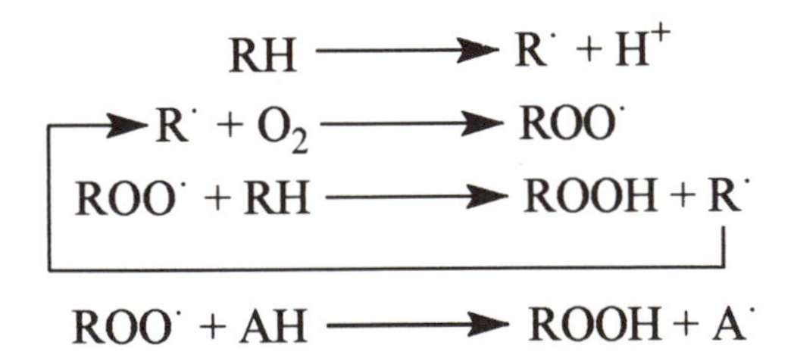

Fig. 14.4. Initiation, propagation and termination of peroxidation chain reaction (RH: PUFA; ROO: peroxide; AH: Vitamin E; ROOH: hydroxy peroxide; A$^{\cdot}$: oxidized vitamin E).

either by replacing vitamin E (AH) or by depleting PUFA. Vitamin E breaks the chain of free radical formation by reacting with the free radicals and converts them to a non-harmful form (A·).

The fatty acyl hydroperoxides formed by the chain reactions as shown in Fig. 14.4 are degraded in the presence of transition metals (Cu^{2+}, Fe^{2+}, Fe^{3+}), haem and haem proteins (cytochromes, haemoglobin, myoglobin) to release radicals that can continue the chain of peroxidation as well as other chain cleavage compounds – malonyldialdehyde, pentane and ethane. The volatile alkanes, pentane and ethane, are excreted by the lungs and can be detected in the breath of vitamin E-deficient subjects.

An intricate series of interrelated mechanisms operate to prevent excessive accumulation of free radicals in tissues; these include enzymes such as superoxide dismutase, and glutathione peroxidase and catalase. Vitamin E plays a vital part in protecting against free radical injury.

It is as yet unclear whether all the actions of vitamin E can be explained in terms of its antioxidant properties and whether the variety of symptoms found in vitamin E-deficient animals and humans can all be attributed to excessive lipoperoxidation and membrane dysfunction. Be this as it may, it is noteworthy that symptoms occur most frequently in those tissues in which there is a high concentration of PUFAs in membrane phospholipids (e.g. the central nervous system) or where oxygen tensions are high (e.g. erythrocytes).

Deficiency

In the 1920s it was shown that feeding rats a semi-purified diet containing vitamins A, B, C and D resulted in malformed embryos in pregnant females and testicular atrophy in male animals. These defects were prevented by fresh lettuce, wheat germ oil and wheat chaff. The common dietary factor in these foods was termed 'vitamin E' and its actions were initially assumed to be confined to the maintenance of reproductive function (Sure, 1924). Later a variety of deficiency symptoms was described in experimental animals. These included necrotizing myopathy in rats, rabbits and guinea pigs, a nutritional encephalomalacia in chicks and a defect in blood vessel permeability termed 'exudative diathesis' (Nelson and Fischer, 1980). In a wide range of animals vitamin E deficiency causes an increase in the tendency for erythrocytes to lyse in a solution of hydrogen peroxide. Of the effects of vitamin E deficiency reported in experimental animals, this effect in erythrocytes is the only feature of deficiency which occurs definitely in man (Horwitt, 1960) and first suggested the possible role of vitamin E in maintenance of membrane stability.

The defect in erythrocytes causing them to lyse in hydrogen peroxide has been shown to be present in erythrocytes from newborn and premature infants and in volunteers receiving a vitamin E-deficient diet. The concentrations of the vitamin in serum of premature infants have been reported to be similar to that

Table 14.5. Serum α-tocopherol concentrations in humans.

Group	α-Tocopherol (mg dl^{-1})
Infants	
Premature	0.23
Premature at 1 month	0.13
Full-term	0.22
2 months breast-fed	0.71
2 months bottle-fed	0.33
5 months old	0.42
2 years old	0.58
Children	
2–12 years	0.72
Adults	
Healthy	0.85
Post-partum mothers	1.33
Diseased cystic fibrotics	
1–19 years	0.15
Biliary atresiacs	
3–15 months	0.10

Adapted from Combs (1992).

in the serum of full-term infants (Table 14.5). However, at one month of age the level reported in premature babies had fallen whereas in full-term babies that had been breast-fed for two months the level had trebled. In the bottle-fed babies whose serum was analysed in the source paper published in 1958 and quoted by Combs (1992), the concentration of vitamin E was less than half that in the babies that had been breast-fed. It might be inferred from these values that the amount of vitamin E incorporated into infant feeds at the time was barely sufficient to meet the needs of rapidly growing infants. In fact, in 1966, Hassan and Hashim reported a syndrome in premature infants associated with oedema, skin lesions and haemolytic anaemia that was reversed by vitamin E. This syndrome has been attributed to feeding a formula which was not only deficient in vitamin E but also abundant in PUFAs (Bell and Filer, 1981).

Serum vitamin E levels in postpartum mothers are higher than in non-pregnant females. In fact, serum vitamin E levels increase during pregnancy whilst fetal levels remain low at about 25% of the maternal level. This difference suggests that there is a transplacental barrier to the transport of the vitamin.

The newborn infant has a greater susceptibility to enhanced free radical generation because of the relatively oxygen-rich environment into which it is born compared with conditions in the uterus. Evidence supporting increased lipoperoxidation in infants compared with adults has been obtained from measurements of breath hydrocarbon (ethane and pentane) (Wispe *et al.*, 1985) and serum malonaldehyde (McCarthy *et al.*, 1981). The situation in premature infants is exacerbated if 100% oxygen is used for respiratory distress syndrome.

It has been suggested that vitamin E supplements should be given prophylactically in the prevention of intracranial haemorrhage (ICH), bronchopulmonary dysplasia (BD) and retrolental fibroplasia (RLF) since these conditions are associated with oxygen therapy in premature infants (Bell, 1987). However, strongest evidence for beneficial effects has been found for RLF (Muller, 1987). Supplements reduce the severity of the condition but not its incidence. Vitamin E should be used with caution in infants (Muller, 1987) and the form in which it is administered carefully considered (Bell, 1987).

Since circulating vitamin E is associated with lipoproteins, and particularly with β-lipoprotein, it follows that the serum tocopherol level varies not only with tocopherol status of the individual but also with the plasma lipid level. It has therefore been suggested that a more meaningful way of expressing serum tocopherol is as a ratio to the total lipid (Horwitt *et al.*, 1972). A further consequence of the vitamin E/tocopherol association is that low serum levels are found consistently in patients with severe malabsorption, of which cystic fibrosis is a good example (Table 14.5). Low values are also found in infants with biliary atresia, small bowel disease, pancreatic disease, gastric surgery, alcoholism, liver disease and obstructive jaundice (Losowsky, 1979). Conversely, high plasma vitamin E levels occur in patients with hyperlipidaemia conditions such as hypothyroidism, diabetes and hypercholesterolaemia.

Patients with cystic fibrosis have been reported to show degeneration changes in the posterior column of the spinal cord which were similar to the cord lesions observed in vitamin E-deficient rats. Children with chronic cholestasis have been reported to show neurological dysfunction which was closely correlated with vitamin E status (Sokol *et al.*, 1985a) and which responded to vitamin E supplements (Sokol *et al.*, 1985b). This evidence, combined with that derived from studies in experimental animals, supports a role for lipoperoxides in the pathogenesis of vitamin E induced CNS disease. Myopathies, particularly exercise-induced myopathy, may also be the result of free-radical generation, secondary to impaired vitamin E function (Jackson, 1987).

Vitamin E deficiency can result from dietary deficiency or impaired absorption of the vitamin. In addition, there is evidence that several other dietary factors affect the need for the vitamin. From the above discussion it is clear that the requirement for vitamin E increases with the amount of dietary PUFAs. Other factors which increase the requirement are deficiencies of sulphur-containing amino acids, deficiencies of copper, zinc and/or manganese and deficiency of riboflavin (Combs, 1992). By contrast, selenium spares vitamin E and decreases the requirement. Vitamin E can be replaced by several lipid-soluble synthetic antioxidants (e.g. butylated hydroxytoluene, butylated hydroxyanisole and *N,N′*-diphenyl-*p*-phenylenediamine, abbreviated to BHT, BHA and DPPD respectively). To some extent, vitamin E can be replaced by the water-soluble antioxidant, vitamin C. The sparing effect of the latter nutrient may be related to its function in the reductive recycling of vitamin E.

In experimental animals vitamin E deficiency has been shown to result in a variety of conditions affecting the neuromuscular, vascular and reproductive systems. Some conditions such as the more general ones of loss of appetite, reduced growth and fetal death, besides responding to vitamin E also respond to selenium and antioxidants. Others, such as myopathy in striated and smooth muscle, liver necrosis, testicular degeneration and exudative diathesis in chicks, respond also to selenium. Conditions responding only to vitamin E include encephalomalacia in chicks, cataract in rats and anaemia and intraventricular haemorrhage in premature human infants.

Requirements

The assessment of a dietary requirement for vitamin E is complicated by a number of factors. There is a clear direct relationship between the dietary intake of PUFAs and the requirement for the vitamin and there is an equally clear relationship between the tissue content of PUFAs and the dietary requirement. In foods it is commonly found that those which contain large amounts of PUFAs also contain large amounts of vitamin E. Thus, individuals who consume such PUFA-rich foods will also receive a commensurate amount of vitamin E. The mixed diets of individuals in western societies may vary greatly in their content of PUFAs. A survey in the UK (Gregory *et al.*, 1990) gave values for the 2.5 and 97.5 centile intakes of *n*-6 PUFA by men of 5.1 and 29 g day^{-1}. The amount of vitamin E required by those with the higher PUFA intakes would be quite large.

Because of these widely differing requirements based solely on the PUFA intake it has been generally considered more realistic to give ranges of acceptable intake rather than a fixed level. Of 1629 subjects included in the survey by Gregory *et al.* (1990), the 2.5 and 97.5 centiles of intakes from 7-day weighed records were 3.5 and 19.5 (median 9.3) mg α-tocopherol equivalent day^{-1} for men and 2.5 and 15.2 (median 6.7) mg day^{-1} for women. Intakes of 3.8–6.2 mg day^{-1} appear to be satisfactory for pregnant and lactating women. Daily intakes of 4 mg and 3 mg day^{-1} α-tocopherol can be adequate for men and women respectively (Department of Health, 1991). Corresponding RDAs in the US are 10 and 8 mg day^{-1} α-tocopherol equivalents.

Some authorities choose to calculate requirements for vitamin E from the PUFA content of the diet. There is no general agreement on what factor should be used but 0.4 mg α-tocopherol g^{-1} dietary PUFA in a normal animal diet appears to be satisfactory (Bieri and Evarts, 1973). If PUFA supplements are taken they should always be accompanied by additional vitamin E.

Infants that are breast-fed have an average daily intake of 2.7 mg α-tocopherol equivalents and in the UK it has been recommended that the vitamin E content of milk formulas shall not be less than 0.3 mg tocopherol equivalents/100 ml of reconstituted feed and not less than 0.4 mg g^{-1} of PUFA (Department of Health, 1991).

Other factors known to affect the requirement for vitamin E in the diet include the content of selenium. This trace element is thought to function as an integral part of glutathione (GSH) peroxidase, an enzyme that reduces toxic lipid peroxides to hydroxy acids (Combs and Combs, 1984). Selenium and vitamin E are complementary in terms of their ability to scavenge free radicals (Fig. 14.5). Dietary selenium can thus compensate for vitamin E requirements.

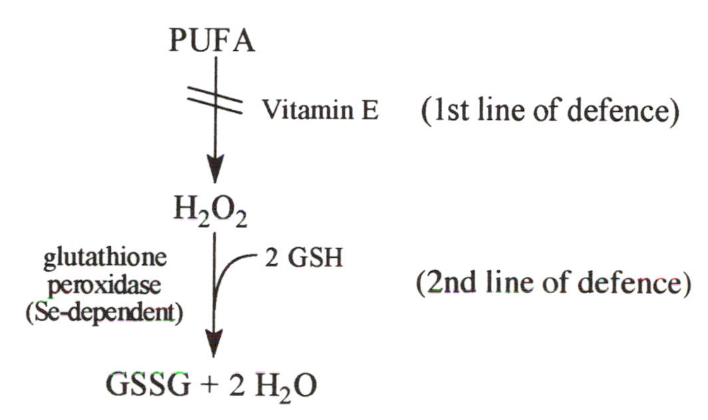

Fig. 14.5. Complementary effects of vitamin E and selenium in reducing free-radical generation from polyunsaturated fatty acid (PUFA).

Assessment of Vitamin E Status

Measurement of serum of plasma concentration of α-tocopherol provides the simplest and most direct evidence of vitamin E deficiency. Values of 5–20 μg ml^{-1} for children of 12 years and older and adults and values of 3–15 μg ml^{-1} for children under 12 years of age are acceptable. If expressed in relation to serum lipid concentration, then acceptable values are 0.8 mg g^{-1} and 0.6 mg g^{-1} respectively.

Another widely used indicator of vitamin E status, to which reference has already been made, is the extent of haemolysis of red cells in the presence of hydrogen peroxide. A high degree of haemolysis accompanies vitamin E deficiency (i.e > 20%). Meticulous care is essential in carrying out this test (Losowsky, 1979) and high peroxide is not specific for vitamin E deficiency. In addition, there are two functional tests which are thought to be useful in determining vitamin E status in humans. One is an *in vivo* test in which pentane, a hydrocarbon gas, is measured as a peroxidized product of PUFA in the body. The second is an *in vitro* test in which peroxidation of PUFA of erythrocytes exposed to hydrogen peroxide is measured by determining generated malonaldehyde.

References

Bell, E.F. (1987) History of vitamin E in infant nutrition. *American Journal of Clinical Nutrition* 46, 183–186.

Bell, E.F. and Filer, L.J., Jr. (1981) The role of vitamin E in the nutrition of premature infants. *American Journal of Clinical Nutrition* 34, 414–422.

Bender, A.E. (1978) Vitamins. In: *Food Processing and Nutrition*. Academic Press, London, pp. 27–57.

Bieri, J.G. and Evarts, R.P. (1973) Tocopherols and fatty acids in American diets. *Journal of American Dietetics Association* 62, 147–150.

Bjorneboe, A., Bjorneboe, G-E.Aa, Bodd, E., Hagen, B.F., Kveseth, N. and Drevon, C.A. (1986) Transport and distribution of alpha-tocopherol in lymph, serum and liver cells in rats. *Biochimica Biophysica Acta* 889, 310–315.

Bjorneboe, A., Bjorneboe, G-E.Aa. and Drevon, C.A. (1990) Absorption, transport and distribution of vitamin E. *Journal of Nutrition* 120, 233–242.

Bunnell, R.H., Keating, J., Quaresimo, A. and Parman, G.K. (1965) Alpha-tocopherol contents of foods. *American Journal of Clinical Nutrition* 17, 1–10.

Chow, C.K. (1985) Vitamin E and blood. *World Review of Nutrition and Dietetics* 45, 133–166.

Combs, G.F. (1992) Vitamin E. In: *The Vitamins*. Academic Press, London, pp. 179–203.

Combs, G.F. and Combs, S.B. (1984) The nutritional biochemistry of selenium. *Annual Review of Nutrition* 4, 257–280.

Cort, W.M., Vicente, T.S., Waysek, E.H. and Williams, B.D. (1983) Vitamin E content of foodstuffs determined by high pressure liquid chromatography fluorescence. *Journal of Agricultural and Food Chemistry* 31, 1330–1333.

Department of Health (1991) *Dietary Reference Values for Food Energy and Nutrients for the United Kingdom*. Report on Health and Social Subjects 41. Her Majesty's Stationery Office, London.

Dickerson, J.W.T. and Williams, C.M. (1990) Vitamin-related disorders. In: Cohen, R.D., Lewis, B., Alberti, K.G.M.M. and Denman, A.M. (eds) The Metabolic and Molecular Basis of Acquired Disease, Volume 1. Baillière Tindall, London, pp. 634–669.

Drevon, C.A. (1991) Absorption, transport and metabolism of vitamin E. *Free Radical Research Communication* 14, 229–246.

Gallo-Torres, H.E. (1980) Absorption, transport and metabolism. In: Machlin, L.J. (ed.) *Vitamin E: A Comprehensive Treatise*. Marcel Dekker, New York, pp. 170–267.

Gregory, J., Foster, K., Tyler, H. and Wiseman, M. (1990) Dietary and Nutritional Survey of British Adults. Her Majesty's Stationery Office, London.

Halliwell, B. and Gutteridge, J.M.C. (1989) *Free Radicals in Biology and Medicine*, 2nd edn. Clarendon Press, London.

Hassan, H. and Hashim, S.A. (1966) Premature infant syndrome in premature infants associated with low plasma Vitamin E levels and high polyunsaturated fatty acid diet. *American Journal of Clinical Nutrition* 19, 147–157.

Horwitt, M.K. (1960) Vitamin E and lipid metabolism in man. *American Journal of Clinical Nutrition* 8, 451–461.

Horwitt, M.K., Harvey, C.C., Dahm, C.H., Jr. and Searcy, M.T. (1972) Relationship between tocopherol and serum lipid levels for determination of nutritional adequacy. *Annals of New York Academy of Science* 203–236.

Jackson, M.J. (1987) Muscle damage during exercise; possible role of free radicals and protective effect of vitamin E. *Proceedings of Nutritional Science* 46, 77–480.

Jacob, H.S. and Lux, S.E. (1968) Degradation of membrane phospholipids and thiols in peroxide hemolysis: studies in vitamin E deficiency. *Blood* 32, 549–568.

Losowsky, M.S. (1979) Vitamin E in human nutrition. In: Taylor, T.G. (ed.) *The Importance of Vitamins to Human Health*. MTP Press, Lancaster, pp. 101–110.

Lucy, J.A. (1972) Functional and structural aspects of biological membranes: A suggested role for vitamin E in the control of membrane permeability and stability. *Annals of New York Academy of Science* 203, 4–11.

McCarthy, K., Bhogal, M., Narch, M. and Hart, D. (1981) Pathogenic factors in bronchopulmonary dysplasia. *Pediatric Research* 18, 483–488.

Muller, D.P.R. (1987) Free radical problems of the newborn. *Proceedings of Nutritional Science* 46, 69–75.

Nelson, J.S. and Fischer, V.W. (1980) Vitamin E. In: Barker, B.M. and Bender, D.A. (eds) *Vitamins in Medicine*, 4th edn, Vol. 1. William Heinemann, London, pp. 147–171.

Olcott, H.S. and Mattill, H.A. (1931) The unsaponifiable lipids of lettuce. II Fractionation. *Journal of Biological Chemistry* 93, 59–64.

Sokol, R.J., Guggenheim, M.A. and Heubi, J.E. (1985a) Frequency and clinical progression of the vitamin E deficiency neurologic disorder in children with prolonged neonatal cholestasis. *American Journal of Diseased Children* 139, 1211–1215.

Sokol, R.J., Guggenheim, M.A. and Iannaccone, S.T. (1985b) Improved neurologic function after long term correction of vitamin E deficiency in children with chronic cholestasis. *New England Journal of Medicine* 313, 1580–1586.

Sure, B. (1924) Dietary requirements for production. II. The existence of a specific vitamin for reproduction. *Journal of Biological Chemistry* 58, 693–709.

Traber, M.G., Kayden, H.J., Balmer-Green, J. and Green, M.H. (1986) Absorption of water-miscible forms of vitamin E in a patient with cholestasis and in thoracic duct-cannulated rats. *American Journal of Clinical Nutrition* 44, 914–923.

Traber, M.G., Ingold, K.U., Burton, G.W. and Kayden, H.J. (1988) Absorption and transport of deuterium-substituted 2R, 4′R, 8′R α-tocopherol in human lipoproteins. *Lipids* 23, 791–797.

Wispe, J.R., Bell, E.F. and Roberts, R.J. (1985) Assessment of lipid peroxidation in newborn infants and rabbits by measurement of expired ethane and pentane: Influence of parenteral lipid infusion. *Pediatric Research* 19, 374–379.

15 Vitamin K

Vitamin K was first discovered in 1935 by a Danish scientist (Dam, 1935) who identified it as the fat-soluble factor necessary for the coagulation of blood. Since the Danish word for the process is spelled Koagulation, the vitamin was later designated as vitamin K, using the first letter of the word. Chemically, vitamin K is not a single substance but a homologous group of fat-soluble vitamins consisting of 2-methyl-1,4-naphthoquinone derivatives.

Phylloquinone, identified as 2-methyl-3-phytyl-1,4-naphthoquinone, is designated as vitamin K_1 (Fig. 15.1). It is the only naturally occurring homologue of vitamin K, synthesized by plants. There is a second series of vitamin K, synthesized by various Gram-positive bacteria, which are called menaquinones, and collectively designated as vitamin K_2. The menaquinone family of K_2 homologues is a large series of vitamins containing unsaturated side chains in the 3-position of the 2-methyl-1,4-naphthoquinone nucleus; the side chains differ in the number of isoprenyl units (Fig. 15.1). Most of the menaquinones contain 6–10 isoprenoid units. In addition to these natural types of vitamin K (i.e. K_1 and K_2), several synthetic compounds containing the 2-methyl-1,4-naphthoquinone structure without the side chain exhibit vitamin K activity (Fig. 15.1). These include menadione or vitamin K_3 and menadiol or vitamin K_4. Menadione, the most potent vitamin K, is available in water-soluble form as its sodium bisulphide derivative, while menadiol, a reduced form of menadione, is available as its sodium diphosphate salt. Menadiol, menadiol sodium diphosphate and menadione sodium bisulphite are all converted *in vivo* to menadione.

Sources

Phylloquinone is the only form of vitamin K present in green and leafy vegetables, while animal tissues contain a mixture of homologues including phylloquinone and menaquinones. The distribution of vitamin K homologues in

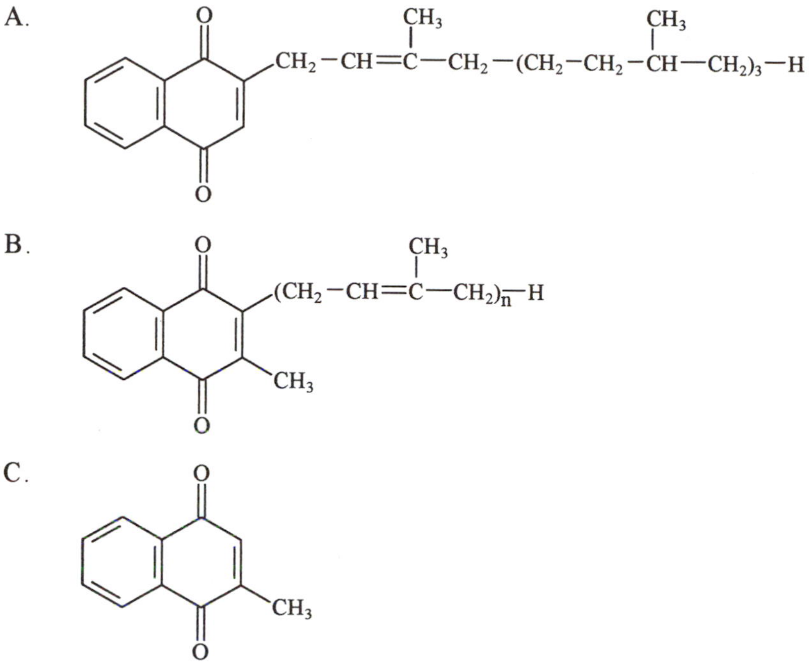

Fig. 15.1. Structural formulas of (A) phylloquinone, K_1; (B) menaquinone, K_2; and (C) menadione, K_3.

animal tissues generally reflects dietary and bacterial sources of the vitamin present in the intestinal tract. The amount synthesized in the gut contributes significantly towards the daily requirement of the vitamin. Thus in conventional rats, the vitamin K requirement is about 10 $\mu g\ kg^{-1}\ day^{-1}$, whereas in germ-free rats the requirement is more than doubled to about 25 $\mu g\ kg^{-1}\ day^{-1}$ (Lefevere *et al.*, 1982).

Vitamin K is widely distributed, the best sources being alfalfa, cabbage, and leafy vegetables, such as spinach. Meats, especially liver, eggs, and cheese provide some vitamin K. Generally speaking, plant foods account for most of the vitamin obtained from the diet.

Absorption and Metabolism

According to animal studies, phylloquinone, the predominant form of dietary vitamin K, is absorbed in the proximal small intestine by a saturable, energy dependent process (Olson, 1984). On the other hand, menaquinones as well as menadione are believed to be absorbed by passive diffusion in the distal intestine and colon. As in the case of other fat-soluble vitamins, absorption of vitamin K is enhanced by the presence of both bile salts and pancreatic juice. Efficiency of absorption in humans may vary from 10 to 70% depending upon these factors as well as the extent of the enterohepatic circulation generally characteristic of isoprenoid lipids (Olson, 1987). Vitamin K along with other lipids, is absorbed as micelles into the luteals.

Unlike vitamins K_1 and K_2, menadione (K_3) and menadiol (K_4) are commercially available in their water-soluble forms as sodium diphosphate and sodium bisulphate derivatives, respectively. These water-soluble forms of vitamin K are readily absorbed when administered orally.

Absorbed vitamin K becomes part of chylomicrons and is carried by chylomicron remnants to the plasma where it is associated with lipoproteins, and subsequently transported to the liver and other organs such as adrenal glands, lungs, bone marrow, and kidneys. As much as 50% of a parenterally or 20% of an orally administered dose of phylloquinone may appear in the liver within 1–2 hours (Dam *et al.*, 1982). Liver vitamin K is usually half in the form of phylloquinone and half as bacterial menaquinones.

Although vitamin K is rapidly concentrated in the liver, there appears to be a rapid turnover, and hence the average body pools are believed to be very small. Phylloquinone is metabolized to various oxygenated derivatives, yielding carboxylic acids that are conjugated with glucuronic acid. Menadione is believed to be rapidly conjugated with sulphate, phosphate and glucuronide (Olson, 1984). The conjugated products of vitamin K are lost mainly in the bile but also partly in the urine. Overall, phylloquinone is degraded more slowly than menadione.

Functions

The clotting of blood

The best known function of vitamin K is to catabolize the synthesis of prothrombin by the liver. In the absence of vitamin K a hypoprothrombinaemia occurs in which blood clotting time may be greatly prolonged. The effect of vitamin K in alleviation of hypoprothrombinaemia, however, is dependent upon the ability of the hepatic parenchyma to produce prothrombin. Hypoprothrombinaemia due to liver damage cannot be corrected by vitamin K.

Blood coagulation is a highly complex process, the mechanism of which is still not fully understood. The process is thought to involve cells, such as thrombocytes, platelets, and erythrocytes, numerous protein factors, and Ca^{2+} ion (Fig. 15.2). Vitamin K is known to be involved in the hepatic synthesis of at least four of the protein factors, which include prothrombin (factor II), proconvertin (factor VII), thromboplastin (factor IX), and the Stuart–Prower factor (factor X).

The blood clot is formed by fibrinogen (factor I), a soluble protein present in the plasma. This protein is transformed to an insoluble network of fibrous material called fibrin. The change of fibrinogen into fibrin is caused by thrombin which in blood exists as prothrombin. The conversion of prothrombin to thrombin depends on the action of plasma and tissue thromboplastin, Ca^{2+} ions and other clotting factors (Fig. 15.2).

Recent advances indicate that vitamin K is necessary for formation of Ca^{2+} binding sites on prothrombin (Olson, 1984). This is essential for prothrombin to be bound to phospholipids so that it is activated to thrombin for participating in the coagulation process. The Ca^{2+} binding sites of prothrombin are formed by the introduction of a second carboxyl group into the glutamyl side chains located in the amino-terminal region of the protein (Fig. 15.3). Once carboxylated, the glutamates are referred to as γ-carboxyglutamic acid (GLA).

In the presence of dicoumarol (Fig. 15.4), a potent antagonist to vitamin K, the prothrombin that is produced *in vivo* has a very low Ca^{2+} binding capacity. In this inactive prothrombin, glutamate residues exist in place of GLA. When

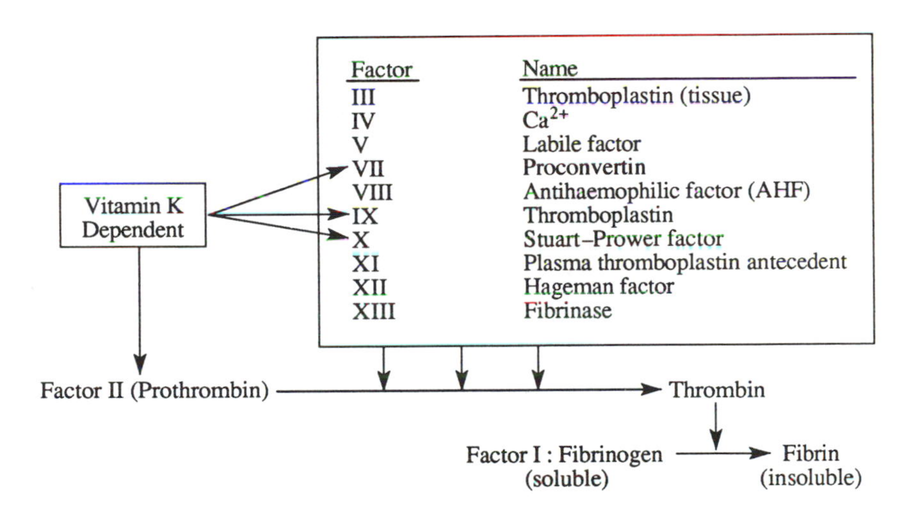

Fig. 15.2. Vitamin K-dependent clotting factors in blood coagulation.

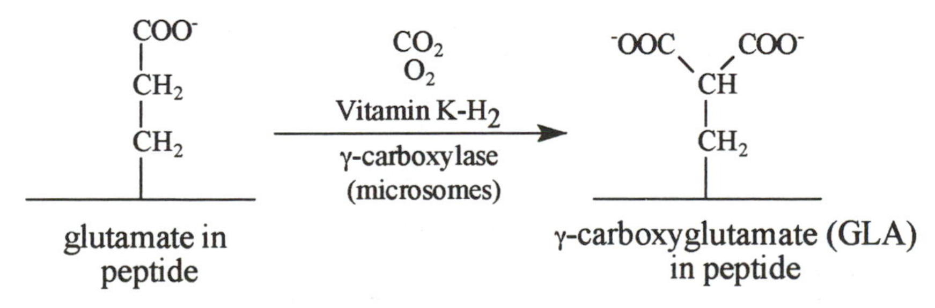

Fig. 15.3. Synthesis of γ-carboxyglutamic acid.

the action of vitamin K is blocked by dicoumarol, calcium ions cannot bind to prothrombin because the protein lacks the added carboxyl groups.

Like prothrombin, factors VII, IX, and X have also been found to contain a series of glutamic acid residues; vitamin K is needed for the carboxylation of these residues. In addition to these coagulation factors, four additional γ-carboxy-containing proteins have been isolated, which have the blood clotting activity and the amino acid sequence homologous to that of prothrombin for residues, and all require Ca^{2+} ions for activity (Olson, 1984). These new clotting factors are protein C, protein S, protein M, and protein Z (Table 15.1).

The vitamin K-dependent carboxylation is carried out by a liver microsomal enzyme through a molecular mechanism that is not fully understood. The γ-carboxylation is believed to require reduced vitamin K (or its epoxide) and CO_2.

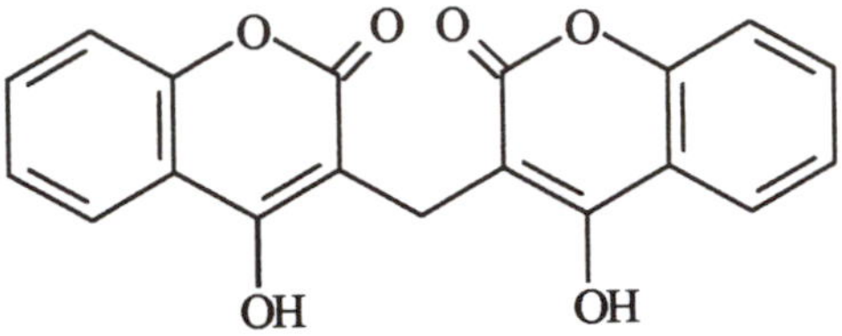

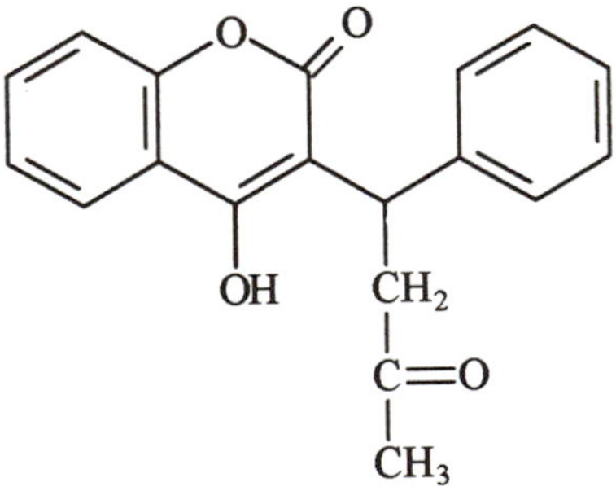

Fig. 15.4. Antagonists of vitamin K. (A) Dicoumarol; (B) warfarin.

Table 15.1. Some characteristics of GLA containing blood clotting proteins.

	Clotting factors							
	II	VII	IX	X	C	S	M	Z
Plasma (μg ml^{-1})	100	1	3	20	10	1	< 1	< 1
Mol. wt × 10)	72	46	55	55	57	69	50	55
Number of GLA residues	10	10	12	12	11	10	(+)	13

Modified from Olson (1984).

This process seems to be coupled with simultaneous epoxidation of vitamin K hydroquinone, the active form of the vitamin. There is an epoxide reductase in liver microsomes that reduces vitamin K epoxide back to the hydroquinone (Suttie, 1985).

The carboxylation of glutamic acid is not potentiated by biotin which is a vitamin known to be involved in carboxylations of many substrates (see Chapter 8). Thus, neither the animals with biotin deficiency exert vitamin K deficiency, nor avidin, a potent inhibitor of biotin-dependent carboxylations, affect carboxylation of glutamic acid. Unlike biotin-catalysed CO_2 fixation, vitamin K-dependent carboxylation does not require ATP.

Other potential roles

Recent reports have indicated that the distribution of GLA is much more widespread than was once realized. Proteins containing GLA have been identified in tissues, such as bone and kidney (Price, 1988). It seems that vitamin K, in addition to its well known involvement in the blood coagulation process, may participate in many other important physiological functions. There appear to be at least two GLA-containing proteins in bone, called bone GLA protein (BGP) and matrix GLA protein (MGP), and one in the cortex of the kidney, which is termed kidney GLA protein (KGP). The function of these proteins has not yet been clearly defined, but there has been an accumulation of evidence suggesting that they may participate in the modulation of bone mineralization.

BGP, also called osteocalcin, contains 49 amino acids including three GLA residues, and has a molecular weight of 5700. Using embryonic chick bones, it has been shown that the appearance of osteocalcin in bones coincides with the beginning of mineralization (Hauschka *et al.*, 1978). Injection of vitamin K antagonists into eggs containing developing embryos has also been shown to result in a reduction in the GLA content of osteocalcin by 20–50%. Furthermore, when rats or chicks are made vitamin K-deficient, the GLA content of isolated osteocalcin appears to drop considerably. All this evidence points to the possibility that the vitamin K-deficient osteocalcin may be an important factor for bone metabolism. Indeed, using cultured osteosarcoma cells, it has been shown that the synthesis of osteocalcin is regulated by 1,25-$(OH)_2D_3$ (Price and Baukol, 1980), the active metabolite of vitamin D (see Chapter 13). Although the physiological role of osteocalcin is equivocal, this GLA-containing protein, which constitutes 1% of total bone protein or 10–20% of non-collagen protein, is considered to modulate calcium deposition in bone matrix.

MGP is another well-characterized GLA protein, which contains about 80 amino acids including five GLA residues. The function of this protein is uncertain but may relate to the action of 1,25-$(OH)_2D_3$ (Price and Baukol, 1980), and therefore the mobilization and deposition of bone calcium. Further searches for GLA-containing protein have led to the discovery of another protein (KGP) in the kidney. This protein is not as well characterized as either BGP or MGP. The kidney protein is thought to contain three to four residues of GLA and have a molecular weight of 1800. KGP has also been suggested to be involved in resorption of Ca^{2+} by the kidney tubules (Lian *et al.*, 1978), a function also related to vitamin D action. This GLA-containing protein has been identified in calcium oxalate renal stones in man (Lian and Prein, 1976). It may solubilize calcium salts in urine.

In recent years, proteins containing GLA residues have been identified in tissues other than the liver, kidney and bone. The distribution of vitamin K-dependent GLA in diverse tissues is clearly an indication of vitamin K having a diversified physiological role. It is likely that we are just beginning to understand the full biological potential of the phyllo- and menaquinones.

Deficiency

A dietary deficiency of vitamin K is not common since the vitamin is fairly well distributed in foods, and intestinal microorganisms synthesize a significant amount of vitamin K in the intestine. Dietary vitamin K is essentially present in the body in its oxidized form, quinone. However, an efficient vitamin K salvage pathway ensures optimal conversion of vitamin-2,3-epoxide back to its active hydroquinone form following the γ-glutamyl carboxylation (Olson, 1984). The formation of the epoxide is an obligatory step in the action of vitamin K in the biosynthesis of prothrombin. A dithiol-dependent quinone reductase is a major pathway for the regeneration of hydroquinone, an active form of vitamin K (Fig. 15.5). The reduced form of the vitamin may also be formed by the action of a number of NAD(P)H-dependent dehydrogenases. The existence of this system for recycling vitamin K also explains the relatively rare occurrence of vitamin K deficiency despite extremely low body stores of the vitamin in man.

Isolated cases of deficiency are seen, but the nature of deficiency appears to be primarily of a secondary nature. Thus, vitamin K deficiency may result from an inadequate absorption or impaired gut synthesis, or as a result of drugs which interfere with its availability. The secondary factors other than drug therapy (discussed in Chapter 18) that could contribute toward vitamin K deficiency, and the situations where the deficiency can be expected, are discussed in the following sections.

Malabsorptive states

Vitamin K requires the presence of bile salts for its absorption from the small intestine. Hence, any disorder that retards the delivery of bile to the small intestine, such as obstructive jaundice or biie fistula, reduces the absorption of vitamin K from the intestine. For this reason any patient being operated on for

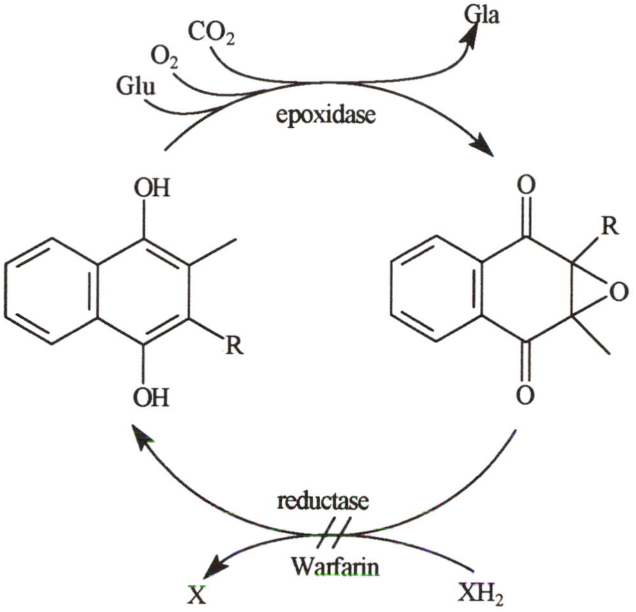

Fig. 15.5. Vitamin K salvage pathway (XH_2:NAD(P)H^+ or RSH–HSR (dithiol)).

obstructive jaundice should receive parenteral vitamin K_1 (5 mg day^{-1}) for three days prior to surgery (Suttie and Olson, 1984). Vitamin K deficiency also has been found to occur in other malabsorptive states such as coeliac disease, Crohn's disease, bowel resection, chronic pancreatic injury, and ulcerative colitis. The malabsorptive states leading to hypoprothrombinaemia can be successfully treated by daily oral administration of 10 mg phylloquinone or menadione.

Hepatic insufficiency

In any form of acute or chronic liver disease, there may be a decreased utilization of vitamin K for the production of the vitamin K-dependent clotting factors, usually as a result of destruction of rough reticulum in the hepatocyte. Patients with hypoprothrombinaemia related to hepatic disorders usually respond to a daily parenteral dose of 10 mg of vitamin K for three days. Serious hepatocellular damage should be suspected if no response to the treatment is noted. Such a situation can be expected in alcohol-induced liver damage.

Vitamin K deficiency in the alcoholic is most frequently manifested by bleeding diathesis accompanied by hypoprothrombinaemia. Reduced prothrombin status of the alcoholic is generally unresponsive to vitamin K because the underlying causes of the manifestation are hepatocellular damage and interference with protein and prothrombin synthesis, rather than the vitamin K deficiency *per se*.

Antagonistic effects of vitamins A and E

According to experimental studies vitamin A, when taken in excess, may have the potential to antagonize vitamin K (Olson, 1984). In rats, hypervitaminosis A appears to precipitate hypoprothrombinaemia accompanied by haemorrhages, and these conditions can be prevented by the administration of vitamin K. However, excess vitamin A induces vitamin K deficiency only when vitamin A is administered orally; no effect of parenterally administered vitamin A is noted. It seems, therefore, that the effect of excess vitamin A is mediated through interfering with the absorption of vitamin K from the gut.

Excess vitamin E may also antagonize the action of vitamin K (Olson, 1984). Human subjects taking megadoses of vitamin E become hypersensitive to the coumarin anticoagulant drugs. The antagonistic effect of vitamin E on vitamin K, however, has yet to be determined. It is not clear if the antagonism is exercised at the level of absorption or of metabolism.

Newborn infants

Both term and preterm babies have inadequate stores of vitamin K, as indicated by its low circulatory and hepatic levels, accompanied by reduced concentrations of blood clotting factors. A number of factors may account for this (Sann *et al.*, 1985; Olson, 1987). Vitamin K does not cross the placental barrier effectively enough from the maternal circulation. In addition, at birth, infants do not normally establish vitamin K-producing bacteria in their intestines. It takes only a few days, however, for the gut flora to become established and begin supplying vitamin K to the infants. Infants are also subject to borderline intakes of vitamin K since milk is generally a poor source. Breast-fed infants are at an even greater disadvantage than bottle-fed infants because human milk is often not produced in significant amounts for several days and contains about one-quarter of vitamin K found in cow's milk (60 $\mu g\ l^{-1}$) (Haroon *et al.*, 1982).

Vitamin K-deficiency-associated hypoprothrombinaemia usually appears during the first week of life, manifested by ecchymoses, nasal or gastrointestinal bleeding, or excessive bleeding at circumcision and umbilical stump (Fig. 15.6). The state of hypoprothrombinaemia can be very severe in the presence of factors such as obstructive jaundice, diarrhoea, treatment with antibiotics, and prolonged breast-feeding (Lane and Hathaway, 1985). The severe signs, generally, involve acute intracranial haemorrhage, widespread deep ecchymoses, excessive bleeding at puncture sites or surgical incisions, and sometimes dysfunction

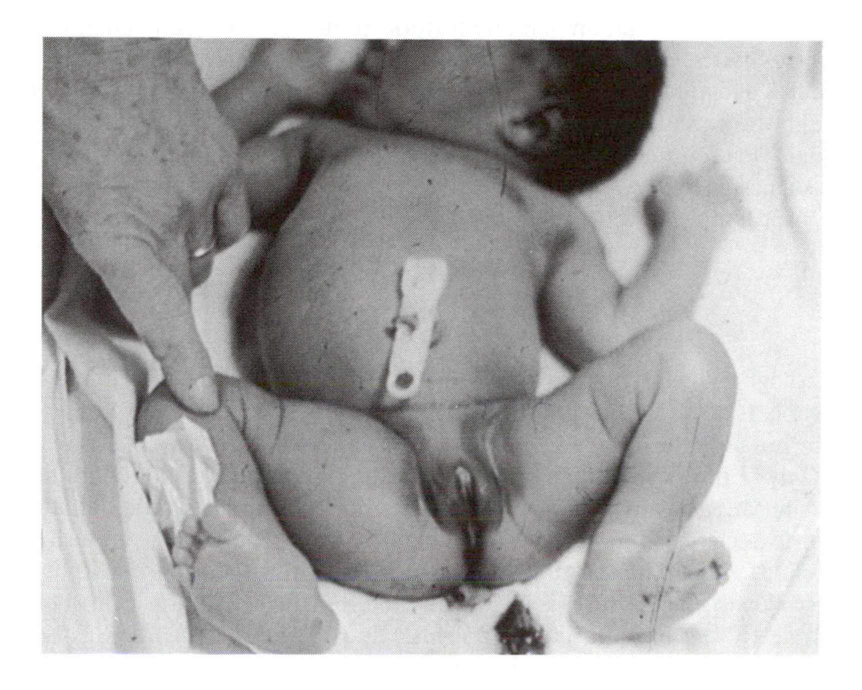

Fig. 15.6. A newborn infant showing bleeding through the umbilical stump and rectum, the most common sites for bleeding in vitamin K deficiency. Source: McLaren, 1981 (with permission).

of the central nervous system with vascular collapse. In order to prevent the occurrence of 'haemorrhagic disease of the newborn', vitamin K is often given either to the mother just before delivery, or to the baby soon after birth. Indeed, according to the American Academy of Pediatrics, there should be a routine parenteral administration of vitamin K at a dose level of 0.5–1.0 mg, to all infants at birth (American Academy of Pediatrics, 1971). Phylloquinone is preferred over water-soluble synthetic vitamin K (menadione) for administration to infants because of a greater safety (see Chapter 21). Since breast milk is severely deficient in vitamin K, it is also recommended for breast-fed infants that either they should receive 1.0 mg phylloquinone intramuscularly once a month for the whole period of breast feeding, or the lactating mother should receive orally 20 mg of phylloquinone twice a week (Shearer *et al.*, 1982).

Hospitalized patients

Vitamin K deficiency with coagulation defects has been reported to exist among patients in hospital (Pineo *et al.*, 1973). The underlying causes for this deficiency appear to be a combination of many factors. Some of these factors include the use of antibiotics, gastrointestinal surgery, and total parenteral nutrition (TPN). Vitamin K does not seem to be metabolized and utilized normally when it is given through the parenteral route at a physiological dose level. Its administration through TPN solutions containing a large dose level (1.0 mg) once a week has been suggested to be reasonably effective in preventing vitamin K deficiency in cases of prolonged TPN (Olson, 1984).

Requirements

It is recognized that vitamin K is a dietary essential, but no recommended allowance is made for the vitamin. This is because the vitamin K requirement is met by not only the dietary intake but also its microbiological synthesis in the gut. Only in newborn infants before establishment of the intestinal flora does there appear to be any need for special attention to vitamin K. The Canadian Pediatric Society recommends that all healthy term infants be given a single dose of 2.0 mg of phylloquinone orally, or 1.0 mg intramuscularly within six hours following birth, and that all preterm, low-birthweight and sick newborns be given 1.0 mg intramuscularly (Health and Welfare, 1990). It has been proposed as a guideline that the dietary intake of infants, especially during their first year of life, should be 10–25 μg day^{-1} (Olson, 1987). For adults, a daily dietary intake of 60–80 μg is believed to be adequate. A normal daily mixed diet in North America contains 300–500 μg vitamin K, an amount exceeding the adequate dietary requirement of the vitamin.

Assessment of Vitamin K Status

Until recently, only functional tests which depend on the measurement of blood clotting time and prothrombin time were used for the assessment of vitamin K status (Simko *et al.*, 1984). A normal prothrombin time is considered to be between 11 and 13 seconds; times greater than 25 seconds are usually associated with severe bleeding. A radioimmunoassay has now been developed which measures prothrombin as well as partially carboxylated prothrombin molecule (des-γ-carboxyglutamyl prothrombin) directly in the plasma (Olson, 1987). The prothrombin/des-γ-carboxyglutamyl prothrombin ratio is believed to be a useful indicator of vitamin K status, because in vitamin K deficiency there is secretion of partially carboxylated prothrombin molecules in the plasma. Another sensitive method for assessing vitamin K nutriture involves high-pressure liquid chromatography (HPLC), which measures plasma phylloquinone (Olson, 1987).

References

American Academy of Pediatrics (1971) Vitamin K supplementation for infants receiving milk substitute infant formulas and for those with fat malabsorption. *Pediatrics* 48, 483–7.

Dam, H. (1935) Antihaemorrhagic vitamin of chick. *Biochemical Journal* 29, 1273–1285.

Dam, H., Sondegaard, E. and Olson, R.E. (1982) Vitamin K. In: Baker, B.M. and Bender, D.H. (eds) *Vitamins in Medicine*, 4th edn, Vol. 2. William Heinemann, London, pp. 92–113.

Haroon, Y., Shearer, M.J., Rahin, S. Gunn, W.G., McEnery, G. and Baskham, P. (1982) The content of phylloquinone (vitamin K-1) in human milk, cow's milk and infant formula foods determined by high-performance liquid chromatography. *Journal of Nutrition* 112, 1105–17.

Hauschka, P.V., Lian, J.B. and Gallop, P.M. (1978) Vitamin K and mineralization. *Trends in Biochemical Sciences* 3, 75–8.

Health and Welfare Canada (1990) In: *Nutrition Recommendations*. The Report of the Scientific Review Committee. Canadian Government Publishing Centre, Ottawa.

Lane, P.A. and Hathaway, W.E. (1985) Vitamin K in infancy. *Journal of Pediatrics* 106, 351–359.

Lefevere, M.F., DeLeenheer, A.P., Claeys, A.E., Claeys, I.V. and Steyaert, H. (1982) Multidimensional liquid chromatography: A breakthrough in the assessment of physiological vitamin K levels. *Journal of Lipid Research* 23, 1068–1072.

Lian, J.B. and Prien, E.L. Jr. (1976) γ-Carboxyglutamic acid in the calcium-binding matrix of certain kidney stones. *Federation Proceedings* 35, 1763.

Olson, J.A. (1987) Recommended dietary intakes (RDI) of vitamin K in humans. *American Journal of Clinical Nutrition* 45, 687–692.

Olson, R.E. (1984) The function and metabolism of vitamin K. *Annual Review of Nutrition* 4, 281–337.

Pineo, G.F., Gallus, A.S. and Hirsh, J. (1973) Unexpected vitamin K deficiency in hospitalized patients. *Canadian Medical Association Journal* 109, 880–883.

Price, P.A. (1988) Role of vitamin K-deficient proteins in bone metabolism. *Annual Review of Nutrition* 8, 565–583.

Price, P.A. and Baukol, S.A. (1980) 1,25-dihydroxyvitamin D increases synthesis of the vitamin K-dependent bone protein by osteosarcoma cells. *Journal of Biological Chemistry* 255, 11660–11663.

Sann, L., Leclercq, M., Guillaumont, M., Trouyez, R., Betherod, M. and Bourgeay-Causse, M. (1985) Serum vitamin K-1 concentrations after oral administration of vitamin K-1 in low birth weight infants. *Journal of Pediatrics* 107, 608–611.

Shearer, M.J., Rahim, S., Barkhan, P. and Stimmler, L. (1982) Plasma vitamin K-1 in mothers and their newborn babies. *Lancet* 2, 460–463.

Simko, M., Cowell, C. and Gilbride, J.A. (1984) *Nutrition Assessment*. Aspen Publishers, Rockville, MD.

Suttie, J.W. (1985) In: Diplock, A.T. (ed.) *Fat Soluble Vitamins*. Technomic Publishing Co., Lancaster, PA, pp. 225–231.

Suttie, J.W. and Olson, R.E. (1984) Vitamin K. In: *Present knowledge in Nutrition*. The Nutrition Foundation, Washington, D.C., pp. 241–259.

Vitamin-like Substances – Pseudovitamins 16

A wide variety of substances, which are widely available in foods or found in unusual foods not normally eaten, are being promoted by the health food industries for their supposedly special health-giving properties. Extravagant claims for the need of these substances are generally made on the basis of deficiency symptoms in species other than humans. Some of these food factors have been given the name 'vitamin' by the proponents, as an attempt to bypass any legal setbacks. None of these factors, however, meet the criteria of 'vitamins', nor are they dietary essentials for humans under normal physiological conditions (Dubick and Rucker, 1983). Most of these substances are widely distributed in a variety of food sources. In addition, they are synthesized in the body in appreciable amounts, and often they can be substituted by other food factors.

Overall, these food substances are biologically active without being dietary essentials for humans. No recommended allowance or estimated levels of safe and adequate intake have been established. Absence of these substances from human diets has not resulted in any deficiency symptoms. There is no consistent evidence of therapeutic efficacy. In fact, some of these factors appear to be potentially toxic. This chapter will discuss the biological potential of some of these substances with no known essential nutrient functions in man.

Amygdalin

Amygdalin, or laetrile as it is commonly called, is a cyanogenic glucoside found in kernels of apricot pits, peach pits, and other stone fruits as well as in nuts like almonds and macademia (Herbert, 1979a). It is composed of two glucose, one benzaldehyde and one cyanide molecules (Fig. 16.1). This cyanide-containing compound has been legalized as a drug in several States of the US. However, in attempts to bypass any legal setbacks its proponents have trade-named it vitamin B_{17}, so that it may sound like an innocuous substance and be made freely

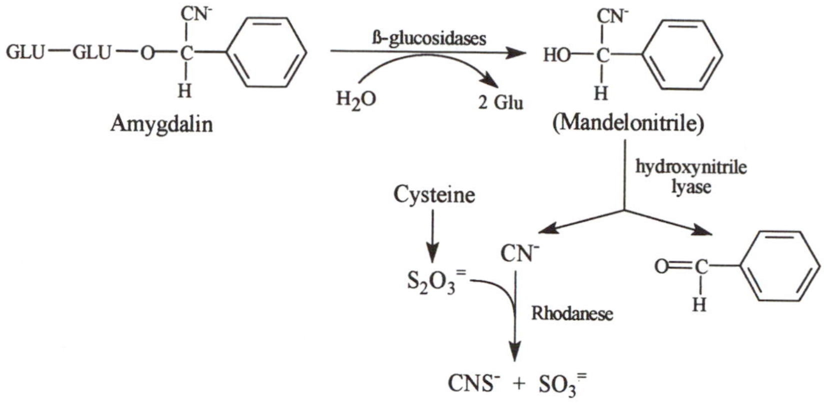

Fig. 16.1. Degradation of amygdalin (laetrile).

available. Although in Canada the use of laetrile is illegal, there is an easy access to laetrile-containing sources, which are available in many health food stores.

Hydrolysis of laetrile releases cyanide (Fig. 16.1), and this can be achieved by treatment with dilute acids and also by specific β-glucosides present in the small intestine. These hydrolytic enzymes are also present in cyanophoric plants, such as celery, bean sprouts, seeds of cherry and apple, and the kernels of peaches and apricots. The enzymes are generally activated when the plant tissue containing them is crushed. The cyanide released from laetrile is normally detoxified to thiocyanate involving the enzyme rhodanese (Fig. 16.1), which is present in the liver, gut, and kidneys (Basu, 1983).

It has been theorized that, unlike normal cells, cancer cells are rich in β-glucosidases while they are lacking in rhodanese enzyme, making laetrile an effective agent for selectively killing cancer cells through cyanide poisoning (Krebs, 1970). Despite the theory that cyanide may be useful in cancer chemotherapy, others have failed to observe any effect of laetrile on experimental tumours (Stock *et al.*, 1978). In a clinical study involving 178 terminally ill cancer patients conducted at four US cancer centres, no differences were observed in terms of cure, disease progression, improvement of symptoms or extension of life span in patients with or without therapy with laetrile (Moertel *et al.*, 1982). Despite the lack of objective evidence to support the use of laetrile in cancer therapy, many patients are taking the substance either alone or in combination with other forms of treatment.

One gram of laetrile yields approximately 60 mg of hydrocyanic acid, which is potentially a toxic level for a healthy individual (Greenberg, 1980). Hence the question of the 'safe' use of laetrile must take as much precedence as the toxic effects of its use. In fact, there is now considerable knowledge concerning the adverse effects of laetrile in both animals and humans. In a study with ten dogs, feeding 1–4 mg of laetrile in tablet or liquid form along with almond paste resulted in the death of six dogs from cyanide poisoning (Schmidt *et al.*, 1978). In another study involving pregnant hamsters, oral administration of laetrile has been reported to cause congenital skeletal malformations among offspring due to increased free cyanide concentration in tissues (Willhite, 1982). One case of human toxicity involved an eleven-month old girl who died following an accidental ingestion of 5 × 500 mg tablets of laetrile (Humbert *et al.*, 1977). Another case was concerned with a 17-year-old girl who committed suicide by overdosing with laetrile (Sadoff *et al.*, 1978). Generally speaking, oral intake of laetrile is more toxic than parenteral administration because of β-glucosidases in gut contents, which release cyanide through hydrolysis of laetrile.

The use of ascorbic acid in large doses may exacerbate the laetrile-mediated cyanide burden in the body (Basu, 1983). Cyanide is detoxified to thiocyanate, requiring cysteine as the sulphate donor (Fig. 16.1). Sulphate formation is also an important pathway in the metabolism of ascorbic acid to ascorbic acid-sulphate (Chapter 10), the sulphate being derived from cysteine. Concomitant intake of laetrile and ascorbate in large doses may, therefore, potentiate the toxicity of cyanide by competing for the cysteine. Individuals taking megadoses of ascorbic acid are expected to be at increased risk of experiencing the adverse side effects from laetrile or other cyanogenic compounds.

Despite ample scientific evidence that laetrile neither prevents nor cures cancer, nor has it any 'vitamin' activities, the manufacturers of this substance continue to thrive. The victims of terminal conditions are often vulnerable to manipulation and exploitation by quacks (Herbert, 1979a).

Pangamic Acid

This compound was first isolated from apricot pits and later from a variety of other sources including liver, blood, rice, and yeast. The chemical structure of pangamic acid (Fig. 16.2) found in these natural sources has been suggested to be D-gluconodimethyl aminoacetic acid, GDA (Stacpoole, 1977). This substance has been widely claimed to be of benefit in diverse conditions such as cancer, asthma, allergies, hypercholesterolaemia, and hypoxia. Like laetrile, pangamic acid is also erroneously promoted as a vitamin, and given the name vitamin B_{15} (Herbert, 1979b). The basis for this is that it is present in nature and that it is useful in a variety of pathological conditions. Scientific evidence, however, does not support either its effectiveness or its lack of adverse effects (Herbert *et al.*, 1980; Gray and Titlow, 1982). Similarly, there is no evidence that physiological or biochemical abnormalities develop when pangamic acid is not included in the diet. Hence the use of the term 'vitamin' is erroneous and misleading.

Analysis of pangamic acid preparations sold in the United States, however, has indicated that they may vary considerably in their chemical composition (Herbert *et al.*, 1980). Thus, they may contain calcium gluconate, *N,N*-

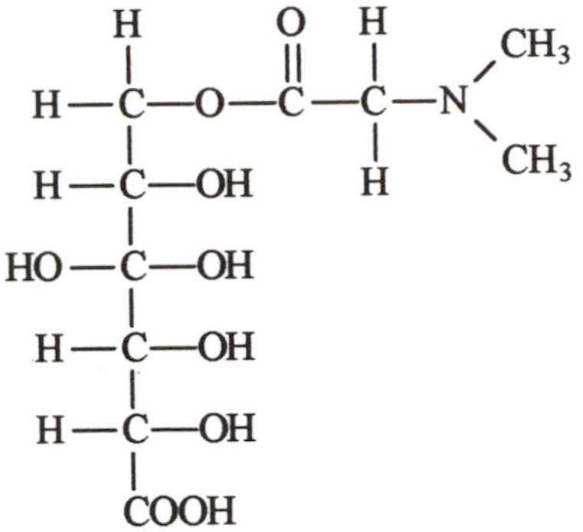

```
     H         O   H
     |         ||  |        CH3
H —  C — O  —  C — C — N <
     |             |        CH3
H —  C — OH        H
     |
HO — C — OH
     |
H —  C — OH
     |
H —  C — OH
     |
    COOH
```

Fig. 16.2. Structure of D-gluconodimethyl aminoacetic acid (pangamic acid in natural sources).

dimethylglycine, *N,N*-diisopropylamine dichloroacetate, or GDA. Since there is no standard of identity for pangamate preparations, their safe use has been seriously questioned. In fact, according to the Ames assay, the *N,N*-diisopropylamine dichloroacetate, one of the components found in the preparations, has been reported to be mutagenic (Gelernt and Herbert, 1982).

Despite pangamic acid's presence in food, there is no evidence that it is a nutrient or vitamin. In fact, it is not a chemically defined substance. According to both the American Food and Drug Administration and the Canada Food and Drug Directorate, this substance has no known values for humans, and may be harmful (Herbert, 1979b).

Bioflavonoids

The bioflavonoids are a large group of phenolic compounds, which were first suggested as dietary factors in 1936 by Albert Szent-Gyorgyi, who later named these substances vitamin P. This name was subsequently withdrawn because of the lack of evidence that they are essential in humans.

Bioflavonoids are widely distributed in the skin and outer layers of fruits and vegetables. Over 800 different flavonoids have been isolated from plant extracts; the typical ones include quercetin, hesperidin, rutin, and naringin (Fig. 16.3). They are colourful antioxidants, found mostly as glycosides, capable of chelating metal ions, and responsible for the colours of fruits (Wollenweber, 1988). It is estimated that approximately 50% of our daily average intake of 1–2 g is absorbed. Flavonoid glycosides are hydrolysed prior to their absorption in the small intestine. Following absorption, they undergo conjugation with glucuronides or sulphates in the liver, and subsequently the conjugated products are excreted essentially in the urine normally within 24 h (Bokkenheuser and Winter, 1987).

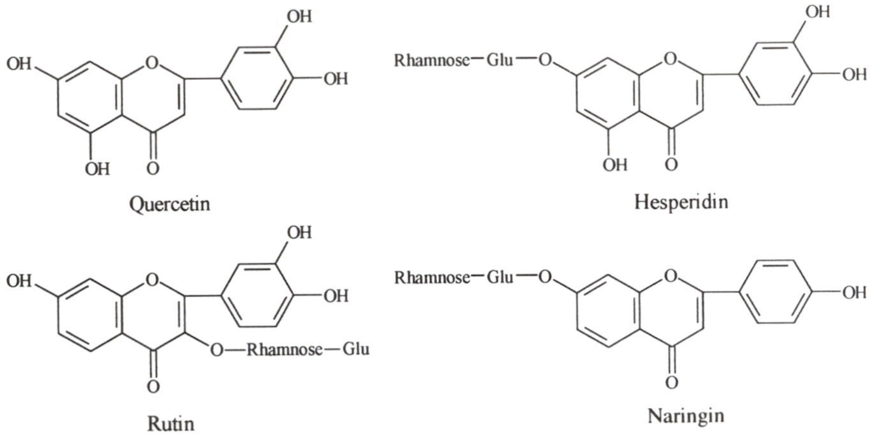

Quercetin

Hesperidin

Rutin

Naringin

Fig. 16.3. Chemical structures of some bioflavonoids.

Numerous biological effects of bioflavonoids have been claimed (Vlietinck *et al.*, 1988). They are believed to reduce red blood cell aggregation and capillary fragility, lower blood cholesterol, and prevent cancer, hypertension, viral infection and allergic reactions. Other potentially important reports of the flavonoid activities include the *in vitro* inhibition of aldose reductase which converts glucose and galactose to their polyols (Varma and Kinoshita, 1976). These metabolic products have been implicated in the neuropathy of diabetes and in the cataract formation in galactosaemia. The overall relevance of this group of compounds to health and disease is, however, inconsistent. So far, scientists have been unable to ascertain any specific biological role of bioflavonoids, although they agree that they may have some pharmacological effect.

Most of the claimed biological potential of the flavonoids as well as their natural sources happen to be similar to those of vitamin C (Chapter 10). Studies are needed to determine if vitamin C or the bioflavonoids cause the reported beneficial effects or if the two agents act synergistically. Until the specific role of bioflavonoids is revealed, there is no justification for their inclusion in nutritional supplements; nor is there a justification in the promotion of certain foods on the basis of a high concentration of the compounds.

Carnitine

Carnitine (3-hydroxy-4-*N*-trimethyl aminobutyrate) was first found in yeast. In 1947, Fraenkel showed that this substance was a dietary essential for the mealworm (*Tenebrio molitor*) larva, and later it was named vitamin B_t (t = tenebrio). Most studies on the requirements for this substance involved insects. This work made no suggestion that carnitine was required for mammals.

Carnitine is not considered to be an essential nutrient in the diet of adult humans and hence a vitamin. This is because its dietary deficiency has never been documented and that it is synthesized from lysine and methionine in kidney and liver by a pathway involving two steps that require Fe^{2+} and ascorbate for hydroxylation (Fig. 16.4; see also Chapter 10). It is truly an amino acid, and found in high concentrations in muscle tissues. This is, in fact, the rationale for the name 'carnitine', which is derived from the Latin word, carnis, meaning flesh or muscle. Carnitine is generally found, at least in significant amounts, exclusively in foods of animal origin, such as red meats and dairy products. However, in a physiological state in adult individuals, its dietary intake appears to have little or no effect on the plasma concentrations of carnitine (Khan-Siddique and Bamji, 1980). This indicates that the amount of this amino acid synthesized in the body may be adequate to meet its requirements.

Carnitine is known to play a role in the β-oxidation of long-chain fatty acids for the production of metabolic energy. The β-oxidation of fatty acids occurs exclusively in the inner region of the mitochondrion, but this inner membrane

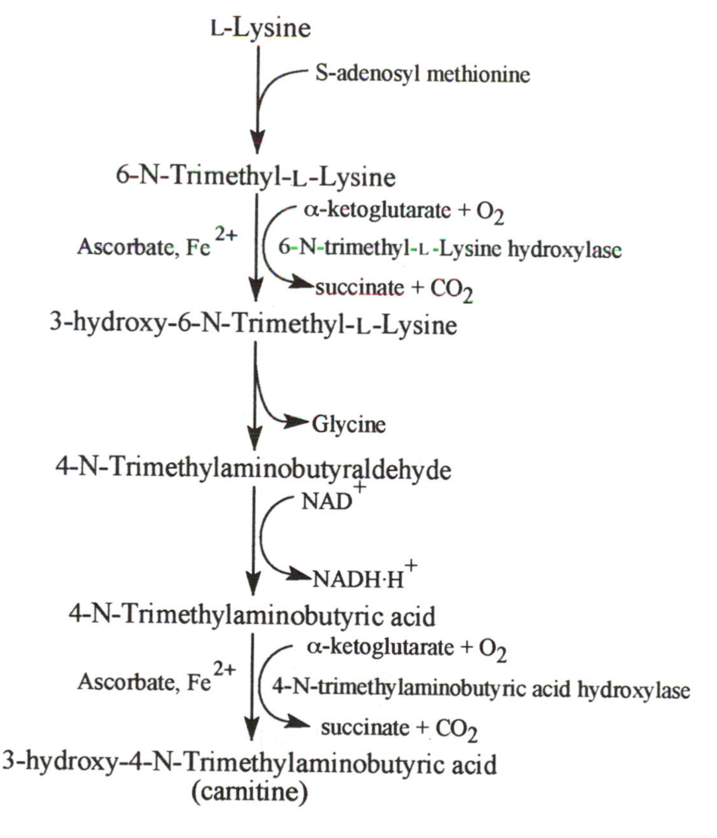

Fig. 16.4. Biosynthesis of carnitine.

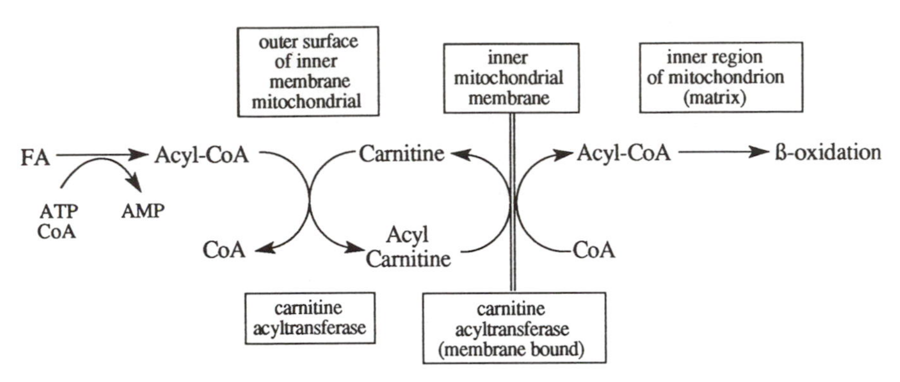

Fig. 16.5. Role of carnitine in β-oxidation of long-chain fatty acids (FA).

is impermeable to long-chain fatty acid CoA moieties. Carnitine facilitates the transport of long-chain fatty acyl CoA into the matrix of the mitochondrion (Fig. 16.5). This transport is catalysed by two enzymes, which include carnitine acyltransferase I, located in the outer surface of the inner mitochondrial membrane, and carnitine acyltransferase II, located in the inner mitochondrial membrane.

Apart from carnitine's important role as being a carrier of long-chain fatty acids into mitochondria for β-oxidation, it is also believed to be involved in carrying acetyl and short-chain acyl units back to the cytosol to form acyl CoA (Bieber, 1988). Research on carnitine continues to explore its further biochemical roles. Its tissue concentrations have been found to be depressed in a variety of conditions which include certain lipid storage diseases of the muscle, muscular dystrophy, hyperthyroidism, diabetes, and during physical exertion (Rebouche and Paulson, 1986). Subjects with these conditions do not generally have a dietary deficiency of carnitine. It is important to clarify if its tissue depletion in these states is the reflection of metabolic difficulties affecting carnitine utilization.

Recent studies have suggested that carnitine may be an essential nutrient for newborns because of their inability to synthesize carnitine in the body (Hahn, 1982; Borum, 1986). This is a concern, since fatty acid oxidation is a major source of metabolic energy for infants. However, milk or infant formula prepared from cow's milk contains appreciable amounts of carnitine which is adequate to meet daily requirements. Soybeans, on the other hand, lack carnitine, and hence infants fed an infant formula prepared from soy protein do not receive any carnitine, unless it is added by the manufacturers. Another situation where infants can become deficient in carnitine is when they are born prematurely. Preterm infants have generally very low tissue stores of carnitine. At the same time they are often maintained on total parenteral nutrition (TPN) containing no carnitine. In summary, carnitine is not a dietary essential and has no use as a supplement for normal adult humans. Evidence exists that infants, especially premature ones, may have limited ability to synthesize this amino acid, but a diet appropriate for their age will supply the amount needed. Supplementation with large doses of D-carnitine is thought to interfere with the body's ability to utilize the physiological L-carnitine. Evidence to date does not justify the use of carnitine supplementation, unless there is medical ground.

Choline

Choline (trimethyl-β-hydroxyethylammonium) is a component of the phospholipids sphingomyelin and phosphatidylcholine (lecithin), both of which are constituents of cell membranes. It is also needed for the synthesis of acetylcholine, a neurotransmitter (Zeisel, 1981). Betaine, an oxidized product of choline, can be a source of labile methyl groups for transmethylation reactions involving methionine synthesis from homocysteine or guanidoacetic acid synthesis from creatine. Another important function of choline is, therefore, that it serves as a source of labile methyl groups. A variety of animal species, such as rats, chicks and turkeys, appear to be most susceptible to inadequate choline intakes; its deficiency in these species is generally associated with fatty deposition in the liver and haemorrhagic kidney disease. In such animals, choline is considered an essential nutrient and perhaps a vitamin.

There are, however, several factors that make choline a non-essential nutrient for humans (McMahon, 1987). Man can synthesize choline from serine provided that enough methionine is present as a methyl donor, and vitamin B_{12} and folacin are present as coenzymes for the transfer of methyl groups (see Chapter 9). Choline is also widely distributed in both animal and plant foods. An average mixed diet for an adult in North America can provide 600–900 mg day^{-1}. Furthermore, the functions of choline, both as a lipotropic factor and as a labile methyl donor, are non-specific. Other dietary lipotropes and methyl donors can substitute for choline, especially in these roles, and thereby may decrease its requirement. The aforementioned factors may explain why no deficiency has yet been produced in humans by an absence of choline in the diet.

Choline or lecithin has been claimed to improve the failing memory of older persons, lower blood cholesterol level, and to be an effective hypertensive agent (Zeisel, 1981). Because of these claims, these substances have been marketed as over-the-counter preparations. To date, however, no acceptable evidence has been offered to support the claims.

Inositol

There are nine isomers of inositol, of which myoinositol is the most important one in nature and the only isomer that is biologically active. Chemically it is acyclic alcohol with six hydroxy radicals and closely related to glucose (Fig. 16.6). It is concentrated in the brain, cerebrospinal fluid, skeletal and heart muscles, and male reproductive organs. The metabolically active form of myoinositol is thought to be phosphatidylinositol. This active form is believed to have many physiologically important functions (Holub, 1986). It may act as an activator of Na^+, K^+-ATPase, an essential constituent of acetyl CoA carboxylase, a membrane-anchor for acetylcholinesterase, and a stimulator of tyrosine hydroxylase, a factor bound to alkaline phosphatase. Myoinositol can thus be an effector of the structure and function of membranes as well as a mediator of cellular responses to external stimuli. The turnover of myoinositol appears to be associated with response of cells to stimuli and subsequent transfer of information (Holub, 1987). It has been hypothesized that this transfer of information may play an important role in signal transmission for various hormones and neurotransmitters. Like choline, inositol is also believed to have a lipotropic action. This action of inositol is thought to be related to its function as a lipoprotein component.

Inositol is quite abundant in foodstuffs. It is distributed in fruits, vegetables, whole grains, meats and milk. The average diet in North America provides about 1 g day^{-1}. In addition, the body may be able to synthesize sufficient amounts to meet its need from glucose. The possibility of intestinal synthesis of inositol along with its wide distribution in animal and plant foods make a deficiency extremely difficult to demonstrate.

There is no known requirement for dietary inositol in man, nor have any signs or symptoms of deficiency been described. It is, however, possible that the significance of inositol in human nutrition has not been sufficiently explored. In recent years, attention has been directed toward the levels and metabolism of inositol-containing phospholipids in relation to many pathological conditions

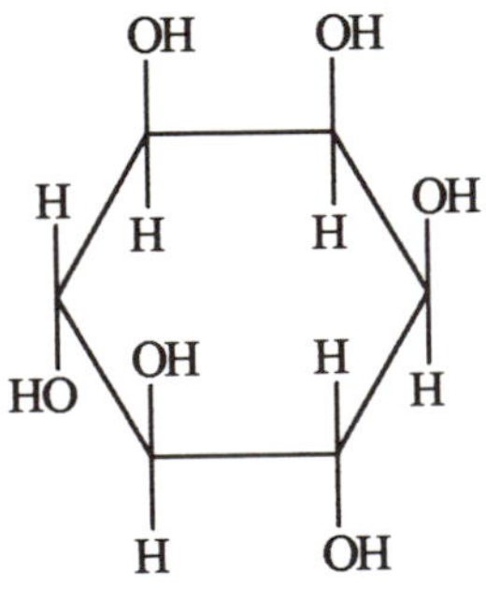

Fig. 16.6. Structure of myoinositol.

(Holub, 1986). Thus, diabetic patients appear to excrete large amounts of myoinositol in urine while its levels in nerve membranes are markedly reduced. This change in myoinositol metabolism may be related to peripheral neuropathies in diabetes. Similar effects have been found in diabetic rats, and the effects appeared to be reversed with inositol supplements (Clements and Reynertson, 1977).

p-Aminobenzoic Acid

p-Aminobenzoic acid (PABA) is a component of folic acid (Chapter 8) and is often classified as a B-vitamin. It is a growth factor for certain microorganisms; its actual role is to provide this component for the synthesis of folic acid by those organisms which do not require a preformed source of folic acid. However, it cannot be substituted for folic acid and is not regarded as a vitamin for animals. Claims, such as PABA retarding the greying of hair and the development of skin cancer, have not been substantiated by scientific evidence. It is an antagonist to the bacteriostatic action of sulphonamide drugs, and hence the ingestion of PABA can worsen infection.

Coenzyme Q (CoQ)

This is a quinone derivative with an isoprenoid tail the length of which varies according to the species. In mammals the tail contains ten isoprene units, and it is called coenzyme Q10. It belongs to a group of compounds known as ubiquinones. CoQ is a lipid-like substance that is somewhat similar in its chemical make-up to vitamin K (Chapter 15). It plays a key role in the respiratory chain as an electron carrier between flavoproteins and cytochromes. In addition, the coenzyme appears to have beneficial effects in certain disease states including muscular dystrophy, periodontal disease, hypertension, and congestive heart failure (Folkers and Yamamura, 1977). A substance, isolated from the phospholipids of soyabeans, was found to aid in blood clotting, and given the name vitamin Q (Quick, 1975).

The ubiquinones are synthesized readily in the body, the ring structure from amino acids such as phenylalanine and the side chain from acetate. They are therefore of little dietary significance and cannot be truly classed as vitamins.

References

Basu, T.K. (1983) High-dose ascorbic acid decreases detoxification of cyanide derived from amygdalin (laetrile): Studies in guinea pigs. *Canadian Journal of Physiology and Pharmacology* 61, 1426–1430.

Bieber, L.L. (1988) Carnitine. *Annual Review of Biochemistry* 57, 162–183.

Bokkenheuser, V.D. and Winter, J. (1987) In: Cody, V., Middeton, E. Jr., Harborne, J.B., and Beretz, A.J. (eds) *Plant flavonoids in Biology and Medicine II*. Alan R. Liss, New York. p. 143.

Borum, P.R. (1986) Carnitine – who needs it? *Nutrition Today* 21, 4–6.

Clements, R.S. Jr. and Reynertson, R. (1977) Myo-inositol metabolism in diabetes mellitus. Effects of insulin treatment. *Diabetes* 26, 215–221.

Dubick, M.A. and Rucker, R.B. (1983) Dietary supplements and health aids – a critical evaluation: I. Vitamins and Minerals. *Journal of Nutrition Education* 15, 47.

Folkers, K. and Yamamura, Y. (1977) Biochemical and clinical aspects of coenzyme Q. In: Folkers, K. and Yamamura, Y. (eds) *Proceedings of the International Symposium on Coenzyme Q*. Elsevier, New York.

Gelernt, M.D. and Herbert, V. (1982) Mutagenicity of diisopropylamine dichloroacetate, the 'active constituent' of vitamin B_{15} (pangamic acid). *Nutrition and Cancer* 3, 129–133.

Gray, M.E. and Titlow, L.W. (1982) The effect of pangamic acid on maximal treadmill performance. *Medical Science, Sports and Exercise* 14, 424–427.

Greenberg, D. (1980) The case against laetrile. *Cancer* 45, 779–787.

Hahn, P. (1982) Carnitine in the perinatal period of mammals. *Nutrition Research* 2, 201–206.

Herbert, V. (1978) The nutritionally unsound 'Nutritional and Metabolic Antineoplastic Diet' of laetrile proponents. *Journal of American Medical Association* 240, 1139–1140.

Herbert, V. (1979a) Laetrile: The cult of cyanide promoting poison for profit. *American Journal of Clinical Nutrition* 32, 1121–1158.

Herbert, V. (1979b) Pangamic acid (Vitamin B_{15}). *American Journal of Clinical Nutrition* 32, 1534–1540.

Herbert, V., Gardner, A. and Colman, N. (1980) Mutagenicity of dichloroacetate, an ingredient of some formulations of pangamic acid (trade-named 'Vitamin B_{15}). *American Journal of Clinical Nutrition* 33, 1179–1182.

Holub, B.J. (1986) Metabolism and function of myoinositol and inositol phospholipids. *Annual Review of Nutrition* 6, 563–597.

Holub, B.J. (1987) The cellular forms and functions of the inositol phospholipids and their metabolic derivatives. *Nutrition Review* 45, 65–71.

Humbert, J.R., Tress, J.H. and Braico, K.T. (1977) Fatal cyanide poisoning: accidental ingestion of amygdalin. *Journal of American Medical Association* 238, 482.

Khan-Siddique, L. and Bamji, M.S. (1980) Plasma carnitine levels in adult males in India: effects of high cereal, low-fat diet, fat supplementation and nutritional status. *American Journal of Clinical Nutrition* 33, 1259–1263.

Krebs, E.T. Jr. (1970) The nitrilosides (Vitamin B_{17}): their nature, occurrence and metabolic significance: antineoplastic vitamin B_{17}. *Journal of Applied Nutrition* 22, 75–86.

McMahon, K.E. (1987) Choline, an essential nutrient? *Nutrition Today* 22, 18–21.

Moertel, C.G., Fleming, T.R., Rubin, J., Kvols, L.K., Sarna, G., Koch, R., Currie, V.E., Young, C.W., Jones, S.E. and Davingnon, J.P. (1982) A clinical trial of amygdalin (laetrile) in the treatment of human cancer. *New England Journal of Medicine* 306, 201–206.

Quick, A.J. (1975) Vitamin Q in control of bleeding. *Wisconsin Medical Journal* 74, 585.

Rebouche, C.J. and Paulson, D.J. (1986) Carnetine metabolism and functions in humans. *Annual Review of Nutrition* 6, 41–66.

Sadoff, L., Fuchs, K. and Hollander, J. (1978) Rapid death associated with laetrile ingestion. *Journal of the American Medical Association* 239, 1532.

Schmidt, E.S., Newton, G.W., Sanders, S.M., Lewis, J.P. and Conn, E.E. (1978) Laetrile toxicity studies in dogs. *Journal of the American Medical Association* 239, 943–947.

Stacpoole, P.W. (1977) Pangamic acid (Vitamin B_{15}). A review. *World Review of Nutrition and Dietetics* 27, 145–163.

Stock, C.C., Tarnowski, G.S., Schmid, F.A., Hutchison, D.J. and Teiker, M.N. (1978) Antitumour tests of amygdalin in transplantable animal tumour system. *Journal of Surgery and Oncology* 10, 81–88.

Willhite, C.C. (1982) Congenital malformations induced by laetrile. *Science* 215, 1513–1515.

Wollenweber, E. (1988) Occurrence of flavonoids aglycones in medicinal plants. *Progress in Clinical Biological Research* 280, 45–55.

Varma, S.D. and Kinoshita, J.H. (1976) Inhibition of lens aldolose reductase by flavonoids – their possible role in the prevention of diabetic cataracts. *Biochemical Pharmacology* 25, 2505–2513.

Vlietinck, A.J., Berghe, D.A.V. and Haemers, A. (1988) In: Cody, V., Middeton, E. Jr., Harborne, J.B. and Beretz, A. (eds) *Plant Flavonoids in Biology and Medicine II*. Alan R. Liss, New York, p. 283.

Zeisel, S.H. (1981) Dietary choline, biochemistry, physiology and pharmacology. *Annual Review of Nutrition* 1, 95–121.

17 Vitamins and Cancer

Interest in the potential role of diet in the aetiology of cancer emerged in the late 1960s when the results of epidemiological investigations highlighted marked international variations in the number of deaths from cancer (Miller, 1985). Studies of disease incidence in migrant populations suggested that these differences were not a function of genetic constitution since, for the most part, cancer incidence in migrant groups shifted from that of the country of origin to that of the host country. Doll and Peto (1981) estimated that 80–90% of human cancers are caused by environmental factors, and that appropriate dietary changes might reduce cancer deaths by as much as 35%, as significant as the contribution from smoking. On the whole, though, associations between the types of food and cancer are rather weak and in most cases the nature of the links between diet and these diseases remain obscure. Only in a very few cases is it possible to identify a particular food component as a causative agent. Aflatoxin, a fungal toxin, is known to be mutagenic and occurs in foods such as nuts and grain stored in tropical conditions. This substance has been reported in high concentrations in populations in Africa at risk of liver cancer, apparently acting synergistically with hepatitis B infection (Wild and Montesano, 1991). By contrast, a high intake of β-carotene is associated with a low incidence of certain cancers (Gerster, 1993).

The role of diet is, in fact, complex. Almost everything natural we eat contains carcinogens, teratogens and clastogens. There are also pyrolysis products from cooking and chemicals produced during smoking or other preservation methods. Caution is necessary in the extrapolation of results obtained in experiments on rodents given near-toxic doses of chemicals in order to assess risk to humans exposed to very low doses. Further, it seems unlikely that geographical patterns of cancer incidence can be explained fully by differences in exposure to dietary carcinogens since plant toxins are so universally distributed. If this is so, it would seem to be more worthwhile to concentrate on the capacity of dietary components to protect against carcinogenesis than to be seeking dietary causes of cancer.

It is beyond the scope of this chapter to discuss the various ways in which diet may influence carcinogenesis. Within the diet considerable attention has focused on the possible benefits to be gained by the consumption of foods that are known to be useful sources of micronutrients and, in particular, vitamins. There are a number of possible mechanisms whereby vitamins may be involved in cancer prevention. It is now generally accepted that the transformation of normal tissue cells into malignant tumour cells is a multi-stage process (Hicks, 1983). The stages include carcinogen metabolism, initiation and promotion, cell differentiation and tumour cell progression, and tumour growth and development; each of these stages may be affected by diet (Poirier, 1987). Xenobiotic chemicals can be transformed into carcinogens by cytochrome P450 and related enzymes whose activity is affected by a number of vitamins (Basu, 1988). Carcinogenesis involves changes in DNA, a not particularly stable molecule (Lindahl, 1993). Spontaneous changes occur and the DNA is prone to damage by environmental agents, especially radiation and chemical mutagens. DNA damage also results from attack by free radicals, or reactive oxygen species, present in cells as a by-product of respiration. The -OH radical is thought to be the most potent and is generated from peroxide with the help of transition metals, for example Fe^{2+} and Cu^{+}. The extent of oxidative damage is limited by the presence of free-radical scavengers, including the dietary antioxidants, vitamin C, vitamin E, carotenoids and possibly flavonoids (Collins *et al.*, 1994). A further mechanism by which carotenoids (especially β-carotene) may be involved in carcinogenesis is via the effect of the provitamin on the immune response (Bendich, 1991).

From what we have considered, it would seem that the involvement of free radicals in the aetiology of many forms of cancer is an attractive possibility because of the availability of dietary (or nutritional) means of prevention by decreasing exposure to oxidative damage. Thus the so-called antioxidant hypothesis (Collins *et al.*, 1994) states: (i) endogenous free radicals cause DNA damage which can lead to cancer; (ii) antioxidants, such as vitamins C and E, carotenoids, and flavonoids, scavenge free radicals before they have a chance to do damage; and (iii) hence, antioxidant vitamins in the diet should decrease the risk of cancer.

We will now consider some of the evidence together with problems in its interpretation for a possible role of vitamin A, β-carotene, folic acid, and vitamins C, D, E, and B_{12} in carcinogenesis.

Vitamin A

Vitamin A has a considerable controlling influence on epithelial differentiation (DeLuca *et al.*, 1994; see also Chapter 11). The tissues that are dependent on vitamin A for normal differentiation include epithelia in the bronchus, trachea, stomach, intestine, kidney, bladder, testes, uterus, breast, prostate, pancreatic duct, and skin. These sites account for well over two-thirds of the total primary cases of cancer in both men and women. Since malignant transformation is fundamentally a process of loss of cellular differentiation, and vitamin A promotes this process, the vitamin has aroused interest in recent years as a chemopreventive and therapeutic agent (Kummet *et al.*, 1983; Lacroix and Bhat, 1988). A link between vitamin A and cancers was suggested as early as 1926 when rats fed a vitamin A-deficient diet developed malignancies (Fujmaki, 1926). Since that time, several studies have shown that vitamin A deficiency enhances the susceptibility of epithelial tissues to chemical-, viral-, and radiation-induced carcinogenesis (Sporn and Roberts, 1983; Moon and Itri, 1984). Conversely high dietary retinoids clearly prevent experimental animal cancers (Moon and Itri, 1984). It has been shown that not only naturally occurring forms of vitamin A, but also synthetic retinoids, can prevent experimental cancers.

In human cancers, the evidence supporting a link between vitamin A deficiency and cancer comes from epidemiological studies and both retrospective and prospective studies involving comparison of serum retinol levels of patients with cancer and controls (Basu *et al.*, 1984; Willett and Hunter, 1994). The epidemiological studies have consistently identified a low intake of vitamin A to be associated with increased risk of developing certain cancers (Kummet *et al.*, 1983). However, the evidence is somewhat confused since many studies have not differentiated between the intake of preformed retinol and carotenoids, which may have anti-cancer effects. Another difficulty in interpreting epidemiological or case-control data involving serum measurements of retinol and its carrier protein (retinol-binding protein, RBP), is that serum levels of these components do not necessarily reflect retinol status or intakes. RBP is a protein with a very short half-life. The synthesis of this protein can be affected by a variety of factors, such as stress, infections, protein–energy malnutrition (PEM), and affected liver (see Chapter 11). These factors may well account for the low serum levels of retinol and RBP found in many case-control studies. It is noteworthy that many of these studies have shown subnormal levels of serum vitamin A in cancer patients compared with those of healthy subjects or unspecified non-malignant patients. Many prospective studies have been carried out establishing retinol status in those individuals who subsequently developed cancer. Many of these studies indicated that those with low serum retinol had a greater risk of developing cancer later, but these findings are not consistent.

Be this as it may, there are theoretical mechanisms to explain cancer risk reduction by vitamin A. It is not only that the vitamin is known to be essential for the maintenance of epithelial tissues in which many cancers occur, immunocompetence, including tumour surveillance, has been shown in animal models to depend on vitamin A status. Further, retinol and its synthetic compounds may directly influence gene expression (Sporn and Roberts, 1983) and have been shown to possess cancer modulating properties in animal models (Hong and Itri, 1994).

Reference has been made to the influence of PEM on serum retinol and RBP levels. Undernutrition is associated with the presence of cancer in some sites but not in others. It is tempting to suggest that compromised nutritional status may account for variations in the findings with respect to serum levels. Patients with breast cancer are not usually malnourished unless the disease is advanced. Patients who remained free of the disease for two years had significantly higher RBP levels than those who suffered a re-appearance of cancer (Mehta *et al.*, 1987). It might be suggested that disappointment and a reduction in psychological status in patients who had a recurrence of the disease resulted in a decreased food intake. The influence of food intake and particularly the intake of vitamin A-rich foods has been found to be significantly related to the development of cancer in a retrospective case-control study (Katsouyanni *et al.*, 1988). In this study the association was stronger for total vitamin A intake (retinol + β-carotene) than for either retinol or β-carotene.

In the genitourinary system vitamin A may have somewhat contrary associations. Thus, vitamin A intake does not appear to reduce the risk of bladder cancer (Tyler *et al.*, 1986). In one study (Michalek *et al.*, 1987) a high intake of total vitamin A was associated with increased risk of recurrence of bladder cancer and in other studies (Heshmat *et al.*, 1985; Kolonel *et al.*, 1987) with increased risk of prostate cancer. In oral leukoplakia, a lesion of the buccal mucosa which can lead to oral cancer, serum levels of vitamin A have been reported to be lower than in controls, with still lower levels in individuals with oral cancer (Wahi *et al.*, 1962). In individuals at high risk of oral cancer due to betel nut and tobacco chewing, high-dose retinol with or without β-carotene may decrease cancer risk (Gaby and Bendich, 1991).

Retinoids have been used to treat a variety of skin disorders in humans including skin cancer (Levine and Meyskens, 1980; Peck, 1982). These compounds have a variety of biological effects and in particular have a striking effect on phenotypic differentiation of the epidermis and the production of keratin and also possess significant antineoplastic activity in experimental systems. Skin carcinogenesis may involve at least three stages, which include initiation, promotion, and malignant transformation (carcinoma). Using 7,12-dimethylbenz[a]anthracine (DMBA) as the initiator and 12-*O*-tetradecanoylphorbol-13-acetate (TPA) as the tumour promotor, treatments with dietary retinoic acid in pharmacological doses (30 $\mu g\ g^{-1}$ diet) have been shown to have a profound inhibitory effect in the conversion of papillomas (benign lesions) to carcinomas in the back skin of female SENCAR mice (DeLuca *et al.*, 1994).

β-Carotene

The possibility that a high intake of β-carotene affords some protection against cancer due to its antioxidant properties has attracted considerable attention (Peto *et al.*, 1981; Temple and Basu, 1988; Gaby and Singh, 1991a). More

Table 17.1. Epidemiological evidence of the protective effect of fruits and vegetables against cancer[1].

Cancer site	Number of studies	Reduced risk	Increased risk
Lung	25	24	0
Larynx	4	4	0
Oral cavity, pharynx	9	9	0
Oesophagus	16	15	0
Stomach	19	17	1
Colorectal	35	20	3
Bladder	5	3	0
Pancreas	11	9	0
Cervix	8	7	0
Ovary	4	3	0
Breast	14	8	0
Prostate	14	4	2

[1]Only studies that reported in terms of relative risk and included 20 or more cancer cases at the site in question are included.
Adapted from Block *et al.* (1992).

than 50 epidemiological studies carried out in the last decade in different parts of the world have consistently shown that a high intake of fruits and vegetables rich in β-carotene is associated with a reduced risk of a variety of cancers, especially of the lung (Table 17.1). In 25 different studies individuals with high intakes or high blood levels of β-carotene had about half the risk of developing lung cancer than those with low intakes or low blood levels (Temple and Basu, 1988).

A number of prospective studies have also been carried out focusing on foods rich in β-carotene or using supplements containing β-carotene. One of the earlier studies involved 1954 men employed by Western Electric Co., Chicago, which was terminated in 1981 after 19 years (Shekelle *et al.*, 1981). This study reported on a distinction between a 'carotene index' based on the consumption of fruits and vegetables and a 'retinol index' based on intake of pre-formed retinol from animal sources. Smokers whose 'carotene index' was in the lowest quartile had a 7-fold greater risk than those in the highest quartile of intake. The 'retinol index' was not associated with the risk of lung cancer. Perhaps the most consistent evidence that carotenes may be protective against lung cancer are the findings from five prospective case-control studies all of which reported that low plasma β-carotene levels were associated with an elevated risk (Connett *et al.*, 1989).

In dietary studies it must always be remembered that carotenoid-containing fruits and vegetables contain other protective nutrients such as vitamin C, fibre and polysaccharides. A recently completed Finnish study involving smokers examined the effects of giving α-tocopherol (50 mg day^{-1}) and β-carotene (20 mg day^{-1}) either separately or together for 5–8 years. The results of this study, however, were disappointing. The group that received β-carotene exhibited a significant 18% relative increase in mortality from lung cancer (The Alpha-Tocopherol Beta-Carotene Cancer Prevention Group, 1994). Other intervention trials in which β-carotene is being tested include the Physicians Health Trial and Women's Health Study in New England, and the Carotene and Retinol Efficiency Trial (CARET) on the west coast of the USA (Hong and Itri, 1994). None of these studies has yet been completed but they should provide valuable information about the usefulness or otherwise of dietary intervention against cancer and other chronic diseases such as cardiovascular disease and degenerative eye diseases.

Such intervention trials with β-carotene have shown promising results against pre-cancerous conditions of the upper air/digestive tract which include head, neck and throat (Stich *et al.*; 1988, Garewal, 1993). Premalignant lesions at other sites, such as cervical dysplasia, are also being studied (Hong and Itri, 1994). The results of prospective studies on the association of dietary carotene intake and cervical cancer are somewhat equivocal. High intakes of β-carotene (LaVecchia *et al.*, 1984) and total carotene (Verreault *et al.*, 1989) have been associated with decreased risk. The negative association reported in this latter study between invasive cervical cancer and total dietary carotene intake did not remain significant when adjusted for other risk factors. In another report, plasma levels were significantly lower in women with mild cervical dysplasia than in controls; still lower levels were associated with severe dysplasia or *in situ* carcinoma, and the lowest levels were in cancer cases (Palan *et al.*, 1988). It is essential in studies of this sort to assess dietary intakes over the period when the plasma levels are measured. Cancer, particularly invasive tumours, may well seriously reduce nutritional status and therefore the blood levels of a number of vitamins.

Patients with oesophageal or gastric cancer are often malnourished. However, in a comparatively well-nourished group in France, in an area with a high incidence of oesophageal cancer, a case-control study of diet showed a decreasd relative risk of the disease associated with a high intake of carotene (Tuyns *et al.*, 1987). The possibility that the foods containing the carotenes may also have contained other nutrients which accounted for the decreased risk is highlighted by the association between total fruit consumption and reduced risk, but no association between risk and a 'carotene index' (Brown *et al.*, 1988). A community intervention trial carried out in Linxian, China, investigated the effects of four dietary interventions on cancer incidence, mainly of gastric and oesophageal cancer (Blot *et al.*, 1993). The only intervention that conferred significant protection compared with placebo was a combination of β-carotene, vitamin E and selenium. It is not known whether β-carotene on its own would have been effective.

Vitamin E

There are a number of reasons why vitamin E might be expected to have protective effects on the development of cancer. The vitamin is an important antioxidant, it plays a role in immunocompetence, inhibits the formation of mutagens and has important effects on the repair of membranes and DNA. Proposed mechanisms are shown in Table 17.2.

There have been few controlled human studies on the effect of vitamin E on cancer. However, the results of epidemiological studies, in accordance with those of animal studies (Gaby and Machlin, 1991), suggest that high intakes may decrease risk of certain cancers (Knekt, 1993). The most consistent associations have been reported for cancers of the lung, oesophagus and colorectum, whereas gynaecological cancers, with the possible exception of cervical cancer, have generally not been found to be associated with vitamin E status.

Animal studies provide stronger evidence for an anticancer effect of vitamin E (Burton, 1994), but generalization from them to effects in humans may be unreliable. Even in humans it is important to recognize that the effects of vitamin E may not be uniformly beneficial and sometimes may, indeed, be harmful. This cautionary note is evident in two studies of chemically-induced (Mitchel and McCann, 1993) and UV-induced (Gerrish and Gensler, 1993) skin cancers in mice. In the case of the chemically-induced skin cancer it was found that all-racemic-tocopherol (a mixture of eight stereoisomers) was a promoter of the cancer. The authors suggested that reduction of cellular oxidant levels may trigger tumour promotion. In the second study, it was shown that although vitamin E can significantly reduce the incidence of photocarcinogenesis in mice, the level of dietary tocopheryl acetate necessary to achieve this effect was ultimately fatal to the animals.

There would seem to be much to learn and understand about the behaviour of vitamin E at a cellular level. Free radicals are fundamentally involved in the

Table 17.2. Possible mechanisms of vitamin E protection against cancer induction.

Blocking nitrosamine formation
Protecting DNA from mutagens
Enhancing immune response in increasing:
humoral antibody production
cell mediated immunity
resistance to bacterial infections
lymphocyte response
tumour necrosis factor
natural killer-cell activity

Modified from Gaby and Machlin, 1991.

regulation of cell growth and development (Rice-Evans and Burdon, 1993). If this is so it can readily be seen that high levels of antioxidants might be detrimental to cells and it is essential that cellular levels of antioxidants would need to be controlled.

Vitamin C

The antitumour potential of vitamin C has attracted considerable attention. Possible mechanisms by which vitamin C may affect carcinogenesis include: acting as an antioxidant; blocking the formation of nitrosamines and faecal mutagens; enhancing immune system response; and increasing the activity of detoxifying hepatic enzymes via an effect on cytochrome P450 (Newberne and Suphakarn, 1984; see also Chapter 10). Vitamin C also plays an important role in interfering with tumour growth by virtue of its inhibiting effect on tumour-cell-induced hyaluronidase and thereby preventing the breakdown of extracellular ground substance.

Patients with cancer are often malnourished (Holmes and Dickerson, 1987) and often have low plasma and tissue concentrations of vitamin C. Some earlier studies (Basu *et al.*, 1974) pointed to an increased requirement for the vitamin because low plasma and leucocyte vitamin C values were found in patients who had been receiving an oral supplement of 100 mg day^{-1}. It has further been suggested that high doses of the vitamin may be of value in increasing the survival time of patients with advanced disease (Cameron and Pauling, 1978) and that the vitamin may also be useful in the prevention of cancer (Block and Menkes, 1989).

It is essential to recognize that 'cancer' is not a single disease and that, as with the other vitamins discussed, a role for vitamin C may be somewhat site-specific depending on the carcinogenetic mechanism involved. A review of this subject (Gaby and Singh, 1991b) contains 49 references to epidemiological investigations of the possible relationship of vitamin C to cancer risk. Case-control studies have shown high dietary vitamin C to be associated with reduced incidence of oesophageal cancer. Similar findings were obtained in a study in areas in Iran that have low incidence, compared with those that have high incidence of the disease.

There appears to be substantial evidence suggesting that vitamin C may reduce the carcinogenic potential of some chemicals, such as nitroso compounds (Ohshima and Bartsch, 1981). The nitroso compounds are produced by the reaction of nitrites with secondary and tertiary amines, amides or ureas (Eq. 10.11). Vitamin C is thought to reduce the amount of available nitrite by reducing nitrous acid to nitrogen oxide (Eq. 10.12). A group at high incidence of oesophageal cancer in China has been reported to have a significantly higher excretion of urinary nitrosamines compared with a group at low risk for the cancer (Lu *et al.*, 1986). The food and water normally consumed by residents of the high risk area are high in nitrite, the precursor of nitrosamines. The urinary nitrosamine concentration in the high risk group was reduced to that of the low risk controls with doses of 100 mg of vitamin C for 3 days. Nitrosamines have also been implicated in a model of gastric cancer development. Gastric cancer follows chronic damage to the mucosa with ulcers, hypochlorhydria and atrophic gastritis representing early stages in this progression (Correa *et al.*, 1975). Supplements of vitamin C (1 g day^{-1}) significantly reduced *N*-nitroso-formation in cases of hypochlorhydria (Reed *et al.*, 1983) and the mutagenicity of gastric juice from patients with duodenal ulcers (O'Connor *et al.*, 1985). Diets rich in ascorbic acid have been found to reduce the risk of atrophic gastritis and stomach cancer. In a British study, measurements of plasma levels of the vitamin in two towns with high and low death rates from stomach cancer, however, showed that plasma concentrations of the vitamin were not related to the incidence of atrophic gastritis, though both plasma vitamin C and fruit intake were lower in the high risk area (Burr *et al.*, 1987).

In Australia, a large-scale study of dietary and other risk factors for colorectal cancer found that dietary vitamin C was protective against the disease at intakes above 230 mg day^{-1} (Kune *et al.*, 1987). This is a high intake when compared with RDAs. A large cohort study has shown that frequent fresh fruit consumption reduces lung cancer risk (Gaby and Singh, 1991b). There is evidence that a protective effect of vitamin C against lung cancer is not seen with intakes of 60 mg day^{-1} or below, but is clearly evident with intakes of 100 mg day^{-1} or more (Kromhout, 1987). It is also noteworthy that cigarette smoking is correlated with both high lung cancer incidence and poor vitamin C status (Hoefel, 1983).

Evidence to date suggests that the major benefit of vitamin C in relation to cancer may be in reducing the risk of developing the disease. Presently available scientific evidence in support of the efficiency of vitamin C in the treatment of patients with advanced cancer is weak. Cameron and Pauling (1978) reported that mean survival times of 100 terminal cancer patients who received 10 mg day^{-1} of vitamin C was 210 days, compared with an average survival time of 50 days for 1000 matched controls; however, this was not a controlled study. Subsequently, a randomized double-blind trial (Creagan *et al.*, 1979) failed to demonstrate any therapeutic value of high doses of vitamin C. However, these patients had all received anti-cancer therapy which could have interfered with the suggested mode of action of vitamin C in promoting immunocompetence. A further study (Moertal *et al.*, 1985) has failed to demonstrate any therapeutic value of high doses (10 g day^{-1}) of vitamin C in 100 patients with advanced colorectal cancer who received no prior chemotherapy. In another study involving patients with breast cancer, administration of 3 g day^{-1} of vitamin C also revealed no appreciable effect on the survival times (Poulter *et al.*, 1984).

Vitamin D

Animal studies have indicated that vitamin D inhibits carcinogenesis and the growth of cancer cells. A 19-year prospective study, known as the 'Western Electric' study (Garland *et al.*, 1985) showed that dietary vitamin D and calcium were inversely related to mortality from colorectal cancer. These observations were confirmed in a second study in which blood levels of 25-OH-D were measured (Garland *et al.*, 1989). These findings are preliminary and further evidence is essential before it can be confidently stated that vitamin D prevents carcinogenesis in humans.

Folic Acid

Folic acid is important for the production of nucleic acids (see Chapter 8). Abnormal cytological changes and formation of abnormal tissue (dysplasia) is recognized as being a pre-cancerous condition in some organs, particularly the cervix and lung. Cervical dysplasia has been reported in folate-deficient monkeys (Mohanty and Das, 1982) and a significantly lower level of plasma folate found in patients with cervical cancer compared with controls (Orr *et al.*, 1985). It is possible that localized folic acid deficiency in certain tissues could lead to chromosomal damage and cancerous growth (Heimburger *et al.*, 1987). In cervical cancer associated with human papillomavirus, it is suggested that folate deficiency facilitates the incorporation of viral genetic material into the host cells. A preliminary trial in smokers with pre-cancerous lesions of the lung indicated that folic acid (10 mg) and vitamin B_{12} (500 µg) supplements may be useful in the treatment of this kind of dysplasia (Heimburger *et al.*, 1988).

Conclusions

There is an increasing amount of evidence that some vitamins, especially those with antioxidant properties, may reduce the risk of developing cancers in certain sites. Further research is necessary to confirm some of the potentially important suggestive findings. The way forward will depend largely on the results of well-planned long-term prospective studies which focus on specific kinds of tumours. The findings presently available are of considerable public health value. Evidence for the therapeutic value of vitamins is at present lacking. However, this is not to say that vitamin supplements should not be given to nutritionally depleted cancer patients for improving their state of well-being and hence their quality of life.

References

Basu, T.K. (1988) *Drug–Nutrient Interactions*. Croom Helm, London.

Basu, T.K., Raven, R.W., Dickerson, J.W.T. and Williams, D.C. (1974) Leucocyte ascorbic acid and urinary hydroxyproline levels in patients bearing breast cancer with skeletal metastases. *European Journal of Cancer* 10, 507–511.

Basu, T.K., Chan, V. and Fields, A. (1984) Vitamin A (retinol) and epithelial cancer in man. In: Prasad, K.N. (ed.) *Vitamins, Nutrition and Cancer*. Karger, Basel. pp. 33–45.

Bendich, A. (1991) β-Carotene and the immune response. *Proceedings of the Nutrition Society* 50, 263–274.

Block, G. and Menkes, M. (1989) Ascorbic acid in cancer prevention. In: Moon, T.E. and Micozzi, M.S. (eds) *Nutrition and Cancer Prevention: Investigating the Role of Micronutrients*. Marcel Dekker, New York, pp. 341–388.

Block, G., Palleison, B. and Suban, A. (1992) Fruit, vegetable and cancer prevention: a review of the epidemiological evidence. *Nutrition and Cancer* 18, 1–29.

Blot, W.J., Li, J.Y., Taylor, P.R. Guo, W., Dawsey, S., Wang, G.Q., Young, C.S., Zheng, S.F., Gail, M. and Li, G.Y. (1993) Nutrition intervention trials in Linxian, China: supplementation with specific vitamin/mineral combinations, cancer incidence and specific mortality in the general population. *Journal of the National Cancer Institute* 85, 1483–1492.

Brown, L.M., Blot, W.J., Schuman, S.H. *et al.* (1988) Environmental factors and high risk of oesophageal cancer among men in coastal south Carolina. *Journal of the National Cancer Institute* 80, 1620–1625.

Burr, M.L., Samloff, I.M., Bates, C.J. and Holliday, R.M. (1987) Atrophic gastritis and vitamin C status in two towns with different stomach cancer deaths. *British Journal of Cancer* 56, 163–167.

Burton, G.W. (1994) Vitamin E: molecular and biological function. *Proceedings of the Nutrition Society* 53, 251–262.

Cameron, E. and Pauling, L. (1978) Supplemental ascorbate in the supportive treatment of cancer: re-evaluation of survival times in terminal cancer. *Proceedings of the National Academy of Sciences, USA* 75, 4538–4542.

Collins, A., Duthie, S. and Ross, M. (1994) Micronutrients and oxidative stress in the aetiology of cancer. *Proceedings of the Nutrition Society* 52, 67–75.

Connett, J.E., Kuller, L.H., Kjelsberg, M.O., Polk, B.F., Collins, G., Rider, A. and Hulley, S.B. (1989) Relationship between carotenoids and cancer: the Multiple Risk Factor Intervention Trial (MRFIT) Study. *Cancer* 64, 126–134.

Correa, P., Haenzel, W., Cuello, C., Tannenbaum, S. and Archer, M. (1975) A model for gastric cancer epidemiology. *Lancet* 2, 58–60.

Creagan, E.T., Moertal, C.G., O'Fallon, J.R., Schutt, A.J., O'Connell, M.J., Rubin, J. and Frytak, S. (1979) Failure of high dose vitamin C (ascorbic acid) therapy to benefit patients with advanced cancer: a controlled trial. *New England Journal of Medicine* 301, 687–690.

DeLuca, L.M., Darwiche, N., Celli, G., Kosa, K., Jones, C., Ross, S. and Chen, L. (1994) Vitamin A in epithelial differentiation and skin carcinogenesis. *Nutrition Reviews* 52, S45–S52.

Doll, R. and Peto, R. (1981) The causes of cancer: qualitative estimates of avoidable risks of cancer in the United States today. *Journal of the National Cancer Institute* 66, 1191–1308.

Fujmaki, Y. (1926) Formation of carcinoma in albino rats fed on deficient diets. *Journal of Cancer Research* 10, 468–471.

Gaby, S.K. and Bendich, A. (1991) Vitamin A. In: Gaby, S.K., Bendich, A., Singh, V.N. and Machlin, L.J. (eds) *Vitamin Intake and Health*. Marcel Dekker, New York, pp. 17–27.

Gaby, S.K. and Machlin, L.J. (1991) Vitamin E. In: *Vitamin Intake and Health*, pp. 71–101.

Gaby, S.K. and Singh, V.N. (1991a) β-Carotene. In: *Vitamin Intake and Health*, pp. 29–57.

Gaby, S.K. and Singh, V.N. (1991b) Vitamin C. In: *Vitamin Intake and Health*, pp. 103–161.

Garewal, H.S. (1993) Beta-carotene and vitamin E in oral cancer prevention. *Journal of Cell Biochemistry* (Suppl. 17F), 262–269.

Garland, C., Shekelle, R.B., Barrett-Connor, E., Crique, M.H., Rossof, A.H. and Paul, O. (1985) Dietary vitamin D and calcium and risk of colorectal cancer: a 19-year prospective study in man. *Lancet* 1, 307–309.

Garland, C.F., Comstock, G.W., Garland, F.C., Sing, K.J., Shaw, E.K. and Graham, E.D. (1989) Serum 25-hydroxy vitamin D and colon cancer: eight-year prospective study. *Lancet* 2, 1176–1178.

Gerrish, K.E. and Gensler, H.L. (1993) Prevention of photocarcino-genesis by dietary vitamin E. *Nutrition and Cancer* 19, 125–133.

Gerster, H. (1993) Anticarcinogenetic effect of common carotenoids. *International Journal of Vitamin and Nutrition Research* 63, 93–121.

Heimburger, D.C., Krumdieck, C.L. and Butterworth, C.E. (1987) Role of folate in prevention of cancers of the lung and cervix. *Journal of the American College of Nutrition* 6, 425–429.

Heimburger, D.C., Alexander, C.B., Birch, R., Butterworth, Jr., C.E., Bailey, W.C. and Krumdieck, L. (1988) Improvement in bronchial squamous metaplasia in smokers treated with folate and vitamin B_{12}: report of a preliminary randomised, double-blind intervention trial. *Journal of the American Medical Association* 259, 1525–1530.

Heshmat, M.Y., Kaul, L., Kovi, J. *et al.* (1985) Nutrition and prostate cancer: a case-control study. *Prostate* 6, 7–17.

Hicks, R.M. (1983) Pathological and biochemical aspects of tumour promotion. *Carcinogenesis* 4, 1209–1214.

Hoefel, O.S. (1983) Smoking: an important factor in vitamin C deficiency. *International Journal of Vitamin and Nutrition Research* 24, 121–124.

Holmes, S. and Dickerson, J.W.T. (1987) Malignant disease: nutritional implications of disease and treatment. *Cancer and Metastasis Reviews* 6, 357–381.

Hong, W.K. and Itri, L.M. (1994) Retinoids and human cancer. In: Sporn, M.B., Roberts, A.B. and Goodman, D.S. (eds) *The Retinoids: Biology, Chemistry and Medicine*. Raven Press, New York, pp. 597–630.

Katsouyanni, K., Willett, W., Trichopoulos, D., Boyle, P., Trichopoulou, A., Vasileros, S., Papadiamentis, J. and Macmahon, B. (1988) Risk of breast cancer among Greek women in relation to nutrient intake. *Cancer* 61, 181–185.

Knekt, P. (1993) Epidemiology of vitamin E: evidence for anti-cancer effects in human. In: Packer, L. and Fuchs, J. (eds) *Vitamin E in Health and Disease*. Marcel Dekker, New York, pp. 513–527.

Kolonel, L.N., Hankin, J.H. and Yoshizawa, C.H. (1987) Vitamin A and prostate cancer in elderly men: enhancement of risk. *Cancer Research* 47, 2982–2985.

Kromhout, D. (1987) Essential micronutrients in relation to carcinogenesis. *American Journal of Clinical Nutrition* 45, 1361–1367.

Kummet, T., Moon, T.E. and Meyskens, F.L. Jr. (1983) Vitamin A: evidence for its preventive role in human cancer. *Nutrition and Cancer* 5, 96–106.

Kune, S., Kune, G.A. and Watson, L.F. (1987) Case-control study of dietary etiological factors: the Melbourne colorectal cancer study. *Nutrition and Cancer* 9, 21–42.

Lacroix, A. and Bhat, P.V. (1988) Vitamin A and carotenoids physiology and role in cancer prevention and treatment: an overview. *Nutrition Quarterly* 12, 20–27.

LaVecchia, C., Franceschi, S., Decarli, A., Gentile, A., Fasoli, M., Pampallonar, S. and Tognoni, G. (1984) Dietary vitamin A and the risk of invasive cervical cancer. *International Journal of Cancer* 34, 319–322.

Levine, N. and Meyskens, F. (1980) Topical vitamin A – acid therapy for cutaneous metastatic metanoma. *Lancet* 2, 224–226.

Lindahl, T. (1993) Instability and decay of primary structure of DNA. *Nature* 362, 709–714.

Lu, S.H., Ohshima, H., Fu, H.M., Tian, Y., Li, F., Blettner, M., Wahrendorf, M. and Bartsch, H. (1986) Urinary excretion of N-nitrosamino acids and nitrite by inhabitants of high and low risk areas for oesophageal cancer in northern China: endogenous formation of nitroso-proline and its inhibition by vitamin C. *Cancer Research* 46, 1485–1491.

Mehta, R.R., Hart, G., Beattie, C.W. and DasGupta, T.K. (1987) Significance of plasma retinol binding protein levels in recurrence of breast tumours in women. *Oncology* 44, 350–355.

Michalek, A.M., Cummings, K.M. and Phelan, J. (1987) Vitamin A and tumour recurrence in bladder cancer. *Nutrition and Cancer* 9, 143–146.

Miller, A.B. (1985) Diet, nutrition and cancer: an epidemiological overview. *Journal of Nutrition, Growth and Cancer* 2, 151–171.

Mitchel, R.E.J. and McCann, R. (1993) Vitamin E is a complete tumour promoter in mouse skin. *Carcinogenesis* 14, 659–662.

Moertel, C.G., Fleming, T.R., Creagan, E.T., Rubin, J., O'Connell, M.J. and Ames, H.M. (1985) High dose vitamin C versus placebo in the treatment of patients with advanced cancer who have had no prior chemotherapy. *New England Journal of Medicine* 312, 137–141.

Mohanty, D. and Das, K.C. (1982) Effect of folate deficiency on the reproductive organs of female rhesus monkeys; a cytomorphologic and cytokinetic study. *Journal of Nutrition* 112, 1565–1576.

Moon, R.C. and Itri, L.M. (1984) Retinoids and cancer. In: Sporn, M.B., Roberts, A.B. and Goodman, D.S. (eds) *The Retinoids*, Vol. 2. Academic Press, Orlando, Vol. 2. pp. 327–371.

Newberne, P.M. and Suphakarn, V. (1984) Influence of the antioxidants vitamin C and E and of selenium on cancer. In: Prasad, K.N. (ed.) *Vitamins, Nutrition and Cancer*. Karger, Basel, pp. 46–67.

O'Connor, H.J., Habibzedah, N., Schorah, C.J. *et al.* (1985) Effect of increased intake of vitamin C on the mutagenic activity of gastric juice and intragastric concentration of ascorbic acid. *Carcinogenesis* 6, 1675–1676.

Ohshima, H. and Bartsch, H. (1981) The influence of vitamin C on the *in vivo* formation of nitrosamines. In: Counsell, J.N. and Hornig, D.H. (eds) *Vitamin C (ascorbic acid).* Applied Science, London, pp. 215–224.

Orr, J.W., Wilson, K., Bobiford, C., Cornwell, A., Soong, SJ., Honea, K.L., Hatch, K.D. and Shingleton, H.M. (1985) Corpus and cervix cancer: a nutritional comparison. *American Journal of Obstetrics and Gynaecology* 153, 775–779.

Palan, P.R., Romney, S.L. and Mikhail, M. (1988) Decreased plasma β-carotene levels in women with uterine cervical dysplasias and cancer. *Journal of the National Cancer Institute* 80, 454–455.

Peck, G.L. (1982) Retinoids: Therapeutic use in dermatology. *Drugs* 24, 341–351.

Peto, R., Doll, R., Buckley, J.D. and Sporn, M.B. (1981) Can dietary beta-carotene materially reduce human cancer rates? *Nature* 290, 201–208.

Poirier, L.A. (1987) Stages in carcinogenesis: alteration by diet. *American Journal of Clinical Nutrition* 45, 185–191.

Poulter, J.M., White, W.F. and Dickerson, J.W.T. (1984) Ascorbic acid supplementation and five year survival rates in women with early breast cancer. *Acta Vitaminologica et Enzymologica* 6, 175.

Reed, P.I., Summers, K., Smith, P.L.R., Welters, C.A., Bartholomew, B.A., Hill, M.J., Vennitt, S., Hornig, D. and Bonjour, J.P. (1983) Effect of ascorbic acid treatment on gastric juice nitrite and *N*-nitroso compound concentrations in achlorhydric subjects. *Gut* 24, 492–493.

Rice-Evans, C. and Burdon, R. (1993) Free radical-lipid interactions and their pathological consequences. *Progress in Lipid Research* 32, 71–110.

Shekelle, R., Liu, S., Raynor, W.J. Jr., Lepper, M., Maliza, C. and Rossof, A.H. (1981) Dietary vitamin A and risk of cancer in the Western Electric study. *Lancet* 2, 1185–1189.

Sporn, M.B. and Roberts, A.B. (1983) Role of retinoids in differentiation and carcinogenesis. *Cancer Research* 43, 3034–3040.

Stich, H.F., Rosin, M.P., Hornby, A.P., Matthew, B., Sankaraharayan, R. and Nair, M.K. (1988) Remission of oral leukoplakias and micronuclei in tobacco/betel Quid chewers treated with β-carotene plus vitamin A. *International Journal of Cancer* 42, 195–199.

Temple, N. and Basu, T.K. (1988) Does beta-carotene prevent cancer? A critical appraisal. *Nutrition Research* 8, 685–701.

The Alpha-tocopherol, Beta-carotene Cancer Prevention Group (1994) The effect of vitamin E and beta-carotene on the incidence of lung cancer and other cancers in male smokers. *New England Journal of Medicine* 330, 1029–1035.

Tuyns, A.J., Rioli, E., Doornbos, G. and Pequignot, G. (1987) Diet and oesophageal cancer in Calandos (France). *Nutrition and Cancer* 9, 81–92.

Tyler, H.A., Barr, L.C., Kissin, M.W., Westbury, G. and Dickerson, J.W.T. (1985) Vitamin A and non-epithelial tumours. *British Journal of Cancer* 51, 425–427.

Tyler, H.A., Notley, R.G., Schweitzer, F.A.W. and Dickerson, J.W.T. (1986) Vitamin A status and bladder cancer. *European Journal of Surgical Oncology* 12, 35–41.

Verreault, R., Chu, J., Mandelson, M. and Shy, K. (1989) A case-control study of diet and invasive cervical cancer. *International Journal of Cancer* 43, 1050–1054.

Wahi, P.N., Bodkhe, R.R., Arora, S. and Srivastava, M.C. (1962) Serum vitamin A studies in leucoplakia and carcinoma of the oral cavity. *Indian Journal of Pathology and Bacteriology* 5, 10–16.

Wild, C.P. and Montesano, R. (1991) Immunological quantitation of human exposure to aflatoxins and N-nitrosamines. In: Vanderlaan, M., Stanker, L.H., Watkins, B.E. and Roberts, D.W. (eds) *Immunoassays for Trace Chemical Analysis*. American Chemical Society, Washington, D.C., pp. 215–228.

Willett, W.C. and Hunter, D.J. (1994) Vitamin A and cancer of the breast, large bowel and prostate: epidemiological evidence. *Nutrition Reviews* 52, S53–S58.

Interactions of Drugs and Vitamins

18

Daily requirements of vitamins for man are often expressed as the optimum amount necessary, or the minimum amount essential for health. The minimum requirements may, however, be insufficient when one is under physiological and pathological stresses. Thus, during periods of pregnancy and lactation, vitamin requirements are raised to compensate for the resources drained by the fetus or infant. Training for athletic competition too imposes its own demands on the body for a more rapid turnover of metabolites. In pathological states, requirements for some or all of the vitamins may also be increased. Infectious disease, hyperthyroidism, fever, and post-operative recuperation are examples of such states demanding supplementation. There appears to be a growing awareness that treatments with drugs of widely differing chemistry also have the potential to cause vitamin deficiencies (Ovesen, 1979; Roe, 1980; Basu, 1988). The drug-induced vitamin deficiency may range from overt depletion with clinical manifestation to subclinical deficiency, depending upon the nutritional status at the onset of exposure, the type, dosage and duration of drug use, the presence of predisposing disease processes, and the age of the patients. Administration of drugs in populations with adequate vitamin intake is usually not a problem, but administration of these drugs in those with borderline intake of vitamins or in patients with low nutritional status can result in symptomatic vitamin deficiency states.

The actual incidence of drug-mediated vitamin deficiency is, indeed, difficult to identify, since cause and effect relationships between an offending drug administered and vitamin deficiency are difficult to prove. However, there is evidence that certain drugs induce deficiencies of specific vitamins, and that these deficiencies occur following long-term, high-dose administration in susceptible patients. The drug-induced vitamin deficiencies may, in turn, have a profound effect on the metabolism of drugs and therefore the body's response to the agents (Roe, 1980; Parke and Ioannides, 1981; Basu, 1988). An understanding of these interrelationships may, therefore, lead to a more rational approach to a reduction in adverse drug reactions and interactions. This chapter will delineate two aspects of drug–vitamin interactions: (i) the effects of drugs on vitamin status, and (ii) the effect of vitamin status on drug metabolism and its consequences.

Drug-induced Vitamin Deficiency

An inadequate dietary vitamin intake is not the only way in which deficiencies occur. A number of secondary mechanisms exist, not the least of which is drug-induced vitamin deficiency. Administration of drugs producing 'relative' deficiencies in otherwise healthy populations may not be much of a problem, but administration of these drugs in patients with superimposed disease processes or in subjects with impaired nutritional status can result in hypovitaminosis with symptoms of avitaminosis. Drugs can affect vitamin status either directly or indirectly. Alterations in absorption, metabolism, and excretion are examples of a drug's direct effect on vitamin status. A drug can also cause vitamin deficiency indirectly by altering appetite and taste, gastrointestinal flora, and the rate of stomach emptying.

Drug-induced malabsorption

A number of drugs listed in Table 18.1 are known to affect vitamin absorption through a variety of mechanisms. These may include a direct effect causing morphological changes in the mucosa of the small intestine, inhibition of mucosal enzymes, binding of bile acids, and interference with gastrointestinal factors, such as intrinsic factor. The absorption of all lipid-soluble vitamins has been reported to be impaired by long-term administrations of a large variety of drugs, which include cholestyramine, laxatives, and broad spectrum antibiotics, such as neomycin. Cholestyramine is an anionic polymeric resin which binds bile acids and prevents the solubilization and absorption of all nutrients that are soluble in fat. This lipid-lowering agent has also been shown to lower the absorption of vitamin B_{12}. It appears to bind the intrinsic factor (IF), and thereby prevents the formation of IF–Vitamin B_{12} complex, the form in which vitamin B_{12} is taken up by the intestinal mucosa. Laxatives, such as mineral oil and phenolphthalein, act as solvents for fat-soluble vitamins. Frequent use of these agents may decrease micelle formation resulting in malabsorption of lipids. In addition, these agents may lead to hyperperistalsis, which may, in turn, affect absorption of nutrients.

Chronic administration of antacid is thought to produce thiamin deficiency due to susceptibility of the vitamin to degradation in an alkaline milieu. Oral administration of neomycin is known to cause a reversible inhibition of vitamin B_{12} absorption. The mechanism of action is not known, but may be due to reduced IF function or to a direct influence on the mucosa of the small intestine.

Table 18.1. Drugs affecting vitamin absorption.

Drug	Pharmacological action	Vitamin	Possible mechanism
Cholestyramine	Hypolipidaemic	Fat-soluble vitamins	Binding of bile acids
		Vitamin B_{12}	Binding of IF
Aluminium hydroxide	Antacid	Thiamin	Destruction
Bisacodyl mineral oil	Laxative	Fat-soluble vitamins	Malabsorption due to a decreased micelle formation
Colchicine	Antigout	Carotenes and vitamin B_{12}	Altered gut mucosa
Sulfasalazine	Anti-inflammatory	Folic acid	Damage to the gut wall
Biguanides	Hypoglycaemic	Vitamin B_{12}	Competitive inhibition of absorption
Neomycin	Antibiotic	Vitamin A	Inhibition of pancreatic lipase; binding of bile acids
		Vitamin B_{12}	Binding of IF
Phenytoin Diphenylhydantoin Phenobarbital	Anticonvulsants	Folic acid	Inhibition of gut peptidase
p-Amino salicylate	Anti-tuberculosis	Vitamin B_{12}	Mucosal block in vitamin B_{12} uptake
Salicylates	Analgesics	Vitamin C	Competitive inhibition of absorption
Potassium chloride	Potassium repletion	Vitamin B_{12}	Decreased ileal pH

IF = Intrinsic factor

Vitamin B_{12} deficiency may also result from treatment with a number of drugs either due to malabsorption of the vitamin or to other unknown mechanisms. Drugs that impair the absorption of vitamin B_{12} without interfering with IF factors include *p*-aminosalicylic acid, colchicin and biguanides.

Folic acid in foodstuffs is a mixture of monoglutamate and polyglutamates of varying chain length (see Chapter 8). For maximal absorption of the vitamin the polyglutamate chain must be split by a peptidase in the mucosa of the small intestine. Long-term administrations of anticonvulsants, and anti-inflammatory agents, such as sulfasalazine, are associated with folic acid deficiency. One of the mechanisms by which these agents induce folate deficiency has been suggested to be mediated by malabsorption of the vitamin through inhibiting the intestinal peptidase enzyme.

The interactions between acetylsalicylic acid (aspirin) and ascorbic acid have been the subject of many studies. Thus, administration of therapeutic doses of aspirin to healthy volunteers has been found to decrease the metabolic availability of ascorbic acid (Basu, 1981). Using (1-^{14}C) ascorbic acid, the rate of exhalation of CO_2 has been shown to be three times higher in guinea pigs receiving the vitamin alone than in those given the vitamin simultaneously with Na-salicylate (Ioannides *et al.*, 1982). The rate of exhalation appeared to reach a peak at about 90 min in the animals given ascorbate alone, while in the presence of salicylate the peak exhalation time was approximately 160 min (Fig. 18.1). The bioavailability of the vitamin during the first 400 min was reduced by half following simultaneous administration of salicylate. It is unlikely that salicylate interfered with the metabolism of ascorbic acid, since a decrease in the rate of CO_2 exhalation following salicylate administration was evident as early as 10 min after the administration of the vitamin when the rate of absorption was much higher than the rate of elimination. Furthermore, the rate of CO_2 exhalation, following the completion of absorption and the

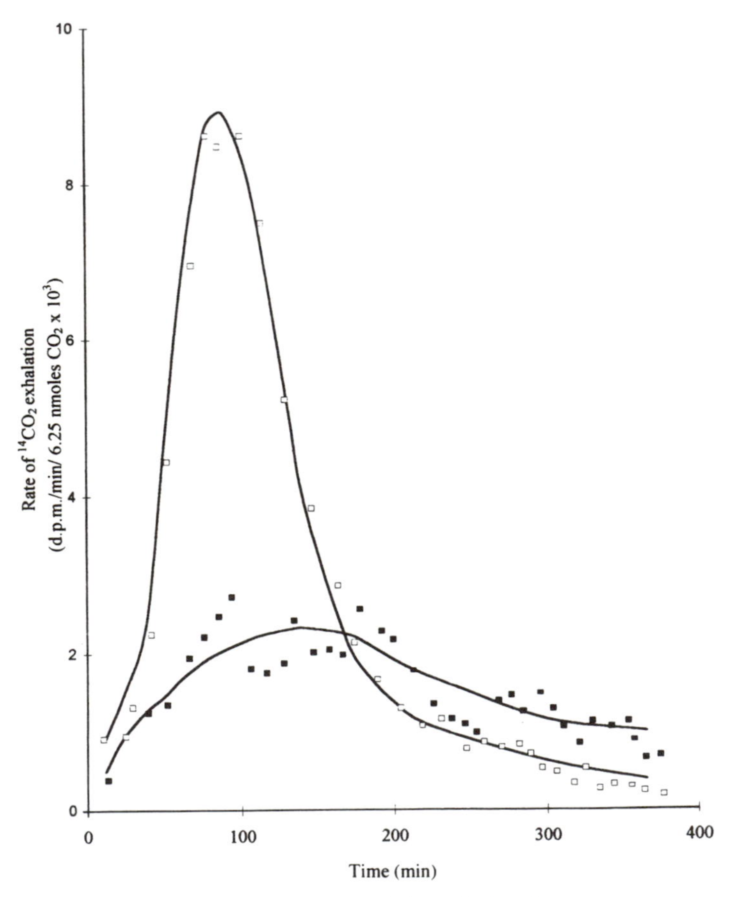

Fig. 18.1. Effect of aspirin on the rate of exhalation of CO_2 following ascorbate administration. □, ascorbate; ■, ascorbate + aspirin. Source: Ioannides *et al.*, 1982.

distribution phases, was the same in animals taking ascorbate alone or ascorbate plus salicylate. These observations indicate that aspirin may impair the gastrointestinal absorption of ascorbic acid. Using gut sac preparations from rats, it has also been shown that both the concentration gradient and the tissue uptake of ascorbic acid are markedly depressed in the presence of acetyl salicylate (Basu, 1981). The fact that aspirin also impedes gastrointestinal absorption of glucose and sodium, and that these decreases are accompanied by a significant fall in intestinal mucosal ATP levels (Arvanitakis *et al.*, 1977), suggests that aspirin may competitively inhibit the Na-dependent active transport of ascorbic acid.

Drugs affecting synthesis

Human requirements for vitamin K are generally satisfied by its availability from dietary sources (phylloquinone) in combination with synthesis by various Gram-positive bacteria in the gut (menaquinone), especially in the ileum where absorption of the vitamin is possible. Oral administration of a variety of broad spectrum antibiotics (Table 18.2) may sterilize the gut and lead to vitamin K deficiency.

Drugs affecting utilization

A variety of drugs are known to interfere with vitamin utilization by blocking or altering transformation of the vitamin to its metabolically active form (Table 18.2). Unlike drugs that interfere with vitamin absorption, agents acting by this mechanism usually produce relatively selective deficiencies. The effect of methotrexate (4-amino-*N*-methyl pteroylglutamic acid, MTX), a widely used cytotoxic agent, on folic acid utilization illustrates this mechanism. In this example, the altered vitamin utilization is of therapeutic significance. The enzymatic activation of folate to dehydrofolate is inhibited by MTX because of its close resemblance in structure to the vitamin (Fig. 18.2). It tends to bind the enzyme dihydrofolate reductase very tightly, inhibiting the synthesis of folic acid coenzymes, which are involved in nucleic acid synthesis (see Chapter 7). The restriction of cellular growth by MTX in patients with malignant disease is principally attained by inducing folate deficiency. A number of other drugs are also known to affect the utilization of folic acid; these include anticonvulsants (phenobarbital, diphenylhydantoin, primidone); primethamine (an antimalarial drug); pentamidine isothionate (a drug used to treat protozoal diseases); and triamterene (a diuretic).

The folate deficiency induced by anticonvulsants in particular received a growing recognition in recent years (Goggin *et al.*, 1987). Prolonged administration of these drugs is associated with low concentrations of folic acid

Table 18.2. Drugs affecting vitamin synthesis and utilization.

Drug	Pharmacological action	Vitamin	Possible mechanism
Neomycin Kanamycin Tetracycline Chloramphenicol Polymyxin Sulphonamides	Antibiotics to sterilize gut	Vitamin K	Inhibition of synthesis by the gut flora
Coumarins	Anti-blood clotting	Vitamin K	Antagonistic effect
Methotrexate	Anti-cancer	Folic acid	Antagonistic effect
Phenobarbital Phenytoin Primidone	Anti-convulsants	Folic acid Vitamin D	Competitive inhibiton of coenzymes
Phenothiazines	Anti-psychotics	Riboflavin	Antagonistic effect
Pyrimethamine	Anti-malarial	Folic acid	Antagonistic effect
Isoniazid Hydralazine Penicillamine	Anti-tuberculosis Anti-hypertensive Anti-rheumatic	Vitamin B_6	Increased urinary excretions of B_6 – drug complex
Oestrogen Progestin	Oral contraceptives	Vitamin B_6	Induction of tryptophan oxygenase
		Vitamin C	Increased ceruloplasmin concentration

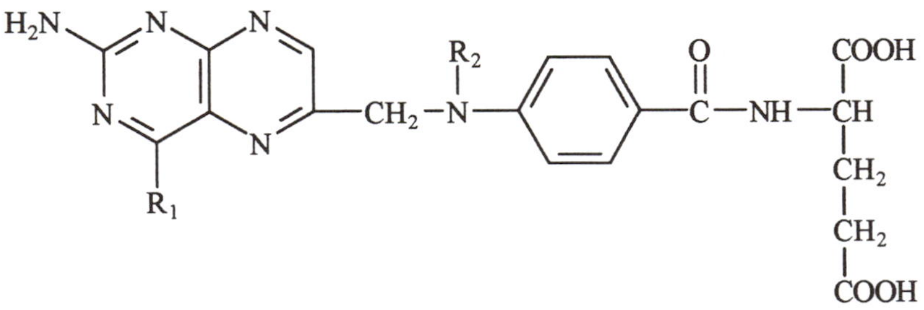

Fig. 18.2. Structural resemblance between folic acid and methotrexate (MTX). Folic acid: R_1 = OH, R_2 = H; MTX : R_1 = NH_2, R_2 = CH_3.

in the serum, red blood cells and cerebrospinal fluid. These biochemical manifestations of folate deficiency are sometimes accompanied by the clinical signs of megaloblastic anaemia. As discussed previously, one of the possible mechanisms by which anticonvulsants precipitate deficiency of folic acid is by affecting its intestinal absorption. There is also evidence that these drugs are potent inducers of NADPH-dependent hepatic drug-metabolizing enzymes (see later). It is therefore possible that the folate-antagonistic effect of anticonvulsants may be mediated through using up NADPH cofactor which is also required for the reduction of folate to its active coenzymes. It remains controversial, however, as to whether folic acid supplements should be given therapeutically or prophylactically to epileptic patients receiving anticonvulsant drugs. The administration of such supplements to epileptics has been reported to exacerbate the seizures (Hartshorn, 1977). On the other hand, there are studies which have failed to demonstrate any difference in seizure frequency between the control and subjects treated with folic acid (Gibberd *et al.*, 1981). In view of the possibility of aggravating seizures in some epileptic patients, folate should always be administered with caution. Its use may be justified if folate deficiency is present. An alternative is to give a rich dietary source, such as liver.

Vitamin D deficiency associated with reduced serum calcium and phosphate levels, and occasional clinical signs of rickets or osteomalacia, has also been recognized as a consequence of long-term treatment with anticonvulsants (Dent *et al.*, 1970; Krause *et al.*, 1982). This side-effect of anticonvulsant therapy is thought to be secondary to an induction in hepatic microsomal enzymes including vitamin D-25-hydroxylase (Morijiri and Sato, 1981), which is responsible for the conversion of vitamin D to 25-OH-D (see Chapter 13). It has been suggested that the drug-induced 25-hydroxylase activity in the liver may be responsible for diverting the metabolism of vitamin D to its more polar and inactive biliary-excreted metabolites rather than to 25-OH-D and subsequently to its active metabolite, 1,25-dihydroxy-D, which is formed in the kidney (Stamp, 1974). This hypothesis is supported by the observations made that phenytoin does not seem to affect the circulatory 1,25-(OH_2)-D, while 25-OH-D concentrations are first increased and then rapidly decreased as the dose of the agent is increased (Fig.18.3). It is generally advisable to supplement the vitamin D intake of epileptics on long-term treatment with anticonvulsants (Christiansen *et al.*, 1973).

Utilization of vitamin K may be competitively blocked by oral anticoagulants of the coumarin or phenindandione groups. One well known antagonist of vitamin K is dicoumarol (Fig. 18.4), the substance found in spoiled sweet clover (*Melilotus*) that causes a fatal haemorrhagic disease in cattle. It is thought that this coumarin antagonist competes directly with vitamin K (Fig. 15.4) at the site where the vitamin exerts its biological activity. The coumarin agents are also thought to block the reduction of oxidized quinone, thus preventing recycling and regeneration of hydroquinone, the active form of vitamin K. Warfarin (Fig. 18.4), a rat poison, and Tromexan are dicoumarol derivatives. Treatment of conditions such as thrombophlebitis with these drugs does deliberately result in inhibition of the normal clotting mechanism in an attempt to prevent additional thrombosis. With increasing use of coumarin drugs, however, haemorrhage can result if the anticoagulant effect is excessive. The anti-

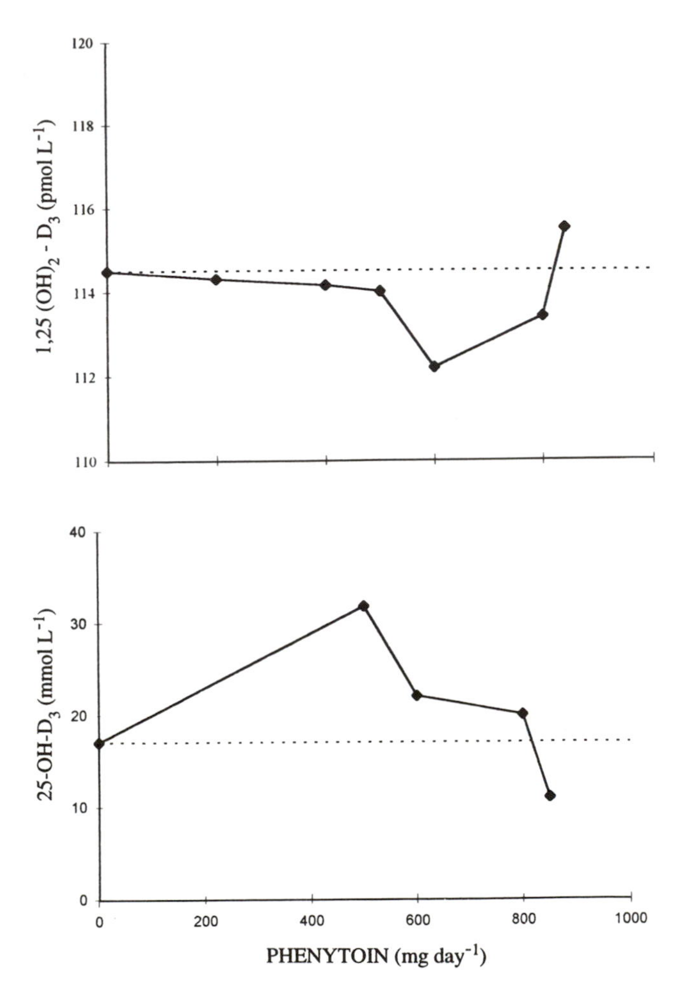

Fig. 18.3. Relationships between the plasma 1,25-$(OH)_2$-D_3 and 25-OH-D_3 concentrations and the dose of phenytoin administered. The patient also received primidone (300 mg day^{-1}) at all doses of phenytoin. The equilibration period for each dose level of phenytoin was 4–5 weeks. Source: Gascon-Barré *et al.* (1984) (with permission).

coagulant effect is usually reversed by phylloquinone or menaquinone but not by menadione. It is thought that the 3-phytyl side chain in phylloquinone or menaquinone is an essential feature which makes the molecules able to reverse the coagulation defect induced by coumarin and its congeners. It is important to restore a normal prothrombin time in patients who are receiving anticoagulants.

Salicylate in doses of 6 g day^{-1} can induce hypoprothrombinaemia in humans. The effect of salicylate differs from that of the 4-hydroxycoumarin

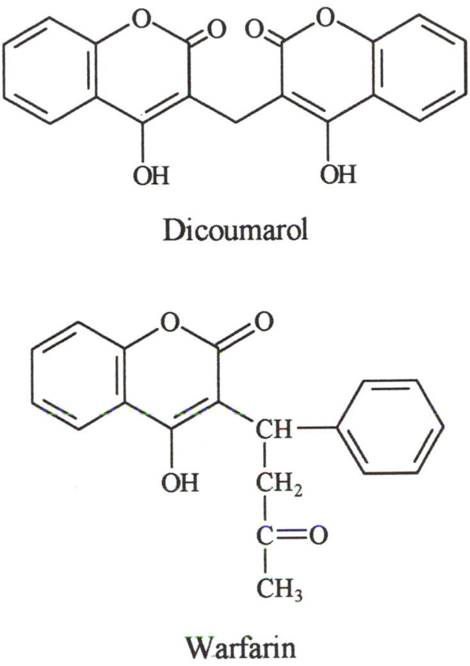

Dicoumarol

Warfarin

Fig. 18.4. Antagonists of vitamin K.

drugs and appears to be directed at the vitamin K reductase enzyme (Hildebrandt and Suttie, 1983). Diphenylhydantoin also reduces prothrombin levels in adults. Infants of mothers on hydantoin therapy should receive prophylactic vitamin K since they appear to be particularly susceptible (Olson, 1982).

Drug-induced oxidation and excretion

A variety of drugs, such as isoniazid, hydralazine and penicillamine, are known to be antagonistic to vitamin B_6 as a result of their structural similarity to the vitamin (Fig. 18.5). Isoniazid and hydralazine are both hydrazides, which form hydrazone complexes with the aldehyde form of vitamin B_6 (pyridoxal). The formation of pyridoxal hydrazone results in enhanced urinary excretion and subsequent depletion of vitamin B_6 from the body. Penicillamine also forms a highly soluble complex with pyridoxal which is then rapidly excreted by the kidney. Vitamin B_6 deficiency associated with clinical signs of peripheral neuropathy and behavioural changes can result after prolonged administration or excessive doses of these drugs. The adverse reactions can usually be treated with judicious doses of vitamin B_6 without compromising the efficacy of the drugs.

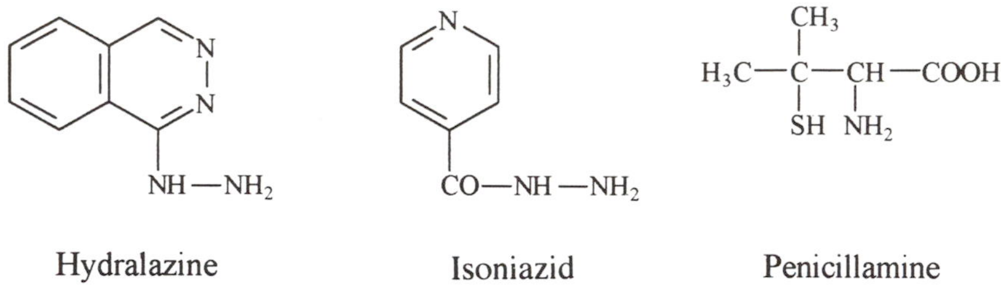

Hydralazine

Isoniazid

Penicillamine

Fig. 18.5. Vitamin B_6 antagonists.

Biochemical evidence of vitamin B_6 deficiency has also been reported among women taking combined oestrogen–progestin oral contraceptives. The biochemical alterations most frequently found were elevated urinary excretion of a number of tryptophan metabolites of the kynurenine pathway following a standard oral tryptophan load; the elevated metabolite excretions could be reversed by supplements of vitamin B_6. The exact mechanism of this effect is not known, but it is thought that a functional deficiency of vitamin B_6 may be due to the induction of tryptophan pyrrolase, which may stimulate the vitamin B_6-dependent NAD pathway of tryptophan metabolism at the expense of the vitamin B_6-dependent serotonin pathway. A proposed scheme outlining the interaction between oral contraceptives and vitamin B_6 is given in Fig. 18.6. The interference of the synthesis of the neurotransmitter may possibly account for the side effects, such as headaches, emotional disturbances and depression, which are associated with the use of oral contraceptives. The users of these agents appear to have low levels of vitamin B_6 during pregnancy and low vitamin B_6 in breast milk after pregnancy (Miller, 1986). Furthermore, long-term users have been found to produce infants with lower Apgar scores than short-term users.

The results of many studies have indicated that ascorbic acid status can also be negatively influenced by administration of oral contraceptives (Rivers, 1975). In a study (Basu, 1986) involving guinea pigs, oral administration of either oestrogen (5 μg) or progestogen (250 μg) in combination with 5 mg of ascorbic acid (minimum requirement) daily for 21 days resulted in significantly lower concentrations of ascorbic acid in plasma, tissues and urine than in animals receiving only 5 mg of the vitamin. None of these animals showed any clinical signs of ascorbic acid deficiency. Clinical manifestations of scurvy were exhibited, however, when animals receiving no ascorbic acid supplement were treated with the steroid hormones for 7 days. These results indicate that the interaction between oral contraceptive hormones and ascorbic acid may be of clinical importance only in cases of border-line intake of the vitamin. The mechanism by which the steroid hormones affect ascorbic acid status is thought

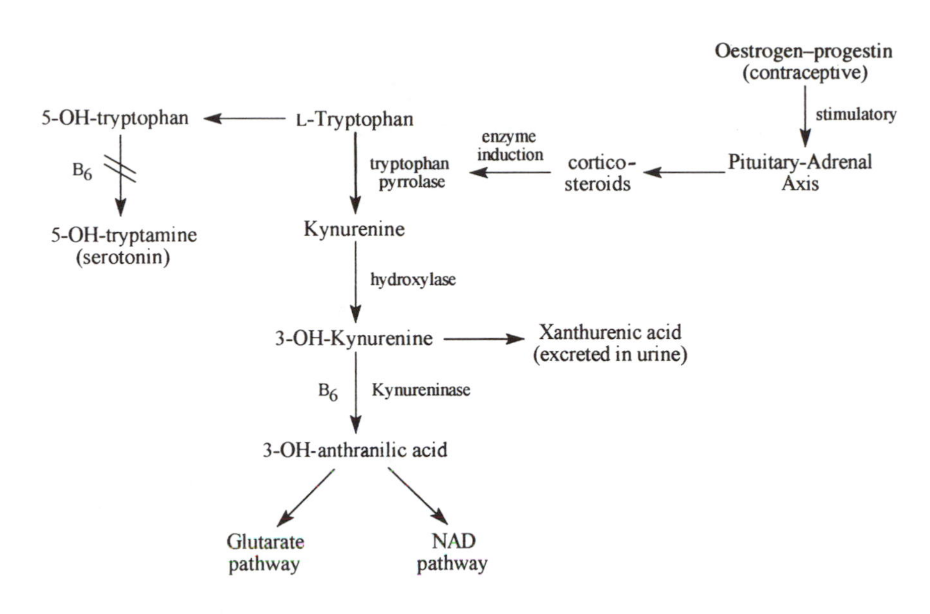

Fig. 18.6. Proposed scheme for oral contraceptive and vitamin B_6 interaction.

to be through its oxidation. *In vitro* studies have revealed a markedly higher rate of oxidation of the vitamin in the presence of either estinyl or progestogen than in untreated controls (Basu, 1986). Furthermore, in animals receiving the hormones there was an associated increased level of serum ceruloplasmin, a copper-containing protein with ascorbate oxidase activity (see Chapter 10).

Alcohol-induced vitamin deficiency

Alcohol (ethanol) causes many pharmacological effects. Alcohol beverages are consumed for their mood-altering actions and pleasing tastes, and in moderate amounts for stimulating appetite. Alcoholic beverages contain no proteins, vitamins or minerals. Although alcohol is generally socially acceptable, when taken in excess, it can affect nutritional status.

Vitamin deficiencies occurring in alcohol abusers are complex in their aetiology (Basu, 1988). Causes of deficiency include dietary insufficiency, malabsorption, hyperexcretion, and a decreased rate of activation. Of all the vitamins, B-vitamin deficiencies are the most common among alcoholics. Deficiencies of these vitamins are often associated with clinical manifestations. In alcohol abusers, thiamin deficiency produces two disorders of the central nervous system: the neurasthenic syndrome characterized by lassitude, irritability and anorexia; and the Wernicke–Korsakoff syndrome which combines the

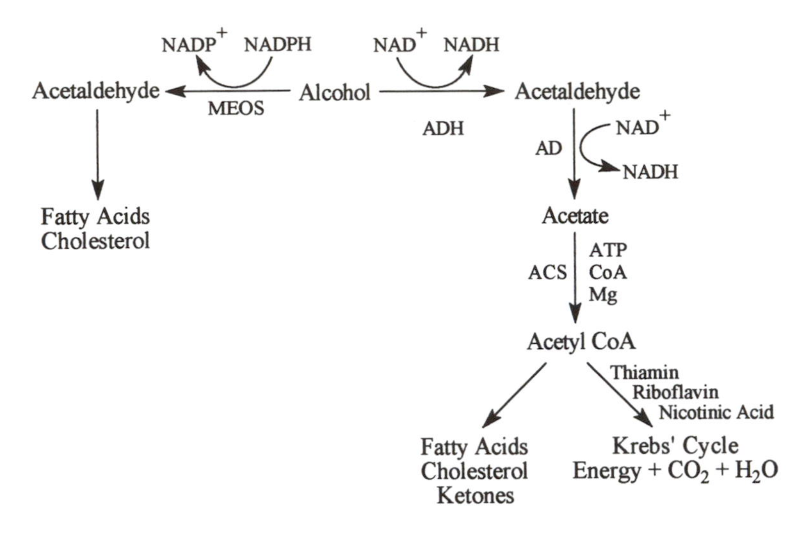

Fig. 18.7. Major metabolic pathways of alcohol (ADH, alcohol dehydrogenase; AD, aldehyde dehydrogenase; ACS, acetyl CoA synthetase; MEOS, microsomal ethanol oxidizing system).

features of Wernicke's encephalopathy (mental confusion, ataxia and ophthalmoplegia) with Korsakoff's psychosis with loss of short-term memory. Folic acid deficiency with clinical manifestations of megaloblastic anaemia occurs in alcoholics but much less commonly than simple macrocytosis. When it does occur the individuals are more frequently malnourished.

Alcohol with empty calories often replaces food in the diet. It can cause inflammation of the stomach, pancreas and intestine, and interferes with the normal process of digestion and absorption. The cytoplasmic alcohol dehydrogenase (ADH) is the rate limiting enzyme for the yield of energy from alcohol (Fig. 18.7). This pathway requires vitamins, such as thiamin, riboflavin, nicotinic acid and pantothenic acid. Prolonged excessive alcohol consumption can therefore increase the body's requirements for these B-vitamins.

The microsomal ethanol oxidizing system (MEOS) is an alternative pathway for the oxidation of alcohol to acetaldehyde (Fig. 18.7), taking place especially when the ADH pathway is exhausted due to the shortage of B-vitamins. The metabolic product, acetaldehyde, can interfere with the activation of vitamins in the liver. The vitamins particularly affected by this mechanism include thiamin, vitamin B_6, folic acid, and vitamin D.

Symptoms of vitamin A deficiency, such as night blindness, accompanied by low circulatory levels of the vitamin, have been found among alcoholics with liver damage (Leo and Lieber, 1982). Retinol is oxidized to retinaldehyde (the active form of vitamin A for vision) by alcohol dehydrogenase (see Chapter 11), the enzyme which also oxidizes ethanol to acetaldehyde. This enzyme has a

50-fold greater affinity for ethanol than retinol. It appears, therefore, that the mechanism for night blindness among alcohol abusers is caused by competitive inhibition of retinal formation by ethanol. Chronic alcoholics have also been reported to be more prone to bone fracture than non-alcoholics (Johnell *et al.*, 1982). The association of alcoholism with bone fracture is attributed to abnormal vitamin D metabolism. The liver is important for both storage and metabolism of vitamin D, and interference with its function by alcohol can cause secondary vitamin D deficiency. The effect of alcohol on the metabolism of the vitamin has been suggested to occur in the kidney where 25-hydroxy-D is converted to either 1,25-dihydroxy-D (active form) or 24,25-dihydroxy-D (inactive form). Thus, alcohol administration to chickens has shown a decrease in the active metabolite and an increase in the inactive one (Kent *et al.*, 1979). Further studies have indicated that the formation of 1,25-dihydroxy-D is under control of the pituitary, the functioning of which is known to be affected by alcohol (MacIntyre, 1979).

Cytotoxic drugs affecting vitamin status

All cytotoxic agents are highly toxic. They not only kill malignant cells, but may also damage normal cells. As a consequence of treatment, side effects such as loss of appetite, nausea, vomiting and gastric irritation may occur when the drugs are administered parenterally. In addition, when the drugs are given orally, further side effects including glossitis, cheilosis and angular stomatitis may also occur. These side effects may often lead to the impairment of nutritional status. One should bear in mind that the nutritional status of a cancer patient may already be impaired due to the disease process. It is, therefore, important that any further nutritional side effects due to the use of drugs are carefully evaluated. A number of anti-cancer agents have been reported to be antagonistic to several vitamins (Basu, 1983a); the antagonistic effect of methotrexate on folic acid has been discussed previously. A list of anti-cancer agents, which are known to be antagonistic to vitamins in animals and humans, is given in Table 18.3.

Vitamin Deficiency and Drug Biotransformation

Drug metabolism

Most drugs are lipid soluble; they are modified prior to excretion. The modification is usually carried out through metabolism in the hepatic and extrahepatic tissues by two distinct phases (Fig. 18.8). The first phase is a synthetic and includes oxygenations, reductions, and hydrolysis. These processes result in the formation of a polar compound, thus making it more suitable for excretion. The

Table 18.3. Cytotoxic agents as antagonists to vitamins.

Agents	Tumours studied	Antagonistic effect
Galactoflavin	Advanced neoplasms	Riboflavin
4-Deoxypyridoxine	Lymphocytic leukaemia	Pyridoxine
6-Aminonicotinamide	Renal cell carcinoma	Nicotinamide
Methotrexate	Lymphocytic leukaemia	Folic acid
Thio-Semicarbazone	Transplanted tumours	Thiamin
5-Fluorouracil	A variety of cancers	Thiamin
Vinblastin	Testicular teratoma	Retinol

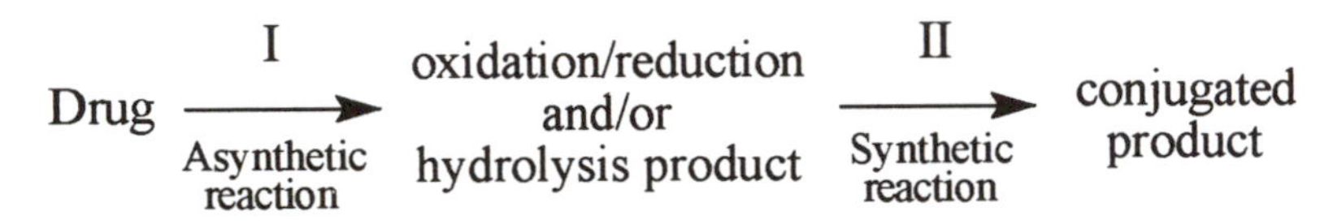

Fig. 18.8. Metabolic transformation of drugs.

compounds formed pass into the second, or synthetic, phase of the process in which conjugates are formed with glucuronic acid, sulphate, or glycine, rendering them even more polar and thus more readily excretable. The enzyme system that brings about the oxygenations of phase I metabolism requires a number of components which include enzymes for hydroxylating and demethylating substrates, flavoproteins, a haemoprotein (cytochrome P450), molecular oxygen and the provision of reducing agents, usually in the form of reduced NADP (Fig. 18.9). The enzymes that catalyse the biotransformations are predominantly the microsomal mixed-function oxidases (MFO), whereas those that catalyse the conjugations and transferases are located in the endoplasmic reticulum or in the cytosol of the cell.

Vitamin deficiency and drug metabolism

There has been increasing experimental evidence suggesting that the hepatic microsomal drug metabolizing system can be affected by deficiency of various vitamins (Table 18.4). Among these vitamins, ascorbic acid has been most extensively studied in relation to drugs. In addition to the evidence linking changes in the activity of drug-metabolizing enzymes with changes in ascorbic acid status (Basu, 1983b) in guinea pigs (one of the species that do not

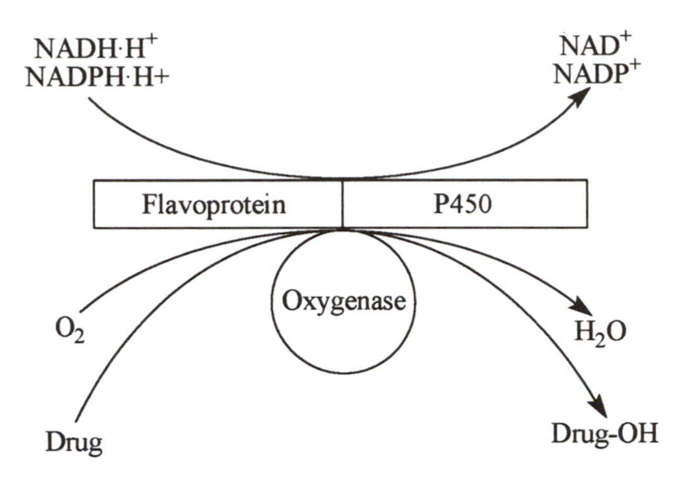

Fig. 18.9. The cytochrome P450 and flavoprotein-dependent oxygenase system.

Table 18.4. Effect of vitamin deficiency on some drug-metabolizing enzyme activities in animals.

Vitamin deficiency	Affected drug metabolism indices
A	Cytochrome P450, oxidations of aminopyrine, ethylmorphine, aniline, benzo(a)pyrene, 7-ethoxycoumarin
B_1	Oxidations of *N*-nitrosodimethylamine, acetaminophen, aniline
B_2	NADPH-P450 reductase, oxidations of aniline, acetanilide, benzo(a)pyrene, aminopyrine
B_3	Oxidation of anaesthetics
C	Cytochrome P450, NADPH-P450 reductase and several mono-oxygenation activities
E	Oxidations of codeine, ethylmorphine, benzo(a)pyrene
Folate	Decrease in induction of cytochrome P450 by barbiturates

Source: Adapted from Zannoni and Sato (1976); Guengerich (1995).

synthesize ascorbic acid), the excretion of the vitamin appears to be markedly increased in rats (animals that synthesize ascorbic acid) receiving a variety of drugs (Wilson, 1974). The drug-induced ascorbic aciduria implies an induced biosynthesis of the vitamin, and this may be due to its increased requirement. It is, therefore, possible that in humans (who, like guinea pigs, are unable to synthesize ascorbate), when undergoing therapy with these drugs, vitamin deficiency may be precipitated if the increased need is not met. However, it should be pointed out that most of this evidence has been derived from experimental animals, and therefore some caution is needed in extrapolation of these results to humans. Only in isolated reports have the modifying effects of vitamins, ascorbic acid in particular, on drug metabolism also been evident in humans (Ginter and Vejmolova, 1981). Thus, in a study involving human volunteers receiving 500 mg ascorbic acid daily for one year, the plasma half-life of antipyrine was decreased, associated with an increased renal clearance of the drug. It is noteworthy that in man, 95% of antipyrine is metabolized by the hepatic MFO system. These results indicate that diets inadequate in ascorbic acid or drug-induced vitamin deficiency may inhibit drug biotransformation in man with consequent alterations in pharmacological activity and manifestation of drug toxicity.

Risk Populations

Drug–vitamin interactions may be of clinical significance, especially in people with a borderline intake and abnormal metabolism of vitamins, such as alcoholics, the ill and the aged, pregnant women, and infants. Apart from alcoholics and the ill, two overriding concerns are: (i) pregnancy outcome in mothers who are deficient in vitamins due to either primary or secondary (drug-induced) causes, and (ii) drug toxicity among the elderly whose nutrient intake is often borderline and who frequently use polypharmacy for various ailments.

Maternal vitamin deficiency and the fetal outcome

Maternal diet and nutritional status are important factors for fetal growth. The degree of rise in birthweight of infants seems to depend upon the nutritional state of the mother (Rush *et al.*, 1980). During the past few decades, ample experimental evidence has been provided showing the involvement of primary or drug-induced vitamin deficiencies in causing congenital abnormalities. The adverse effects of deficiencies of both water- and fat-soluble vitamins upon the fetus and living young have been well documented in the literature (Roe, 1980; Basu, 1983c). The list of pathologies produced in experimental animals by maternal vitamin deficiency is shown in Table 18.5.

Folic acid is required for nucleic acid synthesis and hence growth and development. Its requirement is thought to be much higher in the fetus than in the mother, in order to keep pace with the tremendous mitotic activity that occurs during intrauterine growth. Folic acid deficiency throughout gestation has been shown to produce a number of congenital malformations in rats (Basu, 1983c). Incidence of some of these malformations has been found to be higher among the fetuses of mothers in which folate deficiency is exacerbated

Table 18.5. Avitaminosis and congenital malformations in experimental animals.

Maternal vitamin deficiencies during pregnancy	Congenital malformations among offspring
Thiamin	Stillbirths; low birth weight
Riboflavin	Short long bones; cleft palate; conjoined twins
Niacin	Fetal resorption; hind limb defects
Vitamin B_6	Stillbirths; low birth weight
Folacin	Cleft lip; hydrocephalus; skeletal malformations
Vitamin B_{12}	Hydrocephalus
Ascorbic acid	Impaired collagen synthesis
Vitamin A	Anophthalmia/microphthalmia, hydrocephalus, cardiovascular anomalies, urogenital anomalies
Vitamin D	Skeletal abnormalities
Vitamin K	Brain haemorrhage, hypoprothrombinaemia
Vitamin E	Anencephaly, umbilical hernia, club feet

by prolonged administration of anticonvulsants (Labadarios, 1993). These findings may be of relevance to the congenital abnormalities seen in children born to epileptic mothers on anticonvulsant therapy (Basu, 1983c).

It is important that the use of drugs during lactation is also seriously evaluated. Drugs such as oral contraceptives may diminish breast milk, especially when the agents are taken sooner than six weeks postpartum (Chopra, 1972). Other studies report a decrease in the amount of a variety of nutrients in the breast milk of oral contraceptive users. The oral contraceptive associated reductions in the production of breast milk and its nutrient contents have been reported to be related to a slower growth rate seen in some infants.

Vitamin deficiency and the elderly

There is considerable evidence to suggest that the incidence of adverse drug reaction increases with advancing age (Gillette, 1979). Drugs which have been found to cause adverse reactions include digitalis preparations, antibiotics, corticosteroids, anticoagulants, analgesics, and tranquilizers. Adverse reactions to drugs are frequently encountered in old age when homeostatic mechanisms are impaired leading to subclinical malnutrition, which in turn under conditions of stress, such as infection, trauma, chronic disease, and drug therapy, may precipitate clinical malnutrition.

Consequences of drug–vitamin interactions

Vitamin malnutrition due to either shortage or wrong choice of food appears to be prevalent in a considerable segment of the world's population, especially in the elderly. The deficiency state is exacerbated when these subjects are treated with drugs, resulting in many cases of overt deficiency with clinical manifestations (Fig. 18.10). Such deficiencies may, in turn, result in drug toxicity by impairing the rates of drug metabolism. Thus the study of the interrelation between vitamins and drugs may be viewed as an extension of the factors that modify drug action and dosage. It may be necessary either to increase the intake of vitamins or to decrease the dosage of a drug in order to counteract any detrimental effect of a drug upon vitamin status, and vice versa. Such considerations are of importance, especially in the risk groups such as the elderly, alcoholics, epileptics, pregnant women and infants, where even the normal complex pharmacokinetics of drug metabolism may be further complicated by additional physiological factors.

The level of significance of drug-mediated vitamin deficiency in clinical and subclinical situations is, however, often difficult to evaluate because of variables

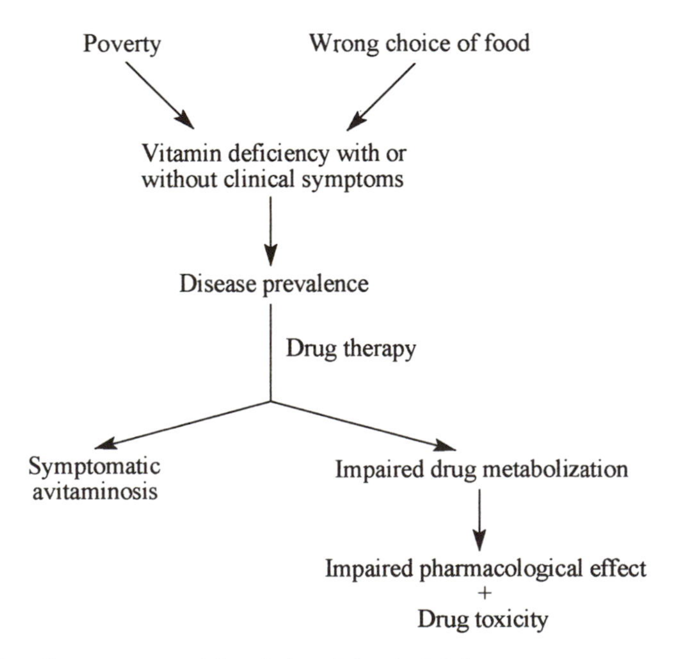

Fig. 18.10. Consequences of drug-induced vitamin deficiency.

such as the nutritional status at the onset of therapy, the type of drug involved, the duration of treatment, and age. As a consequence, drug-induced nutrient deficiency is often an overlooked problem. It is important that the probability of this occurrence, as well as the clinical significance of many interactions, is evaluated in the individual patient rather than by the use of generalized rules.

It is important to realize that many clinical signs which are believed to be 'adverse drug reactions' (Table 18.6) may also be signs of nutrient deficiency. Thus, symptoms such as fatigue and anorexia may be the result of multi-vitamin deficiency. Psychic disturbances may be caused by folate, vitamin B_{12}, and vitamin B_6 deficiencies, and peptic ulcers may be due to vitamin A deficiency. In addition, there are reactions such as anticoagulant-induced haemorrhages which are known to be caused by vitamin K deficiency. Where these clinical symptoms are manifested, supplementation with appropriate vitamins may help to counteract the adverse drug reactions.

There are, however, situations where vitamin supplementation may nullify the beneficial effects of drugs. For instance, the anti-cancer activity of methotrexate is mediated through inducing folic acid deficiency, and supplementation with this vitamin will reverse the therapeutic effect of the agent. Another classic example of this kind is 'the levodopa–vitamin B_6 interaction'. There is evidence that ingestion of vitamin B_6 counteracts the therapeutic effects of levodopa in the control of Parkinson's disease, which is associated with a depression of the dopamine content of the caudate nucleus. Levodopa rather than dopamine is used because it crosses the blood–brain barrier more readily than dopamine. It is metabolized to dopamine by dopa decarboxylase, which requires pyridoxal phosphate as a coenzyme. This reaction takes place in the brain as well as in peripheral tissues. It has been suggested that in the presence of vitamin B_6, much of the levodopa is decarboxylated in extracerebral tissues; as a result an inadequate amount of the drug enters the brain to replenish striatal dopamine. The degree of inhibition of the pharmacological effects of

Table 18.6. Some adverse drug reactions.

Drugs	Reactions
Cardiac glycosides	Fatigue, anorexia, psychic disturbances
Thiazide diuretics	Hypokalaemia, neural dysfunction, arrhythmia
Antihypertensives	Mental depression, dry mouth, diarrhoea
Anticoagulants	Haemorrhage
Hypnotics	Frank psychosis
Tricyclic antidepressants	Confusion
Phenylbutazone	Peptic ulcer, hypertension

levodopa by vitamin B_6 appears to be dose related and can occur with as small a dosage as 5 mg. Patients receiving levodopa therapy are therefore recommended to avoid multiple vitamin preparations and other preparations containing levels of vitamin B_6 greater than 2 mg. It should, however, be pointed out that the concurrent use of carbidopa, a peripheral dopa decarboxylase inhibitor, prevents the levodopa-inhibiting effects of vitamin B_6, and the carbidopa–levodopa combination product is recommended for use in patients receiving vitamin B_6 supplementation.

References

Arvanitakis, C., Chen, G.H., Folscroft, J. and Greenberger, J. (1977) Effect of aspirin on intestinal absorption of glucose, sodium and water in man. *Gut* 18, 187–190.

Basu, T.K. (1981) The influence of drugs with particular reference to aspirin on the bio-availability of vitamin C. In: Counsell, J.N. and Hornig, D.H. (eds) *Vitamin C (Ascorbic Acid)*. Applied Science, London, pp. 273–289.

Basu, T.K. (1983a) Vitamins – cytotoxic drug interaction. *International Journal of Vitamin Nutrition Research* Suppl. 24, 225–233.

Basu, T.K. (1983b) Effects of protein malnutrition and ascorbic acid levels on drug metabolism. *Canadian Journal of Physiology Pharmacology* 61, 295–301.

Basu, T.K. (1983c) Avitaminosis and congenital malformation. *International Journal of Vitamin Nutrition Research* Suppl. 24, 9–14.

Basu, T.K. (1986) Effects of oestrogen and progestogen on the ascorbic acid status of female guinea pigs. *Journal of Nutrition* 116, 570–577.

Basu, T.K. (1988) *Drug-Nutrient Interactions*. London, Croom Helm.

Chopra, J.G. (1972) Effect of steroid contraceptives on lactation. *American Journal of Clinical Nutrition* 25, 1202–1204.

Christiansen, C., Rodbro, P. and Lund, M. (1973) Effect of vitamin D on bone mineral mass in normal subjects and in epileptic patients on anticonvulsants: a controlled therapeutic trial. *British Medical Journal* 2, 208–209.

Dent, C.E., Richens, A., Rowe, D.J.F. and Stamp, T.C.B. (1970) Osteomalacia with long-term anticonvulsant therapy in epilepsy. *British Medical Journal* 4, 69–72.

Gascon-Barré, M., Villeneuve, J. and Lebrun, L. (1984) Effect of increasing doses of phenytoin on the plasma 25-hydroxy vitamin D and 1,25-dihydroxy vitamin D concentrations. *Journal of American College of Nutrition* 3, 45–50.

Gibberd, F.B., Nicholls, A. and Wright, M.G. (1981) The influence of folic acid on the frequency of epileptic attacks. *European Journal of Clinical Pharmacology* 19, 57–60.

Gillette, J.R. (1979) Biotransformation of drugs during ageing. *Federation Proceedings* 38, 1900–1909.

Ginter, E. and Vejmolova, J. (1981) Vitamin C statuş pharmacokinetic profile of antipyrine in Man. *British Journal of Clinical Pharmacology* 12, 256–258.

Goggin, T., Gough, H., Bissessar, A., Crowley, M., Baker, M. and Callaghan, N. (1987) A comparative study of the relative effects of anticonvulsant drugs and dietary folate on the red cell folate status of patients with epilepsy. *Quarterly Journal of Medicine* 65, 911–919.

Guengerich, F.P. (1995) Influence of nutrients and other dietary materials on cytochrome, P450 enzymes. *American Journal of Clinical Nutrition* 61 (Suppl), 651S–658S.

Hartshorn, E.A. (1977) Food and drug interactions. *Journal of the American Dietetics Association* 70, 15–19.

Hildebrandt, E.F. and Suttie, J.W. (1983) The effects of salicylate on enzymes of vitamin K metabolism. *Journal of Pharmacology* 35, 421–426.

Ioannides, C., Stone, A.N., Breacker, P.J. and Basu, T.K. (1982) Impairment of absorption of ascorbic acid following ingestion of aspirin in guinea pigs. *Biochemical Pharmacology* 31, 4035–4038.

Johnell, O., Nilsson, B.E. and Wiklund, P.E. (1982) Bone morphometry in alcoholics. *Clinical Orthopathy* 165, 253–258.

Kent, J.C., Devlin, R.D., Gutteridge, D.H. and Retallack, R.W. (1979) Effect of alcohol on renal vitamin D metabolism in chickens. *Biochemical Biophysical Research Communications* 89, 155–161.

Krause, K.H., Berlit, P. and Bonjour, J.P. (1982) Vitamin status in patients on chronic anticonvulsant therapy. *International Journal of Vitamin Nutrition Research* 52, 375–385.

Labadarios, D. (1993) The effect of chronic drug administration on folate status and drug toxicity. In: Parks, D.V., Ioannides, C. and Walker, R. (eds) *Food, Nutrition and Chemical Toxicity*. Smith-Gordon, London, pp. 71–80.

Leo, M.A. and Lieber, C.S. (1982) Hepatic vitamin A depletion in alcoholic liver injury. *New England Journal of Medicine* 307, 597–601.

MacIntyre, I. (1979) The role of the kidney in vitamin D metabolism. *Advances in Nephrology* 8, 153–163.

Miller, T. (1986) Do oral contraceptive agents affect nutrient requirements – vitamin B_6? *Journal of Nutrition* 116, 1344–1345.

Morijiri, Y. and Sato, T. (1981) Factors causing rickets in institutionalized handicapped children on anti-convulsant therapy. *Archives of Diseases in Childhood* 56, 446–449.

Olson, R.E. (1982) Vitamin K. In: Coleman, R.W. (ed.) *Hemostasis and Thrombosis*. Lippincott, Philadelphia, pp. 582–594.

Ovesen, L. (1979) Drugs and vitamin deficiency. *Drugs* 18, 278–298.

Parke, D.V. and Ioannides, C. (1981) The role of nutrition in toxicology. *Annual Review of Nutrition* 1, 207–234.

Rivers, J.M. (1975) Oral contraceptives and ascorbic acid. *American Journal of Clinical Nutrition* 28, 550–554.

Roe, D.A. (1980) In: *Drug-Induced Nutritional Deficiencies*. The AVI Publishers, Westport, Connecticut.

Rush, D., Stein, Z. and Susser, M. (1980) A randomized controlled trial of prenatal nutritional supplementation in New York City. *Pediatrics* 65, 683–697.

Stamp, T.C.B. (1974) Effects of long-term anticonvulsant therapy on calcium and vitamin D metabolism. *Proceedings of the Royal Society of Medicine* 67, 64–68.

Wilson, C.W.M. (1974) Vitamins and drug metabolism with particular reference to vitamin C. *Proceedings of the Nutrition Society* 33, 231–238.

Zannoni, V.G. and Sato, P.H. (1976) The effect of certain vitamin deficiencies on hepatic drug metabolism. *Federation Proceedings* 35, 2464–2469.

Therapeutic Potential of Vitamins

19

Brief reference was made in the first chapter of this book to the concept of 'Recommended Daily Intake' (RDI), or Recommended Daily Amounts (RDAs) and the fact that many national governments produce tables for the guidance of food providers, dietitians and nutritionists. The values given in such tables are usually arrived at by a panel who make estimates of average requirements that are based on information derived from some, or all, of the following (Department of Health, 1991):

1. The intakes of a nutrient needed to maintain a given circulating level or degree of enzyme saturation or tissue concentration.
2. The intakes of a nutrient by individuals or by groups which are associated with the absence of any sign of deficiency disease.
3. The intake of a nutrient needed to maintain balance noting that the period over which such balance needs to be measured differs for different nutrients and between individuals.
4. The intakes of a nutrient needed to cure clinical signs of deficiency.
5. The intakes of a nutrient associated with an appropriate biological marker of nutritional adequacy.

The recommended daily intakes of vitamins contain a safety margin which, in many countries such as the UK, Canada and the United States, amounts to 2 standard deviations above the estimated average requirement, and it is estimated that such intakes will be sufficient to meet the needs of at least 97.5% of the population. It is recognized that calculation of actual intakes based on food composition tables may leave considerable room for error in the estimation of true amounts of nutrients, and particularly vitamins, ingested. This, together with individual variation in requirements (Table 19.1), combined with the ready availability of a variety of vitamin preparations for purchase over the counter, readily leads to the extension of the idea that 'if a little is good for you, more will be better'. Arguments in support of taking regular vitamin supplements are made seemingly more reasonable by suggestions that what must

Table 19.1. Human vitamin requirements – ranges.

Vitamin	Minimum	Median	Maximum
A (μg retinol equiv.)	360	800	1650
D (μg cholecalciferol)	2.5	5	20
E (mg-tocopherol equiv.)	5	10	50
K (μg phylloquinone)	30	140	3000
C (mg ascorbic acid)	15	60	100
B_1 (mg thiamin)	0.5	1.2	2.2
B_2 (mg riboflavin)	0.8	1.6	3.2
Niacin (mg niacin equiv.)	5.5	1.8	22.5
B_6 (mg pyridoxine)	1	2	4
Folic acid (μg)	100	210	2000
B_{12} (μg cyanocobalamin)	1	2	5
Biotin (μg)	100	200	400
Pantothenic acid (μg)	3	7	14

Adapted from Elias (1993).

always be attempted is 'optimum nutrition'. The definition of this desirable condition is not easy, but it may be considered to be one in which not only is disease prevented, but every system in the body is working with maximum efficiency.

In this chapter we will concentrate on specific medical conditions in which relatively large doses of vitamins have been reported to have some therapeutic potential. The mechanisms by which these effects are brought about are varied as, too, are the large number of metabolic pathways in which vitamins function as coenzymes. The bonding of the apoenzymes to their coenzymes to form the specific halo-enzyme may be by ionic, coordinate covalent or covalent linkages.

Inadequate dietary intake of a vitamin will cause disturbed metabolic activity resulting in the development of a specific disease with clearly recognized, though variable, symptoms. Symptoms of vitamin deficiency can also occur when the dietary intake of a vitamin, in terms of the RDA, is adequate. In these disorders, there is an inherited or induced defect of vitamin utilization such that abnormally large amounts of vitamins are necessary in order to prevent the occurrence of symptoms or to cure an existing disorder. These diseases are described as 'vitamin dependency diseases'. There is often an abnormality in the structure of the apoenzyme such that its affinity for the coenzyme is reduced. A number of these diseases involve vitamin B_6. Defects of vitamin transport and coenzyme synthesis involving vitamin B_{12} and folic acid also exist. Large amounts of other individual vitamins may have therapeutic potential depending upon specific properties of the vitamin, such as the ability to act as a scavenger for free radicals and prevent peroxidation.

This chapter will thus concentrate on a discussion of the potential of comparatively large amounts of vitamins to act as 'drugs'. This use of vitamins is to be distinguished from their use as a public health measure to prevent deficiency states in communities. The relationship of vitamins to the prevention of cancer and the management of cancer patients is discussed elsewhere (Chapter 17).

Vitamin A

Apart from vitamin A deficiency conditions, such as night blindness and xerophthalmia, vitamin A in therapeutic doses has been reported to be beneficial in a number of conditions including skin conditions, such as acne vulgaris, warts and Darier's disease, Down's syndrome, stress-induced ulcers, bronchopulmonary dysplasia (Flodin, 1988), and infectious diseases. Some of these conditions have received more acceptance than others, in terms of their responses to vitamin A therapy.

Acne vulgaris

Acne vulgaris is a condition that is known to respond to a topical application of all-*trans* retinoic acid (tretinoin), which is commercially available either as gel, cream, or solution (Kligman *et al.*, 1969). Topical retinoic acid appears to act by increasing the cellular turnover in the stratum corneum making the epidermis very thin (Matsuoka, 1983). During treatment, the rate of production of loose horny cells in the follicular canal is increased, the formation of comedos, which are keratinous plugs, is prevented, and the existing comedos fall away during washing.

Bronchopulmonary dysplasia

Bronchopulmonary dysplasia (BPD) is another condition that is known to respond to vitamin A treatment. It is a life-threatening condition that occurs in premature infants. It is thought to result from oxygen toxicity, artificial ventilation and from damage caused by other medical interventions such as endotracheal tubes (Anon., 1986). Premature infants commonly have low vitamin A stores and it has been suggested that low vitamin A status interferes with the ability of lung tissue to repair the tissue injury associated with the disorder (Shenai *et al.*, 1985). Plasma retinol values in babies with BPD have been reported to be lower than in those premature babies without BPD. High doses of vitamin A reduced the incidence of BPD from 85% (17/20) in controls to 45% (9/20) in the treated group (Shenai *et al.*, 1987).

Infectious disease

Vitamin A deficiency is associated with increased susceptibility to infections especially of the respiratory and digestive tracts (Sommer, 1994). This is due to a combination effect of immune status and disintegration of mechanical barriers to infections, caused by the loss of epithelial cilia and of mucus secretions (see Chapter 11). Vitamin A deficiency state in general is thought to favour the growth of pathogenic organisms and to impair the host defences against infections.

Circulating levels of vitamin A are usually found to be depressed in children with active infectious diseases (Reddy *et al.*, 1986; Barclay *et al.*, 1987; Coutsoudis *et al.*, 1991). The cause and effect relationship of this is, however, far from being understood. The disease may deplete vitamin A reserves by increasing its requirements at a time when the intake and absorption of nutrients are generally decreased because of the disease. Low plasma vitamin A during active infections could also represent the redistribution of vitamin A from the circulation as an effect of the acute phase response (Thurnham and Singkamann, 1991). The possibility that this internal redistribution during infection might deprive peripheral tissues such as lungs, gut, and skin, of their vitamin A content has been the underlying rationale for a justification of administering vitamin A supplements to children with severe measles, irrespective of their prior vitamin A status (Hussey and Klein, 1990).

In areas where vitamin A deficiency and under-nutrition are public health problems, studies have shown that children with mild vitamin A deficiency are at an increased risk of respiratory disease and diarrhoea (Feachem, 1987; Forman, 1989). Furthermore, children who have recently suffered from measles are much more likely to develop vitamin A deficiency leading to corneal xerophthalmia than those without history of measles (Sommer *et al.*, 1983). It is increasingly being recognized that vitamin A deficiency and infection exist in a synergistic relationship and that lack of vitamin A can affect resistance to infection and childhood mortality even before it is sufficiently severe to cause corneal xerophthalmia. Immunization is the primary means of preventing mortality from infectious disease, such as measles. In developing countries, high infectivity, incomplete vaccine coverage, and vaccine failure due to malnutrition highlight the need for an alternative therapy. Several clinical trials carried out in many developing countries have demonstrated that vitamin A supplementation is effective not only in improving vitamin A nutriture but also in reducing mortality rate from infectious disease. Thus, a dose of 200,000 IU of vitamin A (as retinyl palmitate) in oil administered 1–3 times annually has been shown to reduce mortality by 30–50% or more in pre-school children with infectious disease (Sommer *et al.*, 1983; Milton *et al.*, 1987; Bloem *et al.*, 1990). A randomized double-blind control study in India has revealed a substantial reduction in morbidity and mortality rates among pre-school children with

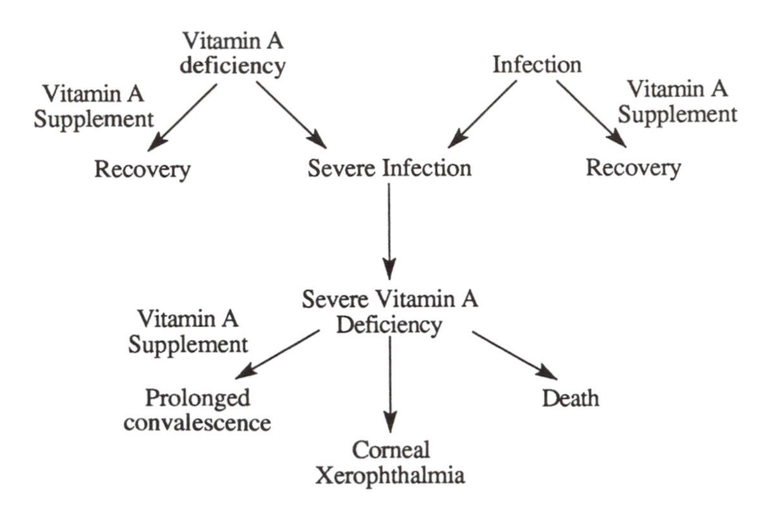

Fig. 19.1. Possible consequences of synergistic effect of vitamin A deficiency and infection.

measles and diarrhoea receiving a small weekly dose (8333 IU) of vitamin A (Rahmathullah *et al.*, 1990)

Normally, vitamin A deficiency or infection alone responds better to vitamin A supplementation than when the two conditions coexist (Fig. 19.1). Because of the available evidence of the dramatic impact of vitamin A status on childhood mortality, the World Health Organization recommends routine vitamin A supplementation for all children with measles in regions where vitamin A deficiency is a recognized problem (WHO/UNICEF, 1987).

β-Carotene

Carotenoids are known to protect plants from ultraviolet light damage (Mathews-Roth, 1986). Observations of this photoprotective effect of carotenoids led to the use of β-carotene in the treatment of patients with light-sensitive skin disorders.

Erythropoietic protoporphyria

Erythropoietic protoporphyria (EPP) is an inherited light-sensitive disease characterized by burning, redness and swelling of skin when exposed to the sun. In a 7-year study involving 133 patients with EPP given β-carotene supplements,

Mathews-Roth *et al.* (1974) found that 84% of the patients increased their ability to tolerate exposure to sunlight without the development of symptoms. Other reports (Thomsen *et al.*, 1979) have confirmed the original studies. Doses up to 180 mg day^{-1} of β-carotene may be needed to achieve results but no serious side effects have been reported. β-Carotene, in effect, remains the drug of choice in erythropoietic protoporphyria. Attempts have been made to use β-carotene for the treatment of other light sensitive skin disorders with varying degrees of success (Gaby and Singh, 1991).

Vitamin D

Vitamin D is a prohormone which plays an important central role in the maintenance of calcium and phosphorus metabolism.

Nutritional rickets

Simple vitamin D-deficiency rickets is conventionally treated with oral vitamin D, 10–50 μg day^{-1}, together with oral calcium supplements (Bachrach *et al.*, 1979; Rudolph *et al.*, 1980). In addition to this condition, there are a number of clinical syndromes associated with abnormal metabolism of vitamin D, where this vitamin or its metabolite is often the therapy of choice. One of these conditions is vitamin D resistant rickets, which is an inherited condition, characterized by rickets or osteomalacia, and short stature. This genetic disorder may be of two types: hypophosphataemic rickets (Chen *et al.*, 1985) and vitamin D-dependent rickets (Delvin *et al.*, 1981).

Congenital rickets

In hypophosphataemic rickets, there is a defect in the renal tubular reabsorption mechanism leading to excessive urinary loss of phosphate and consequently hypophosphataemia. Despite the presence of hypophosphataemia, this condition is associated with inability to hydroxylate 25-OH-vitamin D to 1,25-$(OH)_2$-vitamin D (calcitriol). Oral treatment with either massive doses of vitamin D (0.5–1.0 mg day^{-1}) or low doses of calcitriol (1 μg day^{-1}) plus phosphate (1–4 g day^{-1}) improves intestinal calcium absorption and there is healing of the rachitic epiphyseal lesions. Unlike hypophosphataemic rickets, vitamin D-dependent rickets is characterized by hypocalcaemia with the blood phosphate level normal. In this condition, either a complete lack of the renal 1-α-hydroxylase enzyme or a congenital defect in its regulation is thought to be the underlying biochemical abnormality. Patients with vitamin D-dependent rickets respond well to calcitriol (0.25–2 μg day^{-1}).

Renal osteodystrophy

Osteodystrophy is a common clinical complication in patients with chronic renal disease (Varghese *et al.*, 1979). The complication may take the form of rickets or osteomalacia, or osteitis fibrosa. The osteodystrophy of renal disease is associated with a negative calcium balance, which is a sign of vitamin D deficiency, and yet this condition appears to be resistant to vitamin D when given in its physiological dose levels. Reversal of negative calcium balance and healing of bone lesions can, however, be achieved with massive doses of vitamin D (1 mg day^{-1}).

It is thought that the renal disorder affects the synthesis of 1-α-hydroxylase activity in the kidney, and hence the synthesis of 1,25-$(OH)_2$-vitamin D (Mawer, *et al.*, 1973). Many investigators have attempted to reveal the response of patients with renal insufficiency to various vitamin D metabolites (Massry, 1980). According to these studies, both clinical and biochemical improvement of bone disorders can be effectively achieved in most patients with oral administration of calcitriol at doses of 0.5–3.5 μg day^{-1}.

Hypertension

High blood pressure (BP) is considered an important risk factor for cardiovascular disease (CVD) and a number of epidemiological and intervention trials have shown an inverse relationship between BP and calcium status and/or intake (Belizan *et al.*, 1983; Ljunghall *et al.*, 1987). A number of mechanisms may be involved in causing this effect. These include a reduction in parathyroid hormone production (an effect associated with antihypertensive effects of calcium entry blockers, β-blockers and diuretics), and an alteration in smooth muscle contraction. If vitamin D status is causally related to CVD either positively or negatively, it may be through calcium and/or magnesium (Seelig, 1975) metabolism. However, such a relationship has not been found in subjects who have had myocardial infarcts (Lund *et al.*, 1978) and has not been demonstrated elsewhere. An age-adjusted inverse relationship between dietary vitamin D and systolic BP has been reported in women aged 20–35 years (Sowers *et al.*, 1985). In older women (55–80 years) in the same study, this inverse relationship was maintained only if the subject's consumption of both calcium and vitamin D was below the US RDA. In one study, serum levels of 1,25 $(OH)_2$-D_3 were found to be positively associated with diastolic and systolic BP in a cohort of normotensive and hypertensive older women, when adjusted for age, weight and diuretic use (Sowers *et al.*, 1985). The trend was not significant in premenopausal women.

Osteoporosis

Osteoporosis is a major bone disorder in North America. It is characterized by too little calcified bone, and is caused when bone mineral resorption exceeds new bone formation. The current management of this condition consists primarily of oestrogen therapy and calcium supplements. Treatment of osteoporotic patients with vitamin D or its active metabolite (calcitriol) has been reported to be beneficial in some but without any benefit in others (Brautbar, 1986).

The scope for ingestion of large amounts of vitamin D is limited by its toxicity. In recent years, interest has been centred on the supplementation of individuals, such as the elderly, who are deficient. There is little evidence that supplementation in already replete individuals is likely to have any beneficial effect on the attainment of peak bone mass as a protective measure against osteoporosis. Vitamin D should be supplemented only in cases where dietary intake and sunlight exposure are inadequate.

Vitamin E

Abetalipoproteinaemia

This is a rare autosomal recessive disorder (also called Bassen–Kornzweig disease) in which there is complete absence of production of apolipoprotein B resulting in the absence of circulating chylomicrons and VLDL and LDL (B-lipoprotein). Patients have steatorrhoea because triglycerides cannot be transported out of the mucosal cells, acanthocytosis of the red cells and an ataxic neuropathy and pigmentary retinopathy that develop within the first two decades of life. Absorption of vitamin E is severely impaired in this condition and as a consequence of the lack of plasma lipoproteins to transport the vitamin, a severe deficiency develops. It was suggested (Muller and Lloyd, 1982) that vitamin E deficiency might be the cause of the neurological defects and as a consequence several groups of clinicians started to supplement affected individuals with massive doses of vitamin E. Treatment with 50–200 mg kg^{-1} body weight day^{-1} was associated with prevention of the neuropathy and retinopathy, and the arrest of the progression or reversal of the symptoms (Muller *et al.*, 1983). In addition to vitamin E, vitamin A supplementation (1500–20,000 IU day^{-1}) appears to be necessary for optimal treatment of the retinal degeneration. Because of the very low circulating lipoprotein levels, serum vitamin E measurements do not accurately reflect vitamin E status in this condition. Monitoring of therapy is best done with peroxide haemolysis testing or measurements of vitamin E in adipose tissue biopsies. Measurements of somatosensory or visual evoked responses may be useful in monitoring serial improvement in coordination of affected axonal pathways in response to treatment.

Biliary atresia

Low serum concentrations of α-tocopherol are found in infants with biliary atresia. This is one of a group of 'cholestatic' liver disorders which also includes idiopathic neonatal hepatitis, intrahepatic paucity of bile ducts and familial forms of congenital cholestasis. There is now convincing evidence that vitamin E deficiency is responsible for the neuromuscular disorder in children with chronic cholestasis. Persistent vitamin E deficiency causes a progressive neurological disorder which starts with tendon reflexes at about 18 months of age. Other changes, including truncal and limb ataxia, ophthalmologia and muscle weakness follow within the first decade. Children whose vitamin E status is adequate have normal neurological function. It is interesting to note that a number of the neuromuscular lesions which occur in children with cholestasis resemble those described in animal models of vitamin E deficiency. These include degeneration of the posterior columns of the spinal cord, degeneration and loss of large calibre myelinated fibres in peripheral nerve and neuropathic and myopathic lesions. Repletion of vitamin E by large doses, whether given orally or parenterally, has prevented, reversed or arrested the progression of neurological symptoms (Elias *et al.*, 1981).

Vitamin E status in children with chronic cholestasis should be evaluated every 2–3 months. Elevation of circulating lipids in these patients means that serum vitamin E levels alone do not accurately reflect vitamin E status. It is indeed possible for the children to be vitamin E deficient whilst having normal serum levels. The ratio of serum vitamin E to serum total lipids is measured. A ratio of 0.8 mg g^{-1} for children aged 12 years and older and 0.6 mg g^{-1} for children less than 12 years is indicative of deficiency. Once deficiency is established, treatment should be started with 50 IU kg^{-1} day^{-1} increasing to 150–200 IU kg^{-1} day^{-1} given as a single dose with breakfast several hours before medication that interferes with vitamin E absorption (e.g. cholestyramine).

Cystic fibrosis

The only clinical manifestation of vitamin E deficiency seen in patients with cystic fibrosis is a shortened red blood cell half-life. However, children in whom there is cholestatic liver involvement may have some neurological changes. In the absence of liver involvement, the serum vitamin E levels are usually normalized by supplements of 5–10 IU kg^{-1} day^{-1}.

Intermittent claudication

The use of vitamin E for the management of intermittent claudication caused by arteriosclerosis of peripheral blood vessels was initially suggested by Boyd (see Boyd and Marks, 1963). Since then a number of studies have confirmed that doses of 400–600 mg day^{-1} cause significant improvement in subjective symptoms, distance walked before pain appears and arterial blood flow in calf muscles. A more recent report (Haeger, 1982) emphasized the importance of active exercise in conjunction with 100 mg of D-α-tocopherol acetate three times per day. Unfortunately, many of the studies done on this disorder were not randomized or done in a double-blind manner. Nevertheless, the majority of investigations suggest that vitamin E may benefit some patients.

Platelet function

The aggregation of platelets is a vital function which prevents an organism from bleeding to death after injury. However, an enhancement of this process is potentially dangerous causing the blood to clot more readily than normal and can lead to myocardial infarcts or stroke. In a study in which 47 healthy men and women were given 400, 800 and 1200 IU of vitamin E per day for 2 weeks each, adhesiveness to collagen was significantly decreased (Steiner, 1983). Platelet adhesiveness to collagen is recognized as an important step in the formation of thrombi. The effect of vitamin E on this process would complement the action of agents, such as aspirin, that inhibit aggregation but have no effect on adhesion.

Ischaemic heart disease

Vitamin E status has been identified as a factor in the relative risk of death from ischaemic heart disease. In countries with high mortality rates due to this disease (e.g. Scotland and Finland), plasma vitamin E concentrations have been found to be about 25% lower ($p < 0.01$) than in countries with low to moderate coronary mortality rates (e.g. Italy, Switzerland and Northern Ireland) (Gey *et al.*, 1987).

Cataracts

A number of studies have suggested that the development of cataracts may be related to oxidative stress. Particularly is this so when the ageing eye is not protected by an adequate availability of antioxidants. It seems that the combination of high vitamin E status with high blood levels of β-carotene and/or vitamin C may significantly reduce the risk of developing cataracts (Jacques *et al.*, 1988). The effectiveness of vitamin E in reducing cataract risk was also suggested from a study of people over 55 years of age who had been taking 400 IU of vitamin E per day (Robertson *et al.*, 1989).

Anaemias

Anaemia may be caused by a reduced rate of red cell formation or an accelerated rate of red cell destruction. The lysis of red cells may be caused by oxidative breakdown of red cell membranes. Indeed, increased fragility of RBCs in adults and haemolytic anaemia in premature infants are symptoms of vitamin E deficiency.

A genetic disorder in which there is a deficiency of glucose-6-phosphate dehydrogenase (G6PD) is characterized by a chronic haemolytic anaemia due to reduced antioxidant capacity and red cell survival time. Low serum α-tocopherol levels have been reported in this condition and administration of 800 IU α-tocopherol per day for 1 year has been found to significantly improve red cell life span and haemoglobin concentration. Similar results have been obtained in children with G6PD deficiency (Table 19.2).

Sickle cell anaemia (SCA) is another genetic disorder characterized by red cell haemolysis. In this condition, deformed or sickled cells are easily destroyed with consequent development of anaemia. Administration of 450 IU of vitamin E per day has been found to significantly decrease the percentage of irreversibly sickled cells (Natta *et al.*, 1980). Evidence that blood from SCA subjects forms more oxygen free radicals and that polymorphonuclear leucocytes have a reduced capacity to produce an oxidative burst and that these parameters are normalized by vitamin E supplementation, seems to offer scope for the use of the vitamin in the management of the condition.

Table 19.2. Effect of vitamin E therapy (800 IU day^{-1} for 2 months) on haematological status in patients with glucose-6-phosphate dehydrogenase deficiency.

Parameter	Control	Before treatment	After treatment	*P*
Haemoglobin (g dl^{-1})	12.1	9.0	9.9	< 0.05
Packed cell volume (%)	38.1	28.1	30.5	< 0.05
Reticulocyte count (%)	0.71	3.0	2.3	< 0.05
Red cell half-life (days)	28.0	16.9	22.8	< 0.01
Serum vitamin E (mg dl^{-1})	0.95	0.50	1.15	< 0.001

Adapted from Hafez *et al.* (1986).

Ageing

The possibility that the process of ageing can be retarded by nutrients has attracted considerable attention. Many theories have been proposed to account for the process. One of the potentially most attractive of them is that free-radical damage is involved. Nucleic acids and proteins, as well as unsaturated fatty acids, are known to be susceptible to attack and the products to be injurious to cells. One feature of ageing in tissues which shows the greatest susceptibility to peroxidation (e.g. brain and heart) is the accumulation of a chromolipid or age pigment called lipofuscin. This pigment accumulates more rapidly in vitamin E deficiency. In a life-span study, rats receiving a daily supplement of ethoxyquin (a synthetic antioxidant) or α-tocopherol had a reduced accumulation of lipofuscin in their brains compared with that in pair-fed controls (Rudra *et al.*, 1975). Since it has been suggested that the accumulation of lipofuscin in the brain contributes to the deterioration in the brain with age, it is tempting to entertain the possibility that supplements of vitamin E might slow down the process of ageing and prevent some of the diseases common in later life (Aruoma and Halliwell, 1991).

Vitamin C

Immune function

It has been suggested that vitamin C stimulates the immune system (Anderson, 1981) primarily through enhancement of neutrophil function. *In vitro* studies (summarized by Anderson, 1981) seem to support the view that ascorbic acid stimulates certain aspects of metabolism and activity in neutrophils from normal subjects. These include random motility and chemotaxis, phagocytosis and activity of the hexose monophosphate shunt. The stimulation of neutrophil motility was associated with inhibition of the myeloperoxidase/hydrogen peroxide/halide system. *In vivo* studies, also summarized by Anderson (1981), showed that doses of 1 g or more taken orally stimulated chemotaxis in neutrophils from normal adults. In summary it seems that ascorbic acid may be of value prophylactically and therapeutically in the management of patients with increased susceptibility to infection due to abnormalities of neutrophil function.

Wound healing

The first demonstration that ascorbic acid is essential for wound healing is usually attributed to Crandon (Crandon *et al.* 1940). This surgeon placed himself on a vitamin C-free diet for 6 months and twice during that period had a wound made on his back. Biopsy samples were taken after the wounds had healed. The wound made after 3 months healed normally, while that made after 6 months, when he was scorbutic, superficially appeared to heal, but a biopsy taken after 10 days showed only a blood clot with no evidence of healing. He was given an intramuscular injection of 1 g of ascorbic acid and after a further 10 days there was good healing with normal formation of intercellular material.

If ascorbic acid deficiency prevents the healing of wounds, it seems reasonable to suggest that giving supplements will stimulate healing. In one such study (Taylor *et al.*, 1974) 500 mg of the vitamin was given twice daily to ten surgical patients with pressure sores. The study was carried out in a double-blind manner. Ascorbic acid stimulated the healing of the pressure sores (2.47 cm^2 per month) compared with a placebo (1.45 cm^2 per month). It could be argued from these results that ascorbic acid supplements will accelerate the healing of pressure sores in all patients. However, the patients had been living for some months on a hospital diet and thus could well have been depleted of vitamin C at the start of the study (Schorah, 1979). It is in such patients that vitamin C supplements are likely to have a beneficial effect. From a practical viewpoint, ascorbic acid deficiency should always be suspected in patients with decubitus ulcers and pressure sores. Institutional diets are unlikely to contain sufficient ascorbic acid to maintain the tissue levels of patients who are chronically sick. It is often cheaper to supplement such patients rather than carry out a biochemical assessment of vitamin C status.

The question of whether surgical patients should be given vitamin C supplements is controversial. Decreased blood levels of the vitamin have been consistently reported following surgery. Some researchers have interpreted this fall as evidence of increased need, whereas others suggest that ascorbic acid is redistributed to tissues (Schorah *et al.*, 1986). The fall in blood values may be due to other aspects of treatment such as anaesthesia, blood transfusions or changes in the ratio of the different leucocytes.

Cardiovascular disease

There has been considerable interest in the possibility that latent ascorbic acid deficiency may be implicated as an aetiological factor in the pathogenesis of cardiovascular disease. The possible mechanism(s) by which the vitamin deficiency exerts its atherogenic effect, involving the liver, blood vessels and tissues, is outlined in Fig. 19.2. Atherosclerosis of the walls of arteries results in a narrowing of their lumen thus impeding blood flow. Infarcts and strokes also involve the formation of thrombi, or blood clots, which can completely occlude the atherosclerotic arteries. High blood pressure is also known to be a risk factor and the formation of thrombi is accelerated by increased hyperaggregability of platelets.

It is generally accepted that high serum concentrations of low-density lipoproteins are associated with high risk of coronary heart disease, whilst high

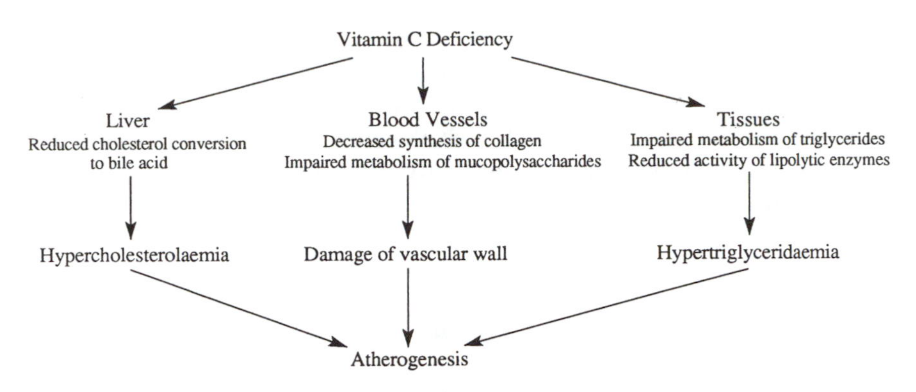

Fig. 19.2. Possible involvement of ascorbic acid deficiency in the cause of atherosclerosis.

levels of the high density lipoproteins (HDL) are protective and decrease risk of heart disease. The influence of ascorbic acid on these serum constituents centres around the fact that cholesterol is associated with both kinds of lipoproteins. Studies in animals suggest that vitamin C stimulates the activity of cholesterol 7-α-hydroxylase, the enzyme regulating the conversion of cholesterol to bile acids. Bile acids are excreted so ascorbic acid could in this way lower serum cholesterol levels. Of interest in this connection is the fact that, in guinea pigs, a dietary deficiency of ascorbic acid has been found to produce atheromatous arterial lesions (Willis, 1953, 1957). In humans, vitamin C does not consistently correlate with total serum cholesterol. The age of the subjects and initial ascorbic status may be complicating variables. Thus, extra dietary ascorbic acid reduces serum cholesterol concentrations in normal people under 25 years (Spittle, 1971) and in hypercholesterolaemic people whose initial ascorbic acid status is low (Ginter *et al.*, 1970) but not if the status is normal (Peterson *et al.*, 1975). Ascorbic acid status in men but not in women is generally correlated positively with HDL-cholesterol concentration (Jacques *et al.*, 1994). Six weeks' treatment with l g ascorbic acid per day, however, has been reported to raise the concentration of HDL-cholesterol in both men and women (Horsey *et al.*, 1981). Furthermore, total serum cholesterol and LDL-cholesterol concentrations are reduced in men with coronary heart disease. Evidence to date suggests that latent ascorbic acid deficiency may be one of several preventable 'risk' factors contributing to coronary heart disease in the western world.

Blood pressure

High blood pressure (hypertension) is a major risk factor for cardiovascular disease. Data obtained from the first National Health and Nutrition Examination Survey (NHANES 1) showed that people with lower vitamin C consumption had higher blood pressure (McCarron *et al.*, 1984). In apparently healthy men in their 30s a strong inverse association was found between serum ascorbic level and high blood pressure (Yoshioka *et al.*, 1984). A similar relationship was found in 54 year old Finnish men (Salonen *et al.*, 1988). A daily 1 g supplement of vitamin C given to mildly hypertensive women was associated with falls in both systolic and diastolic blood pressures as well as in plasma lipids (Koh, 1984).

Platelets

As early as 1980, it was shown that 2 g of vitamin C (in two daily doses) when given to patients with existing coronary heart disease caused a 27% decrease in platelet adhesiveness (Bordia, 1980). Increase in platelet aggregation and adhesiveness induced in coronary artery disease by a high cholesterol meal was prevented by the addition of 1 g ascorbic acid to the food (Bordia and Verma, 1985). The same workers also found that ingestion of 1 g ascorbic acid every 8 hours for 10 days significantly reduced platelet aggregation and adhesion in a group of hyperlipidaemic coronary artery disease patients who had abnormally high measures of these parameters. At present the evidence seems to be that a high threshold of ascorbic acid is required to influence platelet activity.

Stroke

A declining rate of cerebrovascular disease has been associated with an increase in fruit and vegetable consumption in Britain and in the US (Acheson and Williams, 1983) which was suggested to be due to their vitamin C content.

Cataracts

Changes in the lens due to oxidation and polymerization of protein causes cloudiness that interferes with vision. Vitamin supplement use has been studied in 175 people with cataracts and 175 controls (Robertson *et al.*, 1989). Subjects who did not take a supplement had a relative risk of developing cataracts four times higher than did subjects who were taking vitamin C supplements of 300–600 mg day^{-1}.

The 'common cold'

The efficacy of high doses of vitamin C in the prevention or amelioration of the common cold remains controversial. There are reports in the literature showing both benefit and no benefit. The protocols for most of the studies that have looked into this matter are open to criticism. Most of the subjects in them were free living and because of this were exposed to other infections. Moreover, many involved the self-reporting of compliance and symptoms. The original suggested benefit has been brought again into the realm of possibility by the results of a well-controlled, double-blind placebo trial (Mink *et al.*, 1988). Subjects were isolated and exposed to a single type of rhinovirus (rhinovirus type 16) through contact with an infected person. The experimental group received 500 mg vitamin C four times a day and experienced half the degree of symptom severity seen in the placebo group. Serum ascorbic acid levels in the experimental group increased whereas those in the placebo group substantially declined during the study.

The respiratory tract is a major target for oxidative injury from inhaled radicals, such as free oxygen, nitrogen dioxide and those contained in tobacco smoke, and non-radicals, such as ozone and the products of inflammatory cells (Cross *et al.*, 1994). The initial defence against attack by these agents is provided by antioxidants in fluids lining the tract. If these are depleted the defences are correspondingly reduced. Is it possible that antioxidants could ameliorate some of the ill-effects of, say, reactive oxygen radicals and ozone? There is increasing evidence that low micronutrient antioxidant levels may influence asthma; thus supplemental antioxidants, especially ascorbic acid, could have benefits in some forms of airway hyperactivity and increases in ascorbate concentrations have been associated with increased pulmonary function (Schwartz and Weiss, 1994).

Riboflavin

A beneficial effect of therapeutic doses of riboflavin remains yet to be firmly established. However, there are scattered reports of such effects in the literature (reviewed by Matts, 1980). These effects include improvements in psoriasis resulting from intramuscular injection of 5–10 mg day^{-1}, beneficial effects in the treatment of corneal ulcers, photophobia and non-infective conjunctivitis and in the treatment of angular conjunctivitis of the Morax–Axenfeld type.

Niacin

Hyperlipidaemias

Nicotinic acid but not nicotinamide reduces cholesterol as well as triglycerides, and increases HDL-cholesterol concentrations in serum (Altschul *et al.*, 1955; Drood *et al.*, 1991; Schectman *et al.*, 1993). The dose level at which the vitamin is recognized to play a hypolipidaemic effect exceeds 3 g day^{-1} as opposed to its daily physiological requirement of only 20 mg. The effect is, generally, more noticeable in hypercholesterolaemic than in normocholesterolaemic subjects. Proposed mechanisms of lowering the action of nicotinic acid include inhibition of adipose tissue lipolysis, increasing lipoprotein lipase activity and decreasing hepatic lipase activity. A high serum lipoprotein (a) is known to be a risk factor for atherosclerotic and thrombotic cardiovascular disease. The reduction of this lipoprotein level has also been achieved with nicotinic acid treatment, and this effect is thought to be due to a reduction in its synthesis (Seed *et al.*, 1993).

The treatment with nicotinic acid has also been shown to decrease recurrent myocardial infarction without a concurrent mortality increase from cardiovascular disease, as sometimes observed with clofibrate (Coronary Drug Project Research Group, 1975). Furthermore, the lipid-lowering action of nicotinic acid has been clinically tested in conjunction with diet and other drugs including cholestyramine, colestipol and clofibrate. A detectable synergistic effect has been observed in these studies (Kane *et al.*, 1981; Morgan and Cohen, 1987).

Hartnup's disease

This is a genetic disorder characterized by a decreased intestinal absorption and an increased renal excretion of amino acids, including tryptophan (Halvorsen and Halvorsen, 1963). In this condition, the synthesis of niacin from tryptophan is severely affected leading to pellagra-like skin rash with neurological manifestations. Oral administrations of nicotinamide at dose levels of 40–250 mg day^{-1} have been reported to improve the clinical condition.

Pyridoxine

Dependency conditions

Reference has been made already to inherited conditions in which a dependency on large doses of vitamins exists due to abnormalities in the structure of an apoenzyme. Although the number of people affected by these conditions is very small, because of the number of enzymes in the body, the potential for developments of such conditions is large. Table 19.3 lists five conditions in which there is a dependency for pyridoxine.

A condition marked by seizures and convulsions in infancy occurs because of an error in the enzyme, glutamic acid decarboxylase. This enzyme is responsible for the production of γ-aminobutyric acid (GABA) from glutamic acid. Both of these substances act on neurons, with glutamic acid being excitatory and GABA inhibitory. Normal brain function depends on a balance between the

Table 19.3. Inherited diseases showing pyridoxine (B-6) dependency.

Disorder	Clinical manifestations	Apoenzyme affected
Infantile convulsions	Seizures	Glutamic acid decarboxylase
Cystathioninuria	Probably none	Cystathionase
Xanthurenicaciduria	Mental retardation (?)	Kynureninase
Homocystinuria	Ectopic lentis, thrombotic vascular disease, CNS dysfunction	Cystathionine synthase
Hyperoxaluria	Calcium oxalate nephro-lithiosis, renal insufficiency	Glyoxylate α-ketoglutarate carboligase

Adapted from Dickerson and Williams (1990).

action of these two substances. In infants with the dependency syndrome a normal intake of pyridoxine is insufficient to cause production of adequate amounts of GABA with resulting seizures. Giving large amounts of pyridoxine (5–15 mg day^{-1}) normalizes the condition and seizures do not occur.

In homocystinuria there is a functional deficiency of cystathionine synthase (cystathionine β-synthetase), the enzyme in the trans-sulphuration pathway which catalyses the formation of cystathionine from homocysteine. Symptoms include long thin limbs and long thin digits, chest deformities and dislocation of the lenses with tremor of the iris. These symptoms are not present at birth or in early infancy but develop with age. Cystathionine occurs in high concentrations in the human brain. In homocystinuric patients the brain levels are greatly reduced and this affects brain function. Biochemical changes include the presence of homocystine and high levels of methionine in the plasma. The concentration of cystine is low. Homocystinuria leads to alterations in endothelial function initiating changes which ultimately progress to atherosclerosis. Homocystinuria is treated with a low methionine diet. Patients in whom there is a residual activity of cystathionine synthase respond to pharmacological doses of pyridoxine up to several hundred mg per day, both clinically and biochemically (Stern, 1983).

Hyperoxaluria, an increased 24 hour excretion of oxalate, can be due either to a primary inborn error of metabolism or to an acquired condition. The term 'primary hyperoxaluria' is used to include two genetic disorders of glyoxalate metabolism which are characterized by the urinary excretion of large amounts, usually more than 1.0 mmol per 24 hours, of oxalate (Alston, 1983). The conditions are transmitted as autosomal recessive traits and become clinically apparent in early childhood. Clinical manifestations include recurrent calcium oxalate nephrolithiasis or nephrocalcinosis or both. Both conditions are progressive and lead to death in early childhood if left untreated. The two forms of primary hyperoxaluria are known as Type I and Type II. Type I is a glycolic aciduria and Type II an L-glyceric aciduria. It is in Type I that there is a genetic defect in the enzyme, α-ketoglutarate : glyoxylate carboligase. As a result of this defect glyoxylate accumulates and is oxidized to oxalate and reduced to glycolate, both of which are excreted in the urine in excessive amounts. This is a very severe inborn error with no available specific treatment. Large amounts of pyridoxine (200–400 mg per 24 hours) are usually given together with magnesium oxide and a large fluid intake in order to maintain a dilute urine.

Carpal tunnel syndrome

The carpal tunnel is a passage in the wrist surrounded by dense ligaments through which nerve bundles pass. When the passage is narrowed due to inflammation or oedema resulting from chronic repetitive movements, there is pressure on the median nerve, causing pain and/or numbness in the hands. Patients with this syndrome have been reported to be deficient in vitamin B_6 (Ellis *et al.*, 1977; Fuhr *et al.*, 1989). Some (Smith *et al.*, 1984) but not all (Stransky *et al.*, 1989) studies have reported beneficial effects of B_6 supplements. In those studies in which improvement was reported the amounts of B_6 given were 50–150 times the US RDA (that is 100–300 mg day^{-1}).

Premenstual tension syndrome

This condition involves the appearance of symptoms in the premenstruum which disappear in the postmestruum. The symptoms include anxiety, nervous mood swings, abdominal bloating, headache and craving for sweets. There is very considerable confusion about the aetiology of the condition and this is reflected in the variety of treatments that have been suggested. Amongst the nutritional treatments is the use of high doses of pyridoxine (Abraham and Hargrove, 1980). However, pyridoxine status as measured by plasma pyridoxine phosphate has been reported to be normal (Ritchie and Singkaman, 1986) and no evidence of changes in pyridoxine status during the menstrual cycle has been found (Berg *et al.*, 1986). Evidence of benefit from large doses of B_6 is not uniform (Malmgren *et al.*, 1987). The studies that showed the greatest response tended to be those in which doses of up to 500 mg day^{-1} were given, though others (Mattes and Martin, 1982) have reported a good response with a dose of 50 mg day^{-1}.

Conclusions

There is accumulating evidence that amounts of various vitamins in excess of those required to prevent deficiency disorders and given as supplements may be beneficial in the management of a wide range of clinical disorders. For some disorders, such as vitamin dependency conditions, there is some understanding of the possible mechanism of action. In other conditions, such as those due to free radicals and environmental contaminants, an understanding of the mechanism of injury has prompted the use of antioxidant vitamins. The therapeutic use of vitamins, once the province of 'health' devotees, has now become an area for serious research and application by clinical nutritionists.

References

Abraham, G.E. and Hargrove, J.T. (1980) Effect of vitamin B6 on premenstrual symptomatology in women with premenstrual tension syndrome: a double-blind crossover study. *Infertility* 3, 155–165.

Acheson, R.M. and Williams, D.R.R. (1983) Does consumption of fruit and vegetables protect against stroke? *Lancet* 1, 1191–1193.

Alston, W.C. (1983) Urolithiasis. In: Williams, D.L. and Marks, V. (eds) *Biochemistry in Clinical Practice*. Heinemann, London, pp. 290–296.

Altschul, R., Hoffer, O. and Stephen, J.D. (1955) Influence of nicotinic acid on serum cholesterol in man. *Archives of Biochemistry and Biophysics* 54, 556–559.

Anderson, R. (1981) The immunostimulatory and anti-allergic properties of ascorbate. In: Draper, H.H. (ed.) *Advances in Nutritional Research*, Vol. 6. Plenum Press, New York, pp. 19–45.

Anon. (1986) Low plasma vitamin A levels in preterm neonates with bronchopulmonary dysplasia. *Nutrition Reviews* 44, 202–204

Aruoma, O.I. and Halliwell, B. (1991) Oxygen free radicals and human diseases. *Journal of the Royal Society of Health* 111, 172–177.

Bachrach, S., Fisher, J. and Parks, J.S. (1979) An outbreak of vitamin D deficiency rickets in a susceptible population. *Pediatrics* 64, 871–877.

Barclay, A.J., Foster, A. and Sommer, A. (1987) Vitamin A supplements and mortality related to measles: a randomized clinical trial. *British Medical Journal* 295, 294–296.

Belizan, J.M., Villar, J., Pindea, O., Gonzalez, A.F., Sainz, E., Garrera, G. and Sibrian, R. (1983) Reduction in blood pressure with calcium supplementation in young adults. *Journal of the American Medical Association* 249, 1161–1165.

Berg, H. van den, Louwerse, E.S., Brainse, H.W., Thissens, J.T.N.M. and Schrijver, J. (1986) Vitamin B6 status of women suffering from premenstrual syndrome. *Human Nutrition: Clinical Nutrition* 40c, 441–450.

Bloem, M.W., Wedel, M., Van Agfmaa, E.J., Speck, A.J., Saowakontha, S. and Schreurs, S.K.H.P. (1990) Vitamin A intervention: short-term effects of a single oral massive dose on iron metabolism. *American Journal of Clinical Nutrition* 51, 76–79.

Bordia, A.K. (1980) The effect of vitamin C on blood lipids, fibrinolytic activity and platelet adhesiveness in patients with coronary heart disease. *Atherosclerosis* 35, 181–1887.

Bordia, A. and Verma, S.K. (1985) Effects of vitamin C on platelet adhesiveness and platelet aggregation in coronary artery disease patients. *Clinical Cardiology* 8, 552–554.

Boyd, A.M. and Marks, J. (1963) Treatment of intermittent claudication: A reappraisal of the value of α-tocopherol. *Angiology* 14, 198.

Brautbar, N. (1986) Osteoporosis: is 1,25-$(OH)_2D_3$ of value in treatment? *Nephron* 44, 161–166.

Chen, J.C.M., Alon, U. and Hirschman, G.M. (1985) Renal hypophosphatemic rickets. *Journal of Pediatrics* 106, 533–544.

Coronary Drug Project Research Group (1975) Clofibrate and niacin in coronary heart disease. *Journal of the American Medical Association* 231, 360–381.

Coutsoudis, A., Broughton, M. and Coovadia, H.M. (1991) Vitamin A supplementation reduces measles morbidity in young African children: a rondomized, placebo-controlled, double-blind trail. *American Journal of Clinical Nutrition* 54, 890–895.

Crandon, J.H., Lund, C.C. and Dill, D.B. (1940) Experimental human scurvy. *New England Journal of Medicine* 223, 353–369.

Cross C.E., Vliet, A. van den, O'Neill, C.A. and Eiserich, J.P. (1994) Reactive oxygen species and the lung. *Lancet* ii, 930–933.

Delvin, E.E., Glorieux, F.H., Maric, P.J. and Pettifor, J.M. (1981) Vitamin D dependency: Replacement therapy with calcitriol. *Journal of Pediatrics* 99, 26–34.

Department of Health (1991) *Dietary Reference Values for Food Energy and Nutrients for the United Kingdom*. Report on Health and Social Subjects, No. 41. H.M. Stationery Office, London.

Dickerson, J.W.T. and Williams, C.M. (1990) Vitamin-related disorders. In: Cohen, R.D., Alberti, K.G.M.M. and Denman, A.M. (eds) *The Metabolic and Molecular Basis of Acquired Disease*. Bailliére Tindall, London, pp. 634–669.

Drood, J.M., Zimetbaum, P.J. and Frishman, W.H. (1991) Nicotinic acid for the treatment of hyperlipoproteinaemia. *Journal of Clinical Pharmacology* 31, 641–650.

Elias, E., Muller, D.P.R. and Scott, J. (1981) Association of spino-cerebellar disorders with cystic fibrosis or chronic biliary cholestasis and a very low serum vitamin E. *Lancet* ii, 1319–1321.

Elias, P.S. (1993) Vitamins and chemical toxicity – an overview. In: Parke, D.V., Ioannides, C. and Walker R. (eds) *Food Nutrition and Chemical Toxicity*. Smith-Gordon, London, pp. 55–69.

Ellis, J.M., Azuma, J., Watanabe, T. *et al.* (1977) Survey and new data on treatment with pyridoxine of patients having a clinical syndrome including carpal tunnel and other defects. *Research Communications in Chemical Pathology and Pharmacology* 17, 165–177.

Feachem, R.G. (1987) Vitamin A deficiency and diarrhoea: A review of interrelationship and their implications for the control of xerophthalmia and diarrhoea. *Tropical Disease Bulletin* 84, 2–16.

Flodin, N.W. (1988) Vitamin A. In: *Pharmacology of Micronutrients*. Alan R. Liss, Inc. New York, pp. 3–30.

Forman, M.R. (1989) Research priorities and strategies for investigation of the influence of vitamin A supplementation on morbidity. *Food Nutrition Bulletin* 11, 25–35.

Fuhr, J.E., Farrow, A. and Nelson, H.S. Jr. (1989) Vitamin B6 levels in patients with carpal tunnel syndrome. *Archives of Surgery* 124, 1329–1330.

Gaby, S.K. and Machlin, L.J. (1991) Vitamin E. In: Gaby, S.K., Bendich, A., Singh, V.N. and Machlin, L.J. (eds) *Vitamin Intake and Health*. Marcel Dekker, New York, pp. 29–57.

Gaby, S.K. and Singh, V.N. (1991) Beta-carotene. In: Gaby, S.K., Bendich, A., Singh, V.N. and Machlin, L.J. (eds) *Vitamin Intake and Health*. Marcel Dekker, New York.

Gey, K.F., Brubacher, G.B. and Stahlin, H.B. (1987) Plasma levels of antioxidant vitamins in relation to ischaemic heart disease and cancer. *American Journal of Clinical Nutrition* 45, 1368–1377.

Ginter, E., Kajaba, I. and Nizner, O. (1970) The effect of ascorbic acid on cholesterolaemia in healthy subjects with seasonal deficiency of vitamin C. *Nutrition and Metabolism* 12, 76–86.

Haeger, K. (1982) Long-term study of alpha-tocopherol in intermittent claudication. *Annals of New York Academy of Science* 393, 369–375.

Hafez, M., Amer, E.S., Zelan, M., Hammed, H., Soroun, A,H., ed-Desouky, E.S. and Gamil, N. (1986) Improved erythrocyte survival with combined vitamin E and selenium therapy in children with glucose-6-phosphate dehydrogenase deficiency and mild chronic hemolysis. *Journal of Pediatrics* 108, 558–561.

Halvorsen, K. and Halvorsen S. (1963) Hartnup disease. *Pediatrics* 31, 29–38.

Horsey, J., Livesley B. and Dickerson, J.W.T. (1981) Ischaemic heart disease and aged patients: effects of ascorbic acid on lipo-proteins. *Journal of Human Nutrition* 35, 53–58.

Hussey, W.R. and Klein, M.A. (1990) A randomized controlled trial of vitamin A in children with severe measles. *New England Journal of Medicine* 323, 160–164.

Jacques, P.F., Chylack, L.T. Jr., McGandy, R.B. and Hartz, S.C. (1988) Antioxidant status in persons with and without senile cataract. *Archives of Ophthalmology* 106, 337–340.

Jacques, P.F., Sulsky, S.I., Perrone, G.A. and Schaefer, E.J. (1994) Ascorbic acid and plasma lipids. *Epidemiology* 5, 19–26.

Kane, J.B., Malloy, M.J. and Tun, P. (1981) Normalization of low-density lipoprotein levels in heterozygous familial hypercholesterolemia with a combined drug regimen. *New England Journal of Medicine* 304, 251–258.

Kligman, A.M., Fulton, J.E. Jr. and Plewig, G. (1969) Topical vitamin A acid in acne vulgaris. *Archives in Dermatology* 99, 469–476.

Koh, E.T. (1984) Effect of vitamin C on blood parameters of hypertensive subjects. *Journal of Oklahoma State Medical Association* 77, 177–182.

Ljunghall, S., Hvarfner, A., and Lind, L. (1987) Clinical studies of calcium metabolism in essential hypertension. *European Heart Journal* 8, 37–44.

Lund, B., Badskjaer, J., Lund, B. and Soerensen, O.H. (1978) Vitamin D and ischaemic heart disease. *Hormone Metabolism Research* 10, 553–556.

McCarron, D.A., Morris, C.D., Henry, H.J. and Stanton, J.L. (1984) Blood pressure and nutrient intake in the United States. *Science* 224, 1392–1398.

Malmgren, R., Collins, A. and Nilsson, C.G. (1987) Platelet serotonin uptake and effects of vitamin B6 treatment in premenstrual tension. *Neurobiology* 18, 83–86.

Matsuoka, L.Y. (1983) Acne. *Journal of Pediatrics* 103, 849–854.

Massry, S.G. (1980) Requirements of vitamin D metabolites in patients with renal disease. *American Journal of Clinical Nutrition* 33, 1530–1535.

Mathews-Roth, M.M. (1986) Carotenoids quench evolution of excited species in epidermis exposed to UV-B (290–320 nm) light. *Photochemistry and Photobiology* 43, 91–93.

Mathews-Roth, M.M., Pathak, M.A., Fitzpatrick, T.B., Harber, L.C. and Kass, E.H. (1974) β-Carotene as an oral photoprotective agent in erythropoietic protophyria. *Journal of the American Medical Association* 228, 1004–1008.

Mattes, J.A. and Martin, D. (1982) Pyridoxine in premenstrual depression. *Human Nutrition: Applied Nutrition* 36, 131–133.

Matts, S.G.F. (1980) Riboflavin, In: Barker, B.M. and Bender, D.A. (eds) *Vitamins in Medicine* 4th edn, Vol. 1. Heinemann, London, pp. 398–438.

Mawer, E.B., Backhouse, J., Taylor, C.M., Lumb, G.A., Stanbury, S.W. (1973) Failure of formation of 1, 25-$(OH)_2$-D in chronic renal insufficiency. *Lancet* 1, 626–628.

Milton, R.C., Reddy, V. and Naidi, A.N. (1987) Mild vitamin A deficiency and childhood morbidity – an Indian experience. *American Journal of Clinical Nutrition* 46, 827–829.

Mink, K.A., Dick, E.C., Jenings, L.C. and Inhorn, S.L. (1988) Amelioration of rhinovirus colds by vitamin C (ascorbic acid) supplementation. In: DeLaMage, L.M. and Peterson, E.M. (eds) *Medical Virology VII: Proceedings of the 1987 International Symposium on Medical Virology*. Elsevier, New York, p. 356.

Morgan, J. and Cohen, L. (1987) Hyperlipidemia: Further observations on individualized therapy with the combined use of niacin and probucol. *Federation Proceedings* 46, 1470A.

Muller, D.P.R. and Lloyd, J.K. (1982) Effect of large doses of vitamin E on the neurological sequelae of patients with abeta-lipoproteinaemia. *Annals of the New York Academy of Science* 393, 133–144.

Muller, D.P.R., Lloyd, J.K. and Wolff, O.H. (1983) Vitamin E and neurological function. *Lancet* 1, 225–228.

Natta, C.L., Machlin, L.J. and Brin, M. (1980) A decrease in irreversibly sickled erythrocytes in sickle cell anaemia patients given vitamin E. *American Journal of Clinical Nutrition* 33, 968–971.

Nutrition Reviews (1986) Low plasma vitamin A levels in preterm neonates with bronchopulmonary dysplasia. *Nutrition Reviews* 44, 202–204.

Peterson, V.E., Crapo, P.A., Weininger, J., Ginsberg, H. and Olefsky, J. (1975) Quantification of plasma cholesterol and triglyceride levels in hyper-cholesterolaemic subjects receiving ascorbic acid supplements. *American Journal of Clinical Nutrition* 28, 584–587.

Rahmathullah, L., Underwood, B.A., Thulasiraj, R.D., Milton, R.C., Ramaswamy, K., Rahmathullah, R. and Babu, G. (1990) Reduced mortality among children in Southern India receiving a small weekly dose of vitamin A. *New England Journal of Medicine* 323, 929–935.

Reddy, V., Bhaskaram, P., Raghuramulu, N. (1986) Relationship between measles, malnutrition, and blindness: a prospective study in Indian children. *American Journal of Clinical Nutrition* 44, 924–930.

Ritchie, C.D. and Singkamani, R. (1986) Plasma pyridoxal 5-phosphate in women with premenstrual syndrome. *Human Nutrition: Clinical Nutrition* 40c, 75–80.

Robertson, J.M., Donner, A.P. and Trevithink, J.R. (1989) Vitamin E intake and risk of cataract in humans. *Annals of the New York Academy of Science* 570, 372–382.

Rudolph, M., Arulanantham, K. and Greenstein, R.M. (1980) Unsuspected nutritional rickets. *Pediatrics* 66, 72–76.

Rudra, D.N., Dickerson, J.W.T., Walker, R. and Chayen, J. (1975) The effect of some antioxidants on lipofuscin accumulation in rat brain. *Proceedings of the Nutrition Society* 34, 122A.

Salonen, J.T., Salonen, R., Thanainen, M., Parviainen, M., Sephanen, R., Kantola, M., Sephanem, K. and Rauramaa, R. (1988) Blood pressure, dietary fat and antioxidants. *American Journal of Clinical Nutrition* 48, 1053–1056.

Schectman, G., Hyatt, J. and Hartz, A. (1993) Evaluation of the effectiveness of lipid-lowering therapy for treating hypercholesterolaemia in veterans. *American Journal of Cardiology* 71, 759–765.

Schorah, C.J. (1979) Inappropriate vitamin C reserves: their frequency and significance in an urban population. In: Taylor, T.G. (ed.) *The Importance of Vitamins to Human Health.* MTP Press, Lancaster, pp. 61–72.

Schorah, C.J., Habibzadeh, N., Hancock, M. and King, R.F.G.J. (1986) Changes in plasma and buffy layer vitamin C concentrations following major surgery: what do they reflect? *Annals of Clinical Biochemistry* 23, 566–570.

Schwartz, J. and Weiss, S.T. (1994) Relationship between dietary vitamin C and pulmonary function in the First National Health and Nutrition Examination Survey (NHANES*) *American Journal of Clinical Nutrition* 59, 110–114.

Seed, M., O'Connor, B., Peromlaelon, N., O'Donnell, M., Reaveley, D. and Kinoght, G.Z. (1993) The effect of nicotinic acid and acipimax on lipoprotein (a) concentration and turnover. *Atherosclerosis* 101, 61–68.

Seelig, M.S. (1975) Ischaemic heart disease, vitamins D and A and magnesium. *British Medical Journal* 3, 647–648.

Shenai, J.P., Chytil, F. and Stahlman, M.T. (1985) Vitamin A status of neonates with broncho-pulmonary dysplasia. *Pediatric Research* 19, 185–188.

Shenai, J.P., Kennedy, K.A., Chytil, F. and Stahlman, M.T. (1987) Clinical trial of vitamin A supplementation in infants susceptible to bronchopulmonary dysplasia. *Journal of Pediatrics* 111, 269–277.

Smith, G.P., Rudge, P.J. and Peters, T.J. (1984) Biochemical studies of pyridoxal and pyridoxalphosphate status and therapeutic trial of pyridoxine in patients with carpal tunnel syndrome. *Annals of Neurology* 15, 104–107.

Sommer, A. (1994) Vitamin A: Its affect on childhood sight and life. *Nutrition Reviews* 52, 500–566.

Sommer, A., Tarwotio, I., Hussaini, G. and Susento, D. (1983) Increased mortality in children with mild vitamin A deficiency. *Lancet* 2, 585–588.

Sowers, M.R., Wallance, R.B. and Lemke, J.H. (1985) The association of intakes of vitamin D and calcium with blood pressure among women. *American Journal of Clinical Nutrition* 42, 135–142.

Spittle, C.R. (1971) Atherosclerosis and vitamin C. *Lancet* ii, 1280–1281.

Steiner, M. (1983) Effect of alpha-tocopherol administration on platelet function in man. *Thrombosis and Haemostasis* 49, 73–77.

Stern, J. (1983) Hereditary and acquired mental deficiency. In: Williams, D.L. and Marks, V. (eds) *Biochemistry in Clinical Practice.* Heinemann, London, pp. 489–523.

Stransky, M., Rubin, A., Lara, N.S. and Lazaro, R.P. (1989) Treatment of carpal tunnel syndrome with vitamin B6: a double-blind study. *Southern Medical Journal* 82, 841–842.

Taylor, T.V., Rimmer, S., Day, B., Butcher, J. and Dymock, A. (1974) Ascorbic acid supplementation in the treatment of pressure sores. *Lancet* 2, 544–546.

Thomsen, K., Schmidt, H. and Fischer, A. (1979) β-Carotene in erythropoietic protoporphyria: 5 years experience. *Dermatologia* 159, 82–86.

Thurnham, D.I. and Singkamann, R. (1991) The acute phase response and vitamin A status in malaria. *Transactions of the Royal Society of Tropical Medicine and Hygiene* 85, 194–199.

Vargherse, Z., Farrington, K. and Moorhead, J.F. (1979) Renal osteodystrophy: dietary influences and management. *Proceedings of the Nutrition Society* 38, 337–350.

WHO/UNICEF (1987) Expanded program on immunization. Program for the prevention of blindness. *Nutrition Weekly Epidemiological Records* 67, 133–134.

Willis, G.C. (1953) An experimental study of the intimal ground substance in atherosclerosis. *Canadian Medical Association Journal* 69, 17–22.

Willis, G.C. (1957) The reversibility of atherosclerosis. *Canadian Medical Association Journal* 77, 106–109.

Yoshioka, M., Matsushita, T. and Chuman, Y. (1984) Inverse association of serum ascorbic acid level and blood pressure or rate of hypertension in male adults aged 30–39 years. *International Journal of Vitamin and Nutrition Research* 54, 343–347.

Vitamin Abuse 20

The first 50 years of this century have been described as the 'vitamin era' because during this period all the known vitamins were identified. Experimental, clinical and circumstantial evidence have revealed that vitamins are dietary constituents required in very small amounts for normal growth, development and maintenance of health. As discussed in earlier chapters, the amounts meeting the daily needs for adult individuals of different ages range from a few micrograms to a few milligrams, depending on the vitamin. Although the requirements are small, a failure to obtain them from the diet, for periods of time which vary with the age and condition of the individual, and the vitamin, may lead to overt diseases, such as rickets, scurvy, beriberi, pellagra, or xerophthalmia. This relationship points to the fact that vitamins are substances of very high biological activity. Except in vulnerable sections of the population, such as the poor, certain ethnic minorities and the elderly, overt deficiencies of vitamins are now uncommon, especially in the industrialized nations. This is essentially because of the much improved economy, nutrition education, food processing and fortification, and food availability. However, deficiencies at a subclinical level can occur in many situations, some of which are listed in Table 20.1. The amount of each vitamin needed to avoid a marginal deficiency can be readily obtained from a normal mixed diet which is generally comprised of a wide variety of foods including cereals, dairy products, meat or fish, and fruits and vegetables.

Now that clinically manifest vitamin deficiency conditions are not much of a concern in most affluent communities, much attention is increasingly being paid to the potential role of vitamins in the prevention of many chronic diseases and these have been discussed in Chapter 19. Since the mid-1970s, many arguments have been advanced in favour of the prophylactical use of vitamins in large doses against these chronic diseases (Evans and Lacey, 1986). However, it is easy for such arguments (and claims) to be exaggerated, particularly in the minds of the general population who see in them a means of preserving their health in the face of the cost of medical services. The availability of a wide range

Table 20.1. Situations at risk of vitamin deficiencies.

Inadequate intake
Food faddist
Vegetarians
Alcoholics
Drug-induced loss of appetite
Disturbances in absorption
Diseases of:
Gastrointestinal tract
Liver and biliary tract
Prolonged diarrhoea
Pernicious anaemia
Drug-induced malabsorption
Impaired metabolic activation
Hepatic disorders
Renal insufficiency
Alcoholism
Ageing
Increased requirements
Growth
Pregnancy
Lactation
Menstruation

of over-the-counter (OTC) preparations, containing amounts of the various vitamins which are many times the RDA, has encouraged the popular practice in North America and other parts of the western world of consuming high doses of vitamins.

The term 'megadose' is often used in referring to this high dose level of vitamins. There is no clear definition or dose limit to describe 'megadose'. It is generally taken to refer to doses that are ten or more times the RDA. It should be pointed out that 'megavitamin' is actually a misnomer because the vitamins taken in such high doses act more like drugs than nutrients.

Megavitamin Use

It is estimated that about one-third of North Americans use vitamin supplements regularly (Griffith and Innes, 1983), spending 2–2.5 billion dollars each year in the United States alone. The supplemental use of vitamins can be divided into three areas: the prevention or treatment of diseases of deficiency, the treatment of vitamin-responsive inborn errors of metabolism, and the prevention and treatment of many conditions shown in Table 20.2. According to most

Table 20.2. Pathological conditions, claimed to be the consequences of a prolonged marginal deficiency of vitamins.

Vitamins	Conditions
C	Common cold
	Cancer
	Cardiovascular disease
	Bone pain
	Hypersensitivity
B_3	Hypercholesterolaemia
	Schizophrenia
B_6	Premenstrual syndrome
	Schizophrenia
	Impaired memory
	Carpal tunnel syndrome
A (including β-carotene)	Cancer
D	Osteoporosis
E	Haemolytic anaemia
	Claudication
	Cardiovascular disease
	Cancer

of the demographic studies (Block, 1988; Horwath and Worsley, 1989; Donald *et al.*, 1992), women, elderly persons, and individuals with high education and income consume more vitamin supplements. In the United States, more of the white than black population appear to use supplements. It is intriguing that supplement users are generally more nutrition – and diet – conscious than non-users. They pay more attention to the nutritional quality of the foods they eat.

Are Megavitamin Supplements Harmful?

Vitamins which are commonly taken in large doses include vitamin C, pyridoxine, vitamin E, β-carotene, vitamin A, vitamin D, and niacin (Table 20.3; see also Chapter 19). The potential toxicity of large doses of these vitamins varies and has been considered in previous chapters (see also Elias, 1993). Interaction with drugs when taken concomitantly is considered in Chapter 21. Other pertinent facts in relation to this problem are now briefly discussed.

All water soluble vitamins and some of the fat soluble vitamins (e.g. vitamins A and K) are coenzymes, which by virtue of their structural constitutions are able to bind with the cellular enzymes (apoenzymes), forming holoenzymes.

Table 20.3. Use of supplements by the respondents (n = 3000) in the United States.

Vitamins	95th percentile (% of RDA)
A	430
C	2800
D	375
E	6000
B_1	6000
B_2	5000
B_3	1200
B_6	5000
B_{12}	3338
Pantothenic acid	1300
Folacin	200
Biotin	300

It is these holoenzymes that have the capacity to participate in the cellular catalytic functions. The vitamin coenzymes thus appear to function only in the presence of apoenzymes. In fact, one of the rationales of the use of vitamin supplementation is to saturate the apoenzyme and maximize the metabolic processes through forming a maximum level of holoenzyme. It is, however, important to realize that the cellular capacity to synthesis apoenzymes and other proteins per unit of time is limited. The amount of the relevant coenzymes required to meet the daily recommended allowances is thought to be adequate to saturate apoenzymes. An excess coenzyme that is available through megavitamin use may be superfluous for its coenzymatic function in the body.

When water soluble vitamins are taken in excess, neither the plasma nor the tissue levels are generally increased beyond their saturation states, while the urinary concentrations are elevated in a dose-related fashion. For lipid soluble vitamins, on the other hand, the hepatic tissue concentrations are increased with a rise in supplement intake, but without any appreciable changes in their circulatory levels. This plasma level, unresponsive to intakes of megadoses of vitamins, also suggests that the metabolic availability of vitamins cannot be raised by increasing their intakes beyond physiological dose levels.

The Use of High Doses of Vitamins

In Chapter 19 we considered the results of researches that have been directed to the elucidation of the prophylactic and therapeutic potential of the different vitamins. The somewhat preliminary and tentative nature of some of the evidence for this potential does not at present give scientific support to the taking of megadoses of vitamins. However, this is not to say that some individuals may not subjectively 'feel better' as a result of taking them. It must also be conceded that there is considerable scope for biochemical individuality and that, as mentioned previously, this may be due to differences between individuals in the chemical structure and/or physical conformation of certain enzymes (see Chapter 19).

Concluding Comments

Claims for the need of vitamin supplements are often exaggerated. The popularity of megavitamin use is based on erroneous ideas that excessive intakes will help to cope with stress and will prevent diseases ranging from the common cold to cancer. Some individuals may receive little benefit from a selected vitamin supplement, especially if status with respect to the particular vitamin, or vitamins, is of borderline adequacy. The potential for toxicity from large doses of a number of vitamins must be recognized. The taking of large doses of vitamins, besides being of doubtful physiological value may be an expensive way of wasting vitamins and may, in fact, engender some real and potential health hazards (Chapter 21).

The value of vitamin supplements when taken appropriately cannot be disputed. Women of child-bearing age may need a folic acid supplement both before and after conception, though for the same women a vitamin A supplement (or the regular consumption of liver) might be dangerous. Some vegetarians, especially vegans, may need to take vitamin B_{12}. Newborn babies, particularly premature babies, are generally given a single dose of vitamin K to prevent abnormal bleeding. Vitamin D supplements should be given to infants, growing children and pregnant women if they are not having fortified milk or have not adequate exposure to sunlight. Individuals on low-energy diets may have difficulty meeting their vitamin requirements and for this reason need a supplement. In addition, vitamin supplements are also of value in some clinical conditions particularly during rehabilitation after illness when appetite is often poor.

Vitamins C and E, as well as β-carotene, are antioxidants. Their actions both *in vitro* and *in vivo* in preventing free radical damage are well known. However, the claims that because of this property amounts in excess of the RDA will give additional protection against cardiovascular disease, cancer and immunodeficiency may be speculative (Bieri, 1987) and remain to be proven (Meydani, 1995). Normally, in a healthy individual antioxidant supplements are requirements only if the diet is poor. In disease or trauma there may be an increase in metabolism triggered by cytokines and hence increased requirements for these nutrients. There have been suggestions that there should be modest increases of about twofold in the recommended intake for vitamin C, 3–5-fold for vitamin E, and an intake of about 15 mg day^{-1} of β-carotene to provide optimal prophylaxis against the potential harmful effects of free radical processes (Diplock, 1987).

References

Bieri, J.G. (1987) Are the recommended allowances for dietary antioxidants adequate? *Free Radical Biology and Medicine* 3, 193–197.

Block, G. (1988) Vitamin supplement use by demographic characteristics. *American Journal of Epidemiology* 127, 297–309.

Diplock, A.T. (1987) Dietary supplementation with antioxidants: Is there a case for exceeding the recommended dietary allowance? *Free Radical Biology and Medicine* 3, 199–201.

Donald, E.A., Basu, T.K., Hargreaves, J.A., Overton, T.R., Peterson, R.D. and Chao, E. (1992) Dietary intake and biochemical status of a selected group of older Albertans taking or not taking micronutrient supplements. *Journal of the Canadian Dietetics Association* 53, 36–39.

Elias, P.S. (1993) Vitamins and chemical toxicity: an overview. In: Parke, D.V., Ioannides, C. and Walker, R. (eds) *Food, Nutrition and Chemical Toxicity.* Smith-Gordon, London pp. 55–69.

Evans, C.D.H. and Lacey, J.H. (1986) Toxicity of vitamins: complications of a health movement. *British Medical Journal* 292, 509–510.

Griffith, P.R. and Innes, F.C. (1983) The relationship of socioeconomic factors to the use of vitamin supplements in the city of Windsor. *Nutrition Research* 3, 445–455.

Horwath, C.C. and Worsley, A. (1989) Dietary supplement use in a randomly selected group of elderly Australians. *Journal of American Geriatrics Society* 37, 689–696.

Meydani, M. (1995) Vitamin E. *Lancet* 345, 170–175.

Safety Considerations of Excess Vitamin Intakes 21

As discussed in the previous two chapters, the ready availability of vitamins and their many potential prophylactic and therapeutic benefits for various pathological conditions have led to increased consumption of these organic compounds in large doses. The term 'megavitamin' is often used when referring to a vitamin supplement that supplies an amount far in excess of its established recommended intake. It is now increasingly being realized that vitamins, when taken in excess, can be potentially toxic. Safe intake levels may vary widely from vitamin to vitamin. The factors that are generally involved in determining the safety dose limit of vitamins include the sensitivity of the consumer, age, state of health, duration and strength of supplementation, and route of administration. Apart from the direct effects of hypervitaminosis, safety considerations should be given to the interactions of excess vitamins with the normal metabolism of other nutrients, and with therapeutic effects of many pharmacological agents if taken concomitantly.

Lipid-soluble Vitamins

These vitamins, because of their lipid solubility, are liable to accumulate in the body up to toxic levels, if taken in excess for a long period of time. However, not all fat-soluble vitamins are equally toxic. There appears to be more evidence of adverse effects for preformed vitamin A (retinol) and vitamin D than for vitamins E and K. The body distribution and turn-over differences may account for the toxicity differences between the fat-soluble vitamins. Thus, vitamin E is distributed to all tissues including the adipose tissue; this characteristic is unlike that of vitamins A and D, which tend to accumulate predominantly in the liver, making the tissue highly saturated with the vitamins. Vitamin K is rapidly metabolized in the body, making its storage life very short.

Retinoids

Acute toxicity

Adverse effects of vitamin A from its natural dietary sources are rare. The only outbreaks of acute toxicity that have been observed were among Arctic explorers who ingested excessive quantities of the vitamin by eating polar bear liver (containing 6000 μg of vitamin A per gram). The symptoms of acute toxicity due to ingestion of a self-limited single large dose of vitamin A from polar bear liver were first described by Kane, an American Arctic explorer (Knudson and Rothman, 1953). Symptoms include nausea, vomiting, headache, and vertigo. However, there have been many reports of acute hypervitaminosis A, characterized by vomiting, drowsiness, and fontanelle bulging in infants and young children as a result of accidental ingestion of high potency vitamin A preparations (Persson *et al.*, 1965).

Chronic toxicity

The incidence of chronic hypervitaminosis A is relatively more common than acute hypervitaminosis A (Bendich and Langseth, 1989). This is because: (i) supplements containing large amounts of vitamin A are increasingly being given to pre-school children of developing countries where vitamin A deficiency is highly prevalent; (ii) there is an increasing use of vitamin A or its analogues in the treatment of various types of skin disorders such as acne; and (iii) there appear to be an increasing number of people taking excessive doses of vitamin A for reasons of food faddism and health benefits, such as cancer prevention. The chronic form of hypervitaminosis A has been generally attributed to daily ingestions of moderately large doses (7500–15000 μg) over a period of several weeks (Hathock *et al.*, 1990). The population who are particularly sensitive to toxicity at relatively low intakes of vitamin A include infants and young children, and the individuals whose liver function has been compromised by drugs and alcohol abuse, viral hepatitis, or protein–energy malnutrition.

The symptoms of chronic overdoses of vitamin A include a dry, thickening skin with itching, some fissuring and peeling, erythematous eruptions, alopecia, muscular stiffness, enlargement of the liver, spleen, and lymph glands, polydipsis and polyuria. Many of the symptoms of hypervitaminosis A are quite similar to those of a deficiency of the vitamin (Fig. 21.1). These toxic signs are usually reversible and disappear rapidly on cessation of the vitamin intake.

Over the last 40 years, there have been approximately 200 literature reports covering more than 500 individual cases with overt signs of hypervitaminosis A (Table 21.1). Relatively high incidences of vitamin A toxicity were observed during the 1950s. Many of these cases were reported in Europe. This may be

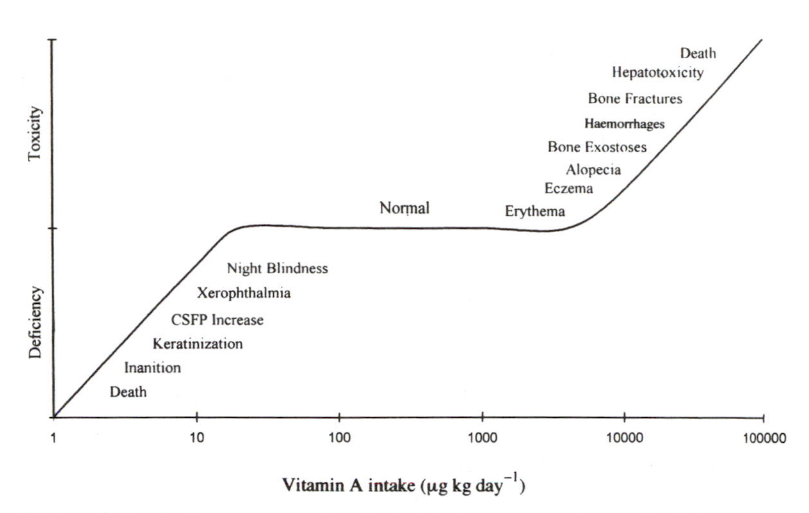

Fig. 21.1. Biological response to vitamin A deficiency, normalcy and toxicity.

Table 21.1. Reported cases of hypervitaminosis A (1950–1987).

Year	Reports[1]	Cases
1950–1959	77	241
1960–1969	37	73
1970–1979	58	169
1980–1987	19	30
Total	191	513

[1]Cited in Bauernfiend, 1980 (1950–1979) and Bendich and Langseth, 1989 (1980–1987).

due to the fact that in the early 1950s (1952–55), two very potent pharmaceutical preparations of vitamin A (containing 105,000–120,000 μg per dose) were available in the market and administered to infants, especially in Spain and France (Bauernfiend, 1980). Another peak period of hypervitaminosis A, as reported in the literature, appeared to be during the 1970s. This may be the reflection of extensive uses of vitamin A for skin disorders, especially during 1970–72.

Teratogenicity

Experimental and isolated clinical evidence suggest that vitamin A, when taken in excess during pregnancy, can be potentially teratogenic. The teratogenic effects of vitamin A are generally shared by all retinoids; some are, however, more toxic (e.g. tretinoin and etretinate) than others (e.g. isotretinoin).

Administrations of an excessive amount of vitamin A (10,500 μg) to pregnant rats from the 2nd, 3rd, or 4th to the 16th day post-coitum, produce a diminished litter rate and gross cranial anomalies among the surviving young (Rosa *et al.*, 1986). The teratogenic action of excess vitamin A has also been demonstrated in other mammalian species including guinea pigs, rabbits, pigs, hamsters, and monkeys. Circumstantial evidence has indicated that maternal excess vitamin A-mediated congenital abnormalities may also occur in humans (Grote *et al.*, 1985). In an attempt to assess the correlation between congenital malformations in infants and vitamin A intake of the mother, Gal and her colleagues (1972) measured the maternal serum vitamin A after delivery. The level of the vitamin was found to be significantly higher in mothers of infants with central nervous system malformations than in those with normal babies. Measurements of vitamin A in the livers of fetuses that were aborted, or of premature infants that died shortly after birth, were also higher in those with malformations. Retinoic acid and its synthetic derivatives have been strongly suggested to be teratogenic in humans especially if ingested during the first 28–70 days of pregnancy (Public Affairs Committee of Teratology Society, 1987). On the basis of this evidence, it is important that a cautious attitude should be taken towards the administration of vitamin A or its analogues to the pregnant woman, for whom the intake of the vitamin has been suggested never to exceed 3000 μg RE day^{-1} as retinyl esters (Marks, 1989). More recently, the American College of Obstetricians and Gynecologists (1993) has recommended a maximum dose of half of this level (1500 μg RE day^{-1}) prior to and during pregnancy. It should, however, be pointed out that both deficiency and excess of vitamin A are equally hazardous with regard to intrauterine life.

Mechanism of action of vitamin A toxicity

The underlying mechanism by which excess vitamin A exerts its toxic effect and produces abnormal development of embryos and fetuses is not fully understood. However, it is thought that some of the effects of hypervitaminosis A may be due to labilization of lysosomal membranes (Zile and Callum, 1983). This may occur through a toxic effect on cell membranes, particularly lysosomal membranes. Normally, vitamin A circulates in serum predominantly as free retinol bound to retinol binding protein (RBP), while in conditions of excessive intake of vitamin A, it is transported in esterified form in association with serum lipoproteins and not bound to RBP. Vitamin A toxicity appears to occur when the capacity of the protein to transport free retinol is exceeded, so that the vitamin in its esterified form is carried to the cell membranes by lipoproteins in large quantities (see Chapter 11). Retinyl esters, because of their amphipathic characteristics, are believed to be more injurious to the membranes via detergent disruption than retinol bound to RBP (Howard and Willhite, 1986).

Carotenoids

β-Carotene in high doses has no known detrimental effect other than yellowing of the skin. Carotene is a yellowish lipochrome, which tends to accumulate in fatty tissues, such as subcutaneous fat, when taken in excess. As a consequence, a yellowish tint of the skin is most clearly seen in the palms and soles of the feet, the chin, behind the ears, over the knuckles and on the abdomen and buttocks (Vakil *et al.*, 1985). Unlike the corneum, the mucosa has no lipid content and hence no affinity for carotene. This may explain the absence of yellow pigmentation in the sclera and oral cavities of individuals with hypercarotenaemia.

In man, ingestion of 30 mg β-carotene over 4 weeks may cause carotenaemia with blood levels above 350 μg dl^{-1}. Carotenaemia is generally considered to be harmless. The slow intestinal absorption over 6–8 hours, and the slower elimination from chylomicrons with a partial shift to other lipoprotein fractions, act as protective mechanisms (Marks, 1989). It should be pointed out, however, that carotenaemia can occur as an inborn error of metabolism of carotenoids, in diabetes and in hypothyroidism (Birkenhead, 1958).

Since β-carotene is a precursor of vitamin A, it might be expected to raise circulating vitamin A levels, but there is no evidence that an excess intake of β-carotene leads to hypervitaminosis A. It appears that the rate of conversion to vitamin A decreases as the amount of β-carotene in the diet increases (see Chapter 12). Therefore, even massive doses of β-carotene will not yield enough retinol to cause systemic vitamin A toxicity. Chronic toxicity studies of β-carotene involving oral doses of up to 1000 mg kg^{-1} day^{-1} for two years in dogs, rats, and mice have shown no evidence of embryotoxicity, carcinogenicity or mutagenicity (Heywood *et al.*, 1985).

In a recent study, β-carotene supplementation was shown to exacerbate the hepatic damage experienced by baboons receiving half of their calories as alcohol (Leo *et al.*, 1992). These results may imply clinical significance in alcohol abusers. Although no evidence of a deleterious interaction between alcohol and β-carotene has, so far, been reported in humans, the interaction needs exploration.

Canthaxanthine (β-carotene-4,4-dione) is a carotenoid without any provitamin A activity. It imparts a red colour and is found in plants, mushrooms, crustaceans, sea trout, and red-feathered birds. Of all the carotenoids, canthaxanthine has been suggested to have an adverse effect in the form of retinopathy, where crystalline deposits identical to synthetic canthaxanthine appear to occur in the retina (Lonn, 1987). Because of this possible side effect, canthaxanthine has not been approved as a drug by the FDA for producing a skin colour similar to a suntan. It is however, approved as a food additive.

Vitamin D

Toxicity

As discussed in Chapter 13, vitamin D (cholecalciferol) is formed in the skin from its precursor, 7-dehydrocholesterol, by UV irradiation. Excessive exposure to sunlight poses no toxicity risk of hypervitaminosis D, although it may be an important contributory factor for skin cancer. Excessive intake of the vitamin, on the other hand, may cause vitamin D toxicity. An excess intake of vitamin D can result in mobilization of calcium in the skeleton and increases the serum calcium level. This calcium is taken up by soft tissues, especially the kidney, resulting in nephrocalcinosis and metastatic calcification of other soft tissues such as blood vessels, myocardium, lungs and skin. The symptoms of hypervitaminosis D include malaise, thirst, polyuria, vomiting, abdominal pain, and gastrointestinal disturbances. The metabolic basis of vitamin D intoxication has recently been discussed by Chapuy and Meunier (1990). Laboratory findings generally include (i) hypercalcaemia and hyperphosphataemia; (ii) hypercholesterolaemia and hypomagnesaemia; and (iii) an abnormal ECG. These symptoms are reversible but, if uncontrolled, may cause death due to kidney failure. Vitamin D toxicity is treated by stopping the vitamin exposure and reducing calcium intake. Calcitonin may help to quickly reduce serum calcium.

The potential toxicity of excessive intakes of vitamin D was evident by a recent incident in Massachusetts in which eight individuals developed vitamin D intoxication after consuming milk that was incorrectly fortified with excessive amounts of the vitamin (5800 μg quart^{-1}); the amount was 580 times the amount stated on the product label (Jacobus *et al.*, 1992). Toxic effects of the vitamin are seen in the adult when doses exceed 1000–3000 IU kg^{-1} day^{-1} for several months. In infants, hypercalcaemia may occur with total daily doses of 3000–4000 IU (Committee on Nutrition, Academy of Pediatrics, 1963). The sensitivity of individuals to an excess of vitamin D is quite variable so that it is not possible to state the minimum safety dose limit. Some individuals show profound toxicity to amounts only slightly above the daily recommended dose. The syndromes of ideopathic hypercalcaemia of infancy and congenital supravalvular aortic stenosis may be examples of this toxicity. It is possible that a little excess vitamin D intake along with large amounts of calcium is sufficient to cause hypercalcaemia. Exposure of patients with sarcoidosis to sunlight causes hypercalcaemia due to the patients having increased sensitivity to vitamin D. The disease is treated with cortisone which decreases the sensitivity.

According to many experimental studies, vitamin D and its synthetic analogues are potentially teratogenic if taken in excess during pregnancy (Nebel and Ornoy, 1972; McClain *et al.*, 1980). Based on this evidence, administration of the vitamin above the recommended intakes is contraindicated for pregnant women.

Mechanism of action of vitamin D toxicity

The homeostasis of blood calcium and phosphate levels determines the rate of production of 1,25-dihydroxy-D (see Chapter 13). The production of this active metabolite of vitamin D in the kidney is thus regulated by a strong feedback mechanism at the physiological level. It is, however, noteworthy that there is no feedback control over the hepatic production of 25-hydroxy-D. There is evidence to suggest that this metabolite may also be active in transporting calcium from the intestinal lumen especially when its blood concentration reaches 200–400 ng ml^{-1} as opposed to its normal level, which is only 40–80 ng ml^{-1}. 25-Hydroxy-D is normally bound to a specific globulin or lipoprotein but its excess is carried to the cell membranes by albumin, and competes with 1,25-dihydroxy-D for receptors in the gut and bone (Brumbaugh and Haussler, 1973). An elevated serum level of 25-hydroxy-D has been found to be associated with raised serum and urinary Ca, possibly due to increased Ca absorption from the intestine and Ca resorption from bone (Davis, 1989). It seems possible that toxicity caused by excess vitamin D administration is mediated through its intermediate, 25-hydroxy-D, rather than its active metabolite, 1,25-dihydroxy-D.

Vitamin E

Vitamin E is essentially non-toxic according to the literature published since 1974 concerning its safety and tolerance (Bendich and Machlin, 1988). Some of the commonly recurring complaints include gastrointestinal disturbances, such as nausea, flatulence, and diarrhoea. These symptoms are generally of a transient nature, and experienced when the intake of vitamin E is as high as 3200 IU day^{-1}. However, concerns have been expressed that vitamin E in excess may interfere with the normal coagulation process (Olson, 1984). It has been suggested that metabolites of this vitamin, such as α-tocopherylquinone or hydroquinone, may act as antimetabolites to vitamin K (March *et al.*, 1973). In a recent clinical trial, patients with retinitis pigmentosa, an hereditary eye disease, experienced a worsening of the retinal function when treated with vitamin E in a dose of 400 IU day^{-1} for 4–6 years (Bersen *et al.*, 1993). In view of the fact that our understanding of the possible subclinical and biochemical effects of excessive intakes of vitamin E is poor, its indiscriminate consumption may be unwise.

Vitamin K

Naturally occurring vitamins K_1 (phylloquinone) and K_2 (menaquinones) are believed to be non-toxic to man even in very large doses, except that there are some isolated reports of idiosyncratic reactions occurring when vitamin K is administered (Miller and Hayes, 1982). Vitamin K is generally taken up rapidly by the liver, distributed widely in body tissues, quickly metabolized into its polar metabolites, and subsequently these metabolic products are excreted in the urine (see Chapter 15); its half-life in the liver is thus less than 24 hours. Therefore, vitamin K is not stored in the body to any appreciable extent as is the case with vitamins A and D. Synthetic vitamin K (menadione) and its water soluble derivatives can, however, be acutely toxic. The toxicity is thought to be due to the unsubstituted carbon 3. The substance binds with tissue sulphydryl groups and thereby it promotes the oxidation of membrane phospholipids (Owen, 1971). This synthetic vitamin K including its derivatives can decrease blood glutathione, interfere with erythrocyte redox systems causing instability, methaemoglobinaemia and haemolysis. Individuals with glucose 6-phosphate dehydrogenase deficiency and premature babies are particularly sensitive to excess vitamin K. Toxic effects reported in infants supplemented with menadione include haemolytic anaemia, hyperbilirubinaemia and kernicterus.

Water-soluble Vitamins

Since the body has the capacity to readily excrete water-soluble vitamins, the use of massive doses of these vitamins has not come into criticism in the same way as has the use of excessive doses of fat-soluble vitamins. There is, however, a general consensus that every biologically active substance has a dose limit beyond which the substance will be potentially toxic. Indeed in recent years, concerns have been expressed with regard to the potential hazard of prolonged and regular ingestion of large amounts of water-soluble vitmins, which were once thought to be innocuous substances (Alhadeff *et al.*, 1984). Of all the water-soluble vitamins, nicotinic acid, vitamin B_6, and ascorbic acid are the most popularly used in megadose quantities for therapeutic and prophylactic purposes (see Chapter 19). These three vitamins have been reported more often to cause overt toxicity in human subjects than the rest of the water soluble vitamins, which include thiamin, riboflavin, pantothenic acid, biotin, folic acid, and cobalamin.

Nicotinic Acid

Nicotinic acid and nicotinamide possess the same vitamin activity with different pharmacological and toxicity profiles. Unlike nicotinamide, an excess intake of nicotinic acid is associated with dilatation of blood vessels, especially capillaries. This action is thought to be due to the release of the vasodilator, histamine. As a consequence, one of the deleterious side effects of nicotinic acid, taken daily in high doses (2–6 g), has been the flushing and warming of the face and neck, usually lasting only for a few minutes (DiPalma and Thayer, 1991). In fact, flushing within fifteen minutes after an oral dose may occur at doses as low as 100 mg day^{-1}. Although uncomfortable, these symptoms are usually tolerated by most subjects. Pretreatment with antihistamine 15 minutes before the nicotinic acid dose, or taking the vitamin after a meal, appear to reduce the severity of such symptoms. However, nicotinic acid, due to its histamine-releasing effect, may be contraindicated in patients with a peptic ulcer and asthma.

Isolated reports have indicated that nicotinic acid in large amounts may be hepatotoxic (Mosher, 1970; Einstein *et al.*, 1975). The signs include clinical cholestatic jaundice associated with elevated serum transaminase activity and serum bilirubin concentration and delayed bromosulphthalein clearance. It is of great interest that slow release nicotinic acid preparations have been reported to be more toxic to the liver than the conventional forms (Knopp *et al.*, 1985; Mullin *et al.*, 1989).

Besides vasodilating and hepatic toxic effects, other side effects of nicotinic acid include decreased glucose tolerance and occasional diarrhoea. Elevated serum uric acid level may also occur. Chronic high doses of nicotinic acid have been reported to be associated with gouty arthritis (The Coronary Drug Project Research Group, 1975). It appears that the vitamin competes with uric acid for excretion, resulting in elevated serum uric acid levels in the presence of an excess intake. Patients with diabetes or gout are generally cautioned against the use of this vitamin in megadose amounts.

Vitamin B_6

Evidence to date indicates that large doses of vitamin B_6 may exert a direct toxicity effect on the peripheral nervous system. According to many experimental studies the adverse reaction appears to be dose related (Bendich and Cohen, 1990). The time course for the development of neuropathy in rats treated with different dose levels of pyridoxine, is shown in Table 21.2. In humans, the minimum toxic dose of vitamin B_6 is believed to be > 500 mg day^{-1} (recommended daily intake, 2 mg). However, women given vitamin B_6 for the treatment of premenstrual tension have been found to vary considerably in their

Table 21.2. Dose–response effect of pyridoxine on the duration for the onset of neuropathy in rats.

Pyridoxine ($mg\ kg^{-1}\ day^{-1}$, i.p.)	Duration (weeks)
200	10–12
300	4
1200	1.3

Adapted from Krinke and Fitzgerald (1988).

sensitivity to the vitamin (Dalton, 1985). Some patients with toxic symptoms were taking only 50 mg day^{-1}.

The clinical manifestations of toxicity consist of a tingling sensation in the neck and legs, and an unsteady gait (Schaumburg *et al.*, 1983). Although most of these symptoms disappear after withdrawal of the vitamin, varying degrees of residual neuropathy tend to remain for a very long time. Interestingly enough, like excess vitamin B_6, a deficiency of this vitamin causes peripheral neuropathy. The neurological disorder probably results from a deficiency of biogenic amines (see Chapter 5). The mechanism by which excess vitamin B_6 intake results in neuropathy is, however, far from being understood. Since pyridines are known to be neurotoxic (Sahenk and Mendell, 1980), the pyridine molecule of vitamin B_6 may be a contributory factor to peripheral neuropathy, especially when the vitamin is taken in excess.

Ascorbic Acid

Ascorbic acid is generally a non-toxic substance for most individuals even if its daily intake is more than 100 times the recommended intake. There are, however, numerous isolated reports describing many adverse reactions, experienced by some individuals, from taking excessive amounts of this vitamin (Basu and Schorah, 1982). Some of these side effects are listed in Table 21.3.

Gastrointestinal disturbances

Massive intakes of ascorbic acid (4–15 g day^{-1}) may cause gastrointestinal discomfort characterized by nausea, abdominal cramps, heartburn and diarrhoea. These symptoms are thought to be caused by either the osmotic activity or acidic effect of the vitamin. The disturbances are perhaps the most consistent abnormalities noted in subjects taking the vitamin in megadose levels. These reactions are, however, of little concern, since the effects may be ameliorated or eliminated by taking the vitamin either with meals or as the sodium salt.

Table 21.3. Some adverse reactions from ingestion of ascorbic acid (AA) in excessive quantities.

Adverse effect	Possible mechanism	Contraindication (to excessive AA)
Gastrointestinal disturbances	An osmotic effect	Sensitive individuals
Formation of renal calculi	Acidifying urine; increased excretions of calcium, urate and oxalate	Individuals with renal insufficiency and who are prone to develop renal stones
Conditioning scurvy	Depletion of tissue ascorbate due to its increased metabolism and excretion	Pregnant mothers
Lysis of erythrocytes	Inhibition of glucose-6-phosphate dehydrogenase (G-6-PD)	Individuals who are genetically deficient in G-6-PD
Iron overload	Induction in iron absorption and storage	Patients with idiopathic haemochromatosis, thalassaemia major, and other haemolytic anaemia
Interference with glucose assay	Interferes with glucose oxidase	Diabetic subjects

Adapted from Basu and Schorah (1982).

Renal stones

Ascorbic acid is partly metabolized to oxalic acid (see Chapter 10). It can also increase urinary excretion of uric acid through inhibiting its binding with serum proteins (Stein *et al.*, 1976). Since the vitamin, when taken in excess, has been reported to mobilize calcium from the skeleton (Ramp and Thornton, 1971), concerns have been expressed that high ascorbic acid intakes might enhance urinary excretions of oxalic acid and uric acid, and increase the risk of calcium oxalate and urate stone formation. Furthermore, the vitamin is known to decrease urinary pH, which could make oxalate or urate even less soluble in urine, and thereby enhance the formation of calcium oxalate and urate crystals in the kidney and bladder.

It has been theorized that the greater the quantity of oxalate excreted, the greater the probability of calcium oxalate calculi formation, and therefore the larger the intake of ascorbic acid the greater will be the chances of developing renal stones. It should be pointed out, however, that in normal subjects oxalic acid formation from ascorbic acid is a saturable process (Schmidt *et al.*, 1981). Thus, the urinary excretion of oxalic acid reaches a plateau after the third dose of 2 g of ascorbic acid, and additional intakes of the vitamin do not appear to increase oxaluria any further. None the less, in view of the isolated reports of renal stones occurring among some mega-ascorbic acid users, ingestion of the vitamin at more than its recommended level is contraindicated for individuals with high intakes of foods containing oxalates (spinach, rhubarb, chocolate), in those who are prone to develop oxalate or urate stones, and in those with a history of calcium urolithiasis.

Conditioning effect

Another frequently expressed concern about high-dose ascorbic acid is that it might be responsible for ascorbate deficiency following a sudden cessation of the extra vitamin intake (Basu and Schorah, 1982). Two cases of infantile scurvy following maternal supplementation with ascorbic acid (400 mg day^{-1}) during pregnancy have been reported. A more recent study examined the withdrawal effects of the vitamin on its changes in plasma and tissue concentrations in guinea pigs exposed to excess ascorbic acid *in utero* and up to weaning (Basu, 1985). Change in ascorbic acid intake from 1.0 to 0.1 mg ml^{-1} drinking water, at the age of 21 days and continued up to 52 days, resulted in significant reductions in ascorbic acid concentrations in plasma, leucocytes and adrenals, compared with those of the pair-fed animals receiving 0.1 mg ml^{-1} drinking water throughout (Table 21.4). These findings indicate that there may be an adaptation to a high intake of ascorbic acid which results in a greater than

Table 21.4. Withdrawal effects of ascorbic acid (AA) on the plasma and tissue concentrations of the vitamin in guinea pigs exposed to excess AA in drinking water during their intrauterine life up to weanling age[1].

Groups	Dose of AA (mg ml^{-1})	Period of exposure	Plasma (mg dl^{-1})	Leucocyte (μg 10^{-8} WBC)	Adrenal (mg g^{-1})
Normal	0.1	From intrauterine life up to 52 days of age	1.5 ± 0.2	51 ± 8	0.09 ± 0.01
High	1.0	From intrauterine life up to 52 days of age	2.2 ± 0.4[1]	97 ± 11[1]	1.16 ± 0.30[1]
High Low	$1.0 \rightarrow 0.1$	High dose from intrauterine life up to 21 days of age (weanling) and then changed to the low dose level from 22 to 52 days of age	0.3 ± 0.03[1]	23 ± 2[1]	0.04 ± 0.002[1]

Adapted from Basu (1985).
[1]Differences between normal and experimental groups are statistically significant.

normal metabolism of the vitamin. It is possible that ascorbic acid continues to degrade to its oxidizing products at a faster rate after the intake has returned to normal (Tsao and Salimi, 1984), so that the intake is no longer adequate.

Impaired copper status

Marked reductions in serum copper and ceruloplasmin levels have been observed in men taking large ascorbic acid supplements (1500 mg day^{-1}). The serum concentrations of these indices, however, appear to increase after the supplementation is discontinued (Finley and Cerklewski, 1983). Since ascorbic acid is oxidized to dehydroascorbate in a reaction catalysed by ascorbate oxidase, a copper containing enzyme (see Chapter 10), it is conceivable that an increased rate of oxidation of the vitamin may account for the conditioning effect.

Lysis of erythrocytes

Fears have been expressed that high ascorbic acid intake might result in increased lysis of erythrocytes. Isolated reports suggest that subjects with glucose-6-phosphate dehydrogenase (G-6-PD) deficiency are more prone to this adverse effect. One of these reports concerns a 60-year-old American black individual (Campbell *et al.*, 1975). As part of treatment for second-degree burns of the hand, he was given 80 g of ascorbic acid intravenously per day, for two consecutive days, presumably with the thought that the vitamin would promote the healing process. Following treatment with ascorbic acid, the subject became oliguric with extensive haemolysis and marked urinary excretion of haemoglobin (4 mg 100 ml). His erythrocytes were found to be deficient in G-6-PD, and he died on the 22nd day following ascorbic acid treatment.

Iron overload

Ascorbic acid is an excellent facilitator of iron absorption from the digestive tract (see Chapter 10). While an increased absorption may be useful for those who are deficient in this nutrient, its accelerated rate of absorption can be detrimental in those who have already abundant iron stores. This effect can lead to haemochromatosis. Since normal individuals can probably regulate their iron intake and absorption effectively, there is no general risk of acquiring excess iron by taking excess ascorbic acid. However, excess ascorbic acid should be contraindicated in individuals with idiopathic haemochromatosis, thalassaemia, and other haemolytic anaemias.

Interference with glucose assay

The presence of excess ascorbic acid in the urine is known to give false-positive or false-negative results on tests for urinary sugar. Because of the strong reducing property of ascorbic acid, it may give a false-positive reaction when 'clinitest' tablets are used for identifying the presence of glucose in urine. On the other hand, a false-negative reaction appears to result when the 'dipstick' type of test is used; this is because the glucose oxidase-horseradish peroxidase system does not function in the presence of ascorbic acid.

Folic Acid

Observations that neural tube defects (NTD) occur more frequently among infants whose mothers are of poor folate status have led to the idea that folic acid supplementation might prevent development of such defects (Cziezel and Dudas, 1992; Werler *et al.*, 1993). A daily folic acid intake of up to 4.0 mg (as opposed to 0.4 mg day^{-1} as RDA) has been recommended for women who are at increased risk of giving birth to a child with NTD (American Academy of Pediatrics, 1993). This dose level can be achieved only through the use of supplements, unless one chooses very selectively the folate-rich foods. In 1993, the Food and Drug Administration (FDA) of the United States proposed fortification of cereal products with folic acid at a level of 0.14 mg/100 g.

Folic acid has, generally, a low acute and chronic toxicity for humans. It should be pointed out, however, although no toxic effects of folate have been clearly established, consuming excessive amounts of this vitamin can mask a vitamin B_{12} deficiency and therefore could delay the diagnosis of pernicious anaemia, with irreversible neurologic damage. Vitamin B_{12} deficiency is rare among women of childbearing age. In older subjects, however, the deficiency is reasonably frequent due to lack of intrinsic factor, gastric atrophy, or intestinal disorders affecting the ileum. Masking B_{12} deficiency by excess folate intakes either through supplementation or consuming folate fortified food may cause a serious problem in this population segment.

Experimental evidence has suggested that excess folate intake may also interfere with the absorption of zinc from the intestine (Keating *et al.*, 1987). Based on the prudent assumption that similar effects might occur in humans, folate intakes in excess of the RDI should be avoided. There is also some question about the safety of high-dose folic acid supplements in patients receiving drugs, such as methotrexate, trimethopterin and anticonvulsants. Folate supplements may affect the efficacy of the pharmacological actions of these drugs (see Chapter 18).

Other Water-soluble Vitamins

These vitamins, which include thiamin, riboflavin, biotin, pantothenate, and vitamin B_{12}, are not generally used by faddists in megadoses. There has been no reported adverse reaction associated with oral intakes of these vitamins in excess of 1000 times their recommended intakes. The toxic effects observed, in isolation of the B-vitamins, appear to be secondary to parenteral administration of the vitamins (Omaye, 1984). Thiamin and vitamin B_{12} given in excess parenterally have been reported in rare cases to produce idiosyncratic reactions, manifested by urticaria, respiratory distress, sweating, vascular collapse, and anaphylaxis. Supplemental intakes of these vitamins are contraindicated in individuals who are known to be sensitive.

Conclusions

The arguments about the beneficial effects of vitamins when taken in large amounts have been countered by claims that most vitamins may be potentially toxic when taken in massive quantities. These toxic effects are essentially mediated through hepatic damage, neurological disorders, interactions with exogenous and endogenous factors, and cause congenital malformations if taken during pregnancy. It is true that some of the evidence of adverse reactions, especially of the water-soluble vitamins, is contradictory and somewhat variable. In view of the seriousness of some of these reactions, it is important to be cautious against taking vitamins in quantities that are many times the recommended intakes. It is also necessary especially for physicians to be aware of the situations in which megadoses of vitamins could be clinically contraindicated.

References

Alhadeff, L., Gualtieri, C.T. and Lipton, M. (1984) Toxic effects of water soluble vitamins. *Nutrition Reviews* 42, 33–40.

American Academy of Pediatrics Committee on Genetics (1993) Folic acid for the prevention of neural tube defects. *Pediatrics* 92, 493–494.

American College of Obstetricians and Gynecologists (ACOG) Committee on Obstetrics: Maternal and Fetal Medicine (1993) Vitamin A supplementation during pregnancy. *International Journal of Gynecology and Obstetrics* 40, 175–178.

Basu, T.K. (1985) The conditioning effect of large doses of ascorbic acid in guinea pigs. *Canadian Journal of Physiology and Pharmacology* 63, 427–430.

Basu, T.K. and Schorah, C.J. (1982) In: *Vitamin C in Health and Disease*. AVI Publishing, Westport, CN, pp. 93–114.

Bauernfiend, J.C. (1980) The safe use of vitamin A. *A Report of the International Vitamin A Consultative Group (IVACG)*. The Nutrition Foundation, Washington, D.C.

Bendich, A. and Cohen, M. (1990) Vitamin B_6 safety issues. *Annals of New York Academy of Science* 585, 321–330.

Bendich, A.B. and Langseth, L. (1989) Safety of vitamin A. *American Journal of Clinical Nutrition* 49, 358–371.

Bendich, A., and Machlin, L.J. (1988) Safety of oral intake of vitamin E. *American Journal of Clinical Nutrition* 48, 612–619.

Bersen, E.L., Rosner, R., Sandberg, M.A., Hayes, K.C., Nicholson, B.W., Weigel-DiFranco, C. and Willelt, W. (1993) A randomized trial of vitamin A and vitamin E supplementation for retinitis pigmentosa. *Archives of Ophthalmology* 111, 761–772.

Birkenhead, T.L.C. (1958) Observations on carotenemia. *Annals of Internal Medicine* 48, 219–227.

Brumbaugh, P.F. and Haussler, M.R. (1973) 1,25-$(OH)_2D$ receptor: competitive binding of vitamin D analogues. *Life Sciences* 13, 1737–1746,.

Campbell, G.D., Steinberg, M.J. and Bower, J.D. (1975) Ascorbic acid-induced hemolysis in G-6-PD deficiency. *Annals of Internal Medicine* 82, 810–813.

Chapuy, M.C. and Meunier, P.J. (1990) Metabolic basis of vitamin D intoxication. In: Cohen, R.D., Lewis B., Alberti, K.G.M.M. and Denman, A.M. (eds) *The Metabolic and Molecular Basis of Acquired Disease*. Bailliére Tindall, London, pp. 1824–1834.

Cook, J.D. and Monsen, E.R. (1977) Vitamin C, the common cold, and iron absorption. *American Journal of Clinical Nutrition* 30, 235–241.

Czeizel, A.E. and Dudas, I. (1992) Preventing of the first occurrence of neural-tube defects by periconceptual vitamin supplementation. *New England Journal of Medicine* 327, 1832–1835.

Dalton, K. (1985) Pyridoxine overdose in premenstrual syndrome. *Lancet* 1, 1168–1169.

Davis, M. (1989) High-dose vitamin D therapy: Indications, benefits and hazards. In: Walker, P., Brubacher, G. and Staehelin, H. (eds) *Elevated dosages of vitamins.* Hans Huber Publishers, Bern, pp. 81–86.

DiPalma, J.R. and Thayer, W.S. (1991) Use of niacin as a drug. *Annual Review of Nutrition* 11, 169–187.

Einstein, N., Baker, A., Galper, J. and Wolfe, H. (1975) Jaundice due to nicotinic acid therapy. *American Journal of Digestive Disease* 20, 282–286.

Finley, E.B. and Cerklewski, F.L. (1983) Influence of ascorbic acid supplementation on copper status in young adult men. *American Journal of Clinical Nutrition* 37, 553–556.

Gal, I., Sharman, I.M. and Pryse-Davies, J. (1972) Vitamin A in relation to human congenital malformations. *Journal of Advances in Teratology* 5, 143–159.

Grote, W.E., Harmus, V., Janig, U., Kietzmann, H., Revens, U. and Schwanze, I. (1985) Malformation of fetus conceived 4 months after termination of maternal etretinate treatment. *Lancet* 1, 1276 (letter).

Hathcock, J.N., Hattan, D.G., Jenkins, M.Y., McDonald, J.T., Sundaresan, P.R. and Wilkening, V.L. (1990) Evaluation of vitamin A toxicity. *American Journal of Clinical Nutrition* 52, 183–202.

Heywood, R., Palmer, A.K., Gregson, R.L. and Hummler, H. (1985) The toxicity of beta-carotene. *Toxicology* 36, 91–100.

Howard, W.B. and Willhite, C.C. (1986) Toxicity of retinoids in humans and animals. *Journal of Toxicology Toxin Review* 5, 55–94.

Jacobus, C.H. *et al.* (1992) Hypervitaminosis D associated with drinking milk. *New England Journal of Medicine* 326, 1173–1177.

Keating, J.N., Wada, L., Stokstad, E.L.R. and King, J.C. (1987) Folic acid: Effect on zinc absorption in humans and in the rat. *American Journal of Clinical Nutrition* 46, 835–839.

Knopp, R.H., Ginsberg, J., Albers, J., Hoff, C. and Ogilvie, J.T. (1985) Contrasting effects of unmodified and time-release forms of niacin on lipoproteins in hyperlipidemic subjects: Clues to mechanism of action of niacin. *Metabolism* 34, 642–650.

Knudson, A.G. and Rothman, P.E. (1953) Hypervitaminosis A: a review with a discussion of vitamin A. *American Journal of Diseases of Children* 85, 316–334.

Krinke, G.J. and Fitzgerald, R.E. (1988) The pattern of pyridoxine-induced lesion: Difference between the high and low toxic level. *Toxicology* 49, 171–178.

Leo, M.A., Kim, C.A., Lowe, N. and Lieber, C.S. (1992) Interaction of ethanol with β-carotene: delayed blood clearance and enhanced hepatoxicity. *Hepatology* 15, 883–891.

Lonn, L.I. (1987) Canthaxanthine retinopathy. *Archives in Ophthalmology* 105, 1590–1591.

McClain, R.M., Langhoff, L. and Hoar, R.M. (1980) Reproduction studies with 1-alpha-25-dihydroxyvitamin D_3 (calcitriol) in rats and rabbits. *Toxicology and Applied Pharmacology* 52, 89–98.

March, B.E., Wong, E., Seier, L., Sim, J. and Biely, J. (1973) Hypervitaminosis E in the chick. *Journal of Nutrition* 103, 371–377.

Marks, J. (1989) The safety of the vitamins: an overview. In: Walker, P., Brubacher, G. and Stachelin, H. (eds) *Elevated Dosages of Vitamins*. Hans Huber Publishers, Bern, pp. 12–20.

Miller, D.R. and Hayes, K.C. (1982) Vitamin excess and toxicity. In: *Nutritional Toxicology*, Vol. 1. Academic Press, New York, pp. 81–133.

Mosher, L.R. (1970) Nicotinic acid side effects and toxicity: A review. *American Journal of Psychology* 126, 1290–1296.

Mullin, G.E., Greenson, J.K. and Mitchell, M.C. (1989) Fulminant hepatic failure after ingestion of sustained-release nicotinic acid. *Annals of Internal Medicine* 111, 253–255.

Nebel, L. and Ornoy, A. (1972) Interdependence of fetal anomalies and placental impairment following maternal hypervitaminosis D and hypercortisonism. In: Klingberg, M.A., Abramovici, A. and Chemke, J. (eds) *Drugs and Fetal Development*. Plenum Press, New York, pp. 251–255.

Olson, R.E. (1984) The function and metabolism of vitamin K. *Annual Review of Nutrition* 4, 281–337

Omaye, S.T. (1984) Safety of megavitamin therapy. In: Friedman, M. (ed.) *Nutritional Toxicological Concepts of Food Safety*. Plenum Press, New York, pp. 169–203.

Owen, C.A. (1971) Vitamin K. Pharmacology and toxicology. In: Friedman, M. (ed.) *Nutritional Toxicological Concepts of Food Safety*. Plenum Press, New York, 2nd edn, Vol. 3.

Persson, B.R., Tunell, E. and Kengren, K. (1965) Chronic vitamin A intoxication during the first half year of life. *Acta Pediatrica Scandinavica* 54, 49–60.

Public Affairs Committee of Teratology Society (1987) Teratology Society Position Paper. Recommendations for vitamin A use during pregnancy. *Teratology* 35, 269–275.

Ramp, W.K. and Thornton, P.A. (1971) Ascorbic acid and the calcium metabolism of embryonic chick tibias. *Proceedings of the Society of Experimental Biology and Medicine* 137, 237.

Rosa, F.W., Wilk, A.L. and Kelsey, F.O. (1986) Teratogen update: Vitamin A congeners. *Teratology* 33, 355–364.

Sahenk, Z. and Mendell, J.R. (1980) In: Spencer, P.S. and Schaumburg, W.H. (eds) *Experimental and Clinical Neurotoxicology*. Williams & Wilkins, Baltimore, pp. 578–597.

Schaumburg, H., Kaplan, J., Windebank, A., Vick, N., Rasmus, S., Pleasure, D. and Brown, M.J. (1983) Sensory neuropathy from pyridoxine abuse: a new megavitamin syndrome. *New England Journal of Medicine*. 309, 445–448.

Schmidt, K.-H., Hagmaier, V., Hornig, D.H., Vuilleumier, J.-P. and Rutishausen, G. (1981) Urinary oxalate excretion after large intake of ascorbic acid in man. *American Journal of Clinical Nutrition* 34, 305–311

Stein, H.B., Hasan, A. and Fox, I.H. (1976) Ascorbic acid-induced uricosuria. *Annals of Internal Medicine* 84, 385–388.

Tsao, C.S. and Salimi, S.L. (1984) Evidence of rebound effect with ascorbic acid. *Medical Hypotheses* 13, 303–310.

Vakil, D.V., Ayiomamitis, A., Nizami, N. and Nizami, R.M. (1985) Hypercarotenemia: A case report and review of the literature. *Nutrition Research* 5, 911–915.

Werler, M.M., Shapiro, S. and Mitchell, A.L. (1993) Periconceptional folic acid exposure and risk of occurrent neural tube defects. *Journal of the American Medical Association* 269, 1257–1261.

Zile, M.H. and Callum, M.E. (1983) The function of vitamin A: current concepts. *Proceedings of the Society of Experimental Biology and Medicine* 172, 139–152.

Index

Abuse of vitamins 313–318
Abetalipoproteinaemia 295
Acetaldehyde 278
Acetyl CoA 15–18, 71, 82–83, 118
Acetyl CoA carboxylase 71
Acetylcholine 18, 43, 114–115
Acne vulgaris 290
Activation of
  folic acid 87
  pyridoxine 52
  thiamin 13
  vitamin D 197
Acyl CoA:retinol acyltransferase (ARAT) 151
Acyl carrier protein (ACP) 80–81
Adenosylcobalamin 113
Adenyl cyclase 133, 136
Ageing and vitamin E 299
Alcohol and
  ascorbic acid 280–282
  B vitamins 277
  folic acid 282
  vitamin A 278
  vitamin D 279
Alcohol dehydrogenase 278
Aldehyde delydrogenase 278
Aluminium hydroxide 269
Amino acids and their interconversions 58–62, 94–95
δ-Aminolevulinic acid synthetase 63
Aminotransferases
  see Transaminases
Amygdalin 240–242
  and ascorbic acid 242
  and cysteine 242
  degradation 241
Anaemias 62–64, 99, 298
Antacid 268
Antioxidants 131–136, 184–188
Arachidonic acid 187
Ariboflavonosis 33
Ascorbic acid 125–144, 328–332
  absorption 129
  biochemical role 131–139
  biosynthesis 125
  body distribution 129
  body pool 129, 140
  as a cofactor 132–136
  deficiency 139–141
  excretion 130
    species variation 130
  as free-radical scavenger 132
  losses 127–129
  metabolism 128, 130
  as nitrite scavenger 138
  redox potential 125
  requirements 141–144
  sources 126
Ascorbic acid assay 146
Ascorbic acid dependent enzymes 132–136
Ascorbic acid risk groups 141
  alcoholics 142
  elderly 142

Ascorbic acid risk groups *contd*
exposure to foreign compounds 144
smokers 143
stress 144
Aspirin 269–271
Assessment 145
Avidin 74
Avitaminosis and congenital malformations 283

B vitamins 326–328, 332, 333
B_{12}–folate interrelationship 92, 111–113, 120
Beriberi 19–21
Biguanides 269
Biliary atresia 296
Bioavailability from cereals 41
absorption 42
deficiency 46
excretion 44, 49
functions 45–48
metabolism 43–45
requirements 48
status 48
tissue storage 43
transport 42
Biochemical analysis of ascorbic acid status 145–146
leucocyte ascorbic acid 145
plasma ascorbic acid 145
urinary ascorbic acid 145
Biochemical assessment for vitamin A 172–173
Biocytin 69
Bioflavonoids 243–244
Biogenic amines 60, 115, 133
Biotin 68–78
absorption 69
activation 69
assessment 77
chemistry 68
deficiency 73–76
metabolic role 70–73, 113
reaction site 70
sources 68
transport 69
Biotransformation of drugs 280
Bisacodyl 269
Bitot spot 165–167
Blood coagulation 230
Blood clotting proteins 232
Branched chain amino acid decarboxylase 14, 18
Bronchopulmunary dysplasia 223, 290
Burning foot syndrome 83–84

Calcium transport 201–203
Calcium binding protein (CaBP) 200–202
Cancer
and β-carotene 255–257
and folic acid 261
and vitamin A 253–255
and vitamin C 259–260
and vitamin D 261
and vitamin E 258–259
and vitamins 252–266
Canthaxanthine and retinopathy 323
Carbon dioxide fixation 70–73
Carboxybiotin 70
γ-Carboxylase 213
γ-Carboxyglutamic acid (GLA) 231
in bone (BGP) 233
in kidney (KGP) 233
in matrix (MGP) 233
Cardiovascular disease and vitamin C 300–301
Carnitine 135, 244–247
and newborns 246
role in β-oxidation 246
synthesis 245
α-Carotene 180
β-Carotene 178–192
absorption 181
assay 190
biosynthesis 178
chemistry 179
properties 181
provitamin activity 150, 184
requirements 188
sources 181
status 188–190
storage 187

transport 181
γ-Carotene 180
Carotenoids 148, 178
Carpal tunnel syndrome 306
Casal's collar *see* Pellagra
Cataracts 297, 302
and riboflavin deficiency 33
Catecholamines and ascorbic acid 133
Cellular retinol-binding protein (CRALBP) 153
Cerebrocortical necrosis 18
Chloramphenicol 272
Cholecalciferol *see* Vitamin D
Cholestyramine 268, 269
Choline 247–248
biosynthesis 113
Chromium *see* Glucose tolerance factor
11-*cis* Retinal *see* Rhodopsin
Clotting factors 231–233
Coenzyme A 15, 79, 82
Coenzyme Q 249
Coffee
NAD 42, 43–49
NADP 42, 43–39
vitamin B_6 40, 44, 61–63
zinc 44
see also Trigonelline
Colchicine 269
Collagen 33, 135
Common cold and vitamin C 302
Conditioning effect of vitamin C 329–331
Copper and vitamin C 134, 331
CoQ *see* Coenzyme Q
Cryptoxanthine 178–180, 188
Cutaneous synthesis of vitamin D *see* Vitamin D, synthesis
Cyanocobalamin *see* Vitamin B_{12}
Cyclic nucleotides 136
Cystathionase 60
Cystic fibrosis 296
Cytochrome P450 280, 281
Cytotoxic drugs 279, 280

D-Gluconodimethyl aminoacetic acid 242–243
Decarboxylations
non-oxidative 60
oxidative 14–18
Dehydroascorbic acid 125, 128
7α-Dehydrocholesterol 194
Dehydrogenase complex 14–17
Dehydrolipoyl dehydrogenase 14–17
Deoxyuridine (dU) suppression test
Desaturases 73
Detoxification of benzoic acid *see* Hippuric acid
Diabetes and vitamin A 160
Dicoumarol 231, 272, 273, 275
Dietary losses of
ascorbic acid 127
β-carotene 187
pyridoxine 53
riboflavin 32
thiamin 12
vitamin A 149
Dietary sources of
biotin 68
β-carotene 181
folic acid 88
niacin 40
pantothenic acid 80
pyridoxine 53
riboflavin 29
thiamin 11
vitamin A 149
vitamin B_{12} 107
vitamin C 126
vitamin D 195
vitamin E 215
vitamin K 228
Dihydrolipoyl transacetylase 14–17
Dioxygenase 134–136, 182, 183
Diphenylhydantoin 269, 271, 272
Diphyllobothrium latum 117, 121
Dopamine β-hydroxylase 133
Drug-induced vitamin 268
Drug-induced vitamin deficiency 268–279, 284
Drugs
interactions with vitamins 267–287
reactions *see* Drugs, toxicity
toxicity 284–286

Elderly and
thiamin 22, 33, 96, 203, 206
vitamin B_{12} 332
Energy metabolism 14–18, 31–32, 45–46, 70, 83, 118
Epithelical differentiation and vitamin A 253
Ergocalciferol 193
Erythropoietic protoporphyria 292
Exudative diathesis 224

Fatty acid synthesis *see* Acetyl CoA carboxylase
FIGLU test 101
Fish tapeworn and vitamin B_{12} *see Diphyllobothrium latum*
Flavoenzymes *see* Flavoproteins
Flavoproteins 31, 35
Folate
malabsorption 97
masking vitamin B_{12} deficiency 332
and methionine synthesis 91
and nucleic acid synthesis 92–94
status 96–99
factors affecting 96–99
and zinc 332
Folate binding proteins (FBP) 90
Folate coenzymes and their interconversions 92
Folic acid 4, 86–105
absorption 89, 91
assessment 100–103
chemistry 86–88
deficiency 95–99
functions 90–95
molecular weight 86
requirements 97–100
sources 88
status 96–99
synthesis 88
transport 89, 91
Formimino glutamic acid 95, 101
Free radicals and vitamin E 220
Functional CaBP 200

Gluconeogenesis 71
see also Pyruvate carboxylase
5-*O*-β-Glucopyranosyl-pyridoxine 55
Glucose-6-phosphate dehydrogenase 298, 326, 331
Glucose assay and vitamin C 332
Glucose tolerance factor 46
see also Niacin–chromium complex
β-Glucosidase 241
γ-Glutamylcarboxypeptidase 90
Glucosidic vitamin B_6 55
Glutathione peroxidase 225
Glutathione reductase 32–36
FAD-stimulating effect 32
Glutathione-reductase activation coefficient 32, 35
Glycogen phosphorylase 61
Glycoprotein synthesis 162
Goat's milk and folate 97

6-Hydroxypyridoxine and ascorbic acid 54
Haem synthesis 62, 83
see also δ-Aminolevulinic acid synthetase
Haemorrhagic diathesis 235, 237
Hartnup's disease 304
Hepatic retinyl ester hydrolase 152
Hepatic alcohol dehydrogenase 159
Hereditary rickets 209, 293
Hippuric acid 83
Holo-RBP 152
Homocysteine 112–115
Homocystinuria 305
Hydralazine 272, 275, 276
Hydrocyanic acid 241
Hydroxylysine 133
Hydroxyproline 133
Hyperlipidaemias and niacin 303
Hyperoxaluria 305, 329
Hypertension
and ascorbic acid 301
and vitamin D 294
Hypervitaminosis A *see* Toxicity of vitamin A
Hypochronic anaemia *see* Microcytic anaemia

Hypophosphataemic rickets *see* Hereditary rickets
Hypoprothrombinaemia 235, 236

Immune function and
  β-carotene 187–189
  vitamin A 163, 291
  vitamin C 135, 199, 302
Infectious disease 291–292
  and vitamin A 291–292
Inositol 248–249
Inter-photoreceptor retinol-binding protein (IRBP) 153
Intermittent claudication 297
Intracranial haemorrhage 223
Intrinsic factor 108, 110, 116–118
Intrinsic factor antibody 117
Intrinsic factor–vitamin B_{12} complex 268, 269
Iron absorption and vitamin C 137
Iron overload and vitamin C 331
Ischaemic heart disease 297, 300
Isoniazid 272, 275, 276

Kanamycin 272
Keratinized epithelial cells 161, 165, 168
Keratomalacia 166
α-Ketoacid decarboxylases 15–18, 24
Kynureninase 61

L-Gulono-γ-lactone oxidase 126
Laennec's cirrhosis 76
Laetrile *see* Amygdalin
LDL-oxidation 185
Lecithin 247–248
Lecithin:retinol acyltransferase (LRAT) 151
Leigh's disease 24
Leucine 48, 73
Leucocyte ascorbic acid 145
Levodope–vitamin B_6 interaction 285
Lipoic acid 15–17
Losses of
  ascorbic acid 127
  β-carotene 181
  pyridoxine 53
  riboflavin 32
  thiamin 12
  vitamin A 149
  vitamin B_{12} 107
Lumisterol 194
Lycopene 178

Macrocytic anaemia 95, 115
Maple syrup urine disease 15
Megadose 314
Megaloblastic anaemia *see* Macrocytic anaemia
Megavitamin use 315–317
  see also Therapeutic potential of vitamins
Menadione 228–230
Menaquinone 228–230
Methotrexate 271, 272
Methylcobalamin 111–113
β-Methylcrotonyl CoA carboxylase
Methylmalonic aciduria 113, 120
Methylmalonyl mutase 113, 118
Microcytic anaemia 64
Microsomal ethanol oxidizing system 278
Mineral oil 269
Mono-oxygenase 133
Myelin sheath 114
Myoinositol 248–249

Necrotizing encephalomyelopathy 18, 24
Neomycine 269, 272
Nervous system 18, 48, 60, 64, 114, 133, 140
Neural tube defect 98, 332
Neuropathy and vitamin B_6 327–328
Niacin 39–50
  chemistry 39
  relationships with tryprophan 40–45, 63
  sources 40–42
    dietary 41

Niacin *contd*
synthesis 40
Niacin–chromium complex 46
Niacin-dependent enzymes 47
Niacytin 41
Night-blindness 165–167
Nitrite scavenger 138
Nomenclatures for
folic acid 4, 87
vitamin K_1 4
Nucleic acids and their biosynthesis 92–94

Oestrogen 272, 273, 277, 293
see also Osteoporosis
Opsin *see* Rhodopsin
Optimum nutrition 289
Oral contraceptives 276, 277, 283
Osteocalcin 233
Osteomalacia 204–208
after gastrectomy 207
drug-induced 207
and intestinal malabsorption 207
and liver disease 207
and renal disease 208, 294
Osteoporosis 208–209, 295
pathogenesis of 208
oestrogen and 208
Oxalic acid 128
see also Ascorbic acid, metabolism
Oxidative decarboxylations 13–18
Oxygen therapy in premature infants 223
Oxygenases 132–136
see also Ascorbic acid, as a cofactor

Pangamic acid 242–243
Pantothenic acid 79–85
absorption 81
chemistry 79
deficiency 83
excretion 81
functions 83
metabolism 82
requirements 84
sources 80

p-Amino salicylate 269
p-Aminobenzoic acid 86–89, 249
Parietal cell antibody 117
Parkinson's disease 285
Pathophysiology of vitamin A
epithelial cell differentiation 161, 168
immune system 163
reproduction 165, 166
vision 163–167
visual cycle 163–164
see also Rhodopsin
Pellagra 4, 47, 84
Penicillamine 272, 275, 276
Pentamidine isothionate 271
Pernicious anaemia 116–118
see also Intrinsic factor
Peroxidation chain reaction and vitamin E 220
Petechial haemorrhages 139
Phenobarbital 269, 272
Phenothiazines 272
Phenytoin 269, 272, 274
Phosphate absorption 201–203
Phosphodiesterase 133
Phototherapy and riboflavin 32
Phylloquinone 228
see also Vitamin K
Phytones 178
Pigeon chest 204, 205
Plaque development 186
Platelet function 297, 302
Potassium chloride 269
Prealbumin *see* Transthyretin
Premenstrual tension syndrome 306
Primethamine 271, 272
Progestin 272
Propionyl CoA carboxylase 72, 113
Prothrombin 231, 274
Provitamin A 182, 183–184
Pteroylglutamic acid *see* Folic acid
Pyridoxic acid 51, 52, 57
Pyridoxyl-ε-aminolysine 53
Pyruvate carboxylase 70
Pyruvate decarboxylase 14–16, 24

R proteins 108, 110

R-binders 111–113
RBC folate 101
Raw egg and biotin deficiency *see* Avidin
Recommended daily intake 8, 288, 289
Reducing activity 136
Relative dose response test 172
Renal stones 329
Renal osteodystrophy 294
Retinol-binding protein 152–160
  body pool size 158
  half-life 158
  relation with age 159
  relation with protein 158
  relation with zinc 159
Retinol function tests 170–171
  conjunctival impression cytology 171
  rapid dark adaption 170
  rose bengal staining test 171
Retinoic acid receptors 153
Retinoids 148
Retinol equivalent (RE) 150
Retrilental fibroplasia 223
Rhodanase 241
Rhodopsin 163–165
Rhodopsin–vitamin A cycle 163–165
Riboflavin 28–38
  absorption 29, 31
  assessment 35
  chemistry 28
  deficiency 32–34
  dietary sources 29
  metabolism 30
  reaction site 29
  requirements 34
  transport 30
Riboflavin relationships with
  cataract 33
  FAD 28, 31
  FMN 28, 31
  folate 34
  iron 34
  vitamin B_6 33, 34
Rickets 203–206, 209–210

Salicylate 269
Schiff base 59
Schilling test 121
Scurvy 139–141
Selenium, its relations with vitamin E 225
Serum retinol-binding protein 172
Serum vitamin B_{12} 119
Serum folate 100, 102
Serum transthyretin 172
Serum vitamin A 172
Sickle cell anaemia 298
Smokers and vitamin C 143
Somatic function of vitamin A *see* Pathophysiology of vitamin A
Squamous metaplasia 161
Streptozotocin-induced vitamin A deficiency 160
Stroke and vitamin C 302
Subacute necrotizing encephalomyclopathy *see* Leigh's disease
Sudden infant death syndrome (SIDS) 76
Sulfasalazine 269
Sulphonamides 272
Supplements 288

Tachysterol 194
Teratogenicity of vitamin A 321–322
Testosterone and vitamin A 166
Tetracycline 271
Tetrahydrofolic acid 88
Therapeutic potential of vitamins 288–312
Thiamin 11–27
  absorption 13
  activation 13–15
  assessment 25
  biosynthesis 11
  coenzyme functions 14–19
  deficiency 19–25
  dietary sources 11
  genetic disorders 24
  losses 12, 18, 23
  metabolism 13
  reaction site 14
  requirements 25
  structure 11

Thiamin antagonists 23
Thiamin pyrophosphate 13, 14–26
 dependent enzymes 13–18
Thiamin status 25–26
 activity coefficient 26
 TPP effect 25
Thiaminase 13, 18, 23
Thioester 15–17
Thymidylate synthetase 94
 see also Folate and nucleic acid synthesis; Folate coenzymes and their interconversions; Amino acids and their interconversions
Thyrotoxicosis 23
α-Tocopherol
 in milk formula 224
 in serum 222
Tocopherols *see* Vitamin E
Tocotrienols *see* Vitamin E
Total parental nutrition 23, 74–76
Toxicity of
 carotenoids 323–324
 vitamin A 320–323
 vitamin D 324–325
TPP effect *see* TPP stimulating effect
TPP stimulating effect 19, 25
Transaminases 58–60
Transcobalamin 110, 118
Transketolase 15, 18–20, 24–26
Transmethylase 111–113
Transthyretin 152–160
Triamterene 271
Trigonelline 41, 44

Uric acid 327, 329

Vegetarians and vitamin B_{12} 116
Vitamin A 148–177
 absorption 150
 assay 173
 biotransformation 154–156
 cellular distribution 153–156
 chemistry 148
 excretion 154–156
 homeostasis 156–158
 losses 149
 requirements 168–170
 sources 149
 status, factors affecting
  age 159
  dietary intake 156
  disease 160
  lipid 158
  protein 158
  stress 160
  zinc 159
 storage 152
 transport 151–160
Vitamin B_6 51–67
 absorption 56
 assessment 65
 bioavailability 53
 chemistry 51
 conditioned deficiency 64
 deficiency 64
 dependency 304–306
 dependent enzymes 58
 excretion 57
 functions 57
 metabolism 56
 relationships with
  cysteine *see* Cystathionase
  glycogen 61
  haem 62
   see also δ-Aminolevulenic acid synthetase
  neuroactive amines *see* Biogenic amines
  niacin 40, 44, 61
   see also Kynureninase
  tryprophan 40–45
   see also Kynureninase
  requirements 65
  sources 53, 54–56
  transport 56
 vitamers and their interconversions 52, 56
Vitamin B_{12} 106–124
 absorption 108, 110, 121
 chemistry 106
 deficiency 114–119
 excretion 111
 half-life 111, 119

metabolic rate 111–113
and methionine biosynthesis 111–113
as methyl donor 247
molecular weight 106
requirements 119
sources 107, 109
status 119
transport 108, 110
Vitamin C and
carnitine synthesis 136
cholesterol degradation 136
collagen synthesis 135
copper interrelationship 134
folate interrelationship 138
immune function 138
non-haem iron absorption 137
synthesis of catecholamines 133
Vitamin D 193–213
assessment 210
chemistry 193–195
functions 199
dietary sources 195
absorption 196
metabolism 197–199
metabolites 197–199
requirements 202
status, seasonal effect 210
synthesis 194–196
Vitamin D-binding protein (DBP) 196
Vitamin D resistant rickets *see* Hereditary rickets
Vitamin E 214–227, 325
absorption 216–218
chemistry 214
deficiency 221
dietary sources 215–217
function 219
its status 225
metabolism 218
requirements 224
and polyunsaturated fatty acid intake 224
transport 218
Vitamin K 228–229 271, 273, 326
absorption 229
and its antagonist vitamin A 236
and its antagonist vitamin E 236
chemistry 228
deficiency 234
in newborns 236, 237
functions 230–234, 249
in obstructive jaundice 234
requirements 238
status 238
synthesis 228, 234
Vitamins 1–8
body pool of 5
coenzyme forms 3
deficiency 1, 2
stages of 1,2
definition 1
history 1–3
recommended allowances 8, 9
requirement 4–8, 25, 34, 48, 65, 76, 99, 119, 141, 168, 188, 202, 224, 238
status 6–8, 25, 36, 48, 65, 77, 96–102, 119–122, 145, 170–173, 180, 210, 225, 238
factors affecting 6–8

Warfarin 232, 235, 275
Wernicke–Korsakoff syndrome 22, 24, 278
Wound healing 299
see also Vitamin C

Xanthurenic aciduria 65
Xerophthalmia 167
Xerosis 165–167

Zinc 44, 159, 332

NORTHERN MICHIGAN UNIVERSITY
3 1854 005 479 723